Human ochratoxicosis and its pathologies
Ochratoxicose humaine et ses pathologies

***Colloques* INSERM**
ISSN 0768-3154

Other *Colloques* published as co-editions by John Libbey Eurotext and INSERM

133 Cardiovascular and Respiratory Physiology in the Fetus and Neonate. *Physiologie Cardiovasculaire et Respiratoire du Fœtus et du Nouveau-né.*
Scientific Committee : P. Karlberg, A. Minkowski, W. Oh and L. Stern;
Managing Editor : M. Monset-Couchard.
ISBN : John Libbey Eurotext 0 86196 086 6
INSERM 2 85598 282 0

134 Porphyrins and Porphyrias. *Porphyrines et Porphyries.*
Edited by Y. Nordmann.
ISBN : John Libbey Eurotext 0 86196 087 4
INSERM 2 85598 281 2

137 Neo-Adjuvant Chemotherapy. *Chimiothérapie Néo-Adjuvante.*
Edited by C. Jacquillat, M. Weil and D. Khayat.
ISBN : John Libbey Eurotext 0 86196 077 7
INSERM 2 85598 283 7

139 Hormones and Cell Regulation (10th European Symposium). *Hormones et Régulation Cellulaire (10e Symposium Européen).*
Edited by J. Nunez, J.E. Dumont and R.J.B. King.
ISBN : John Libbey Eurotext 0 86196 084 X
INSERM 2 85598 284 7

147 Modern Trends in Aging Research. *Nouvelles Perspectives de la Recherche sur le Vieillissement.*
Edited by Y. Courtois, B. Faucheux, B. Forette, D.L. Knook and J.A. Tréton.
ISBN : John Libbey Eurotext 0 86196 103 X
INSERM 2 85598 309 6

149 Binding Proteins of Steroid Hormones. *Protéines de liaison des Hormones Stéroïdes.*
Edited by M.G. Forest and M. Pugeat.
ISBN : John Libbey Eurotext 0 86196 125 0
INSERM 2 85598 310 X

151 Control and Management of Parturition. *La Maîtrise de la Parturition.*
Edited by C. Sureau, P. Blot, D. Cabrol, F. Cavaillé and G. Germain.
ISBN : John Libbey Eurotext 0 86196 096 3
INSERM 2 85598 311 8

Suite page 241
(Continued p. 241)

Human ochratoxicosis and its pathologies

Ochratoxicose humaine et ses pathologies

Proceedings of the International Symposium : Human ochratoxicosis and associated pathologies in Africa and developing countries, held in Bordeaux (France) on July 4-6, 1993

Actes du Symposium International : Ochratoxicose humaine et pathologies associées en Afrique et dans les pays en voie de développement, tenu à Bordeaux du 4 au 6 juillet 1993

Sponsored by the Institut National de la Santé et de la Recherche Médicale (INSERM), the Commission of the European Communities, Directorat-General XII, the United Nations Environment Programme, Nairobi (Kenya), the Conseil Régional d'Aquitaine

Edited by
E.E. Creppy
M. Castegnaro
G. Dirheimer

LES EDITIONS INSERM

John Libbey EUROTEXT

British Library Cataloguing in Publication Data

A catalogue record for this book
is available from the British Library

ISBN 2-7420-0017-8
ISSN 0768-3154

First published in 1993 by

Editions John Libbey Eurotext
6 rue Blanche, 92120 Montrouge, France. (33) (1) 47 35 85 52
ISBN 2-7420-0017-8

John Libbey and Company Ltd
13 Smiths Yard, Summerley Street, London SW18 4HR, England.
(44) (81) 947 27 77

Institut National de la Santé et de la Recherche Médicale
101 rue de Tolbiac, 75654 Paris Cedex 13, France.
(33) (1) 44 23 60 00
ISBN 2-85598-543-9

ISSN 0768-3154

Foreword

Ochratoxin A (OTA) is a mycotoxin produced essentially by fungi of *Aspergillus genera* and by *Penicillium verrucosum*. It was discovered in South Africa by van der Merwe et al. (1965) who established it structure. Its molecular mode of action was determined by Röschenthaler *et al.* for prokaryotes and Creppy and Dirheimer for eukaryotes.

Among the mycotoxins of human health concern, the ochratoxins involvement is increasing since the early seventies, as it is implicated in Balkan endemic nephropathy (BEN). In these areas of central Europe (Bulgaria, Romania and Yugoslavia), OTA was first detected in food and feed (cereal, poultry and pork) and afterwards identified in blood of inhabitants suffering from BEN, an evolutive tubulo-interstitial nephropathy with associated urinary tract tumours (Krogh et al., Petkova-Bocharova et al.).

The carcinogenicity of OTA is established by now in rodents (NTP 1989) and it is regarded as possible carcinogen in human (classified 2 B) by the International Agency for Research on Cancer (IARC). During the last decade, human ochratoxicosis has been identified as a potential risk factor for human in western European countries, northern America, northern Africa and Asia, as ochratoxin has been found in blood of people in Germany, Scandinavia, France, Canada, Tunisia, Algeria, India and Japan.

In addition to its nephrotoxicity in all animal species studied so far including human and its carcinogenicity, OTA, at least in laboratory animals, is immunosupressive, teratogenic and genotoxic. It likely disturbs several metabolic and enzymatic systems. Facing the new situation, the ochratoxin A wordwide specialists in a round-table discussion of the last meeting at IARC (Lyon) arose the question to know whether BEN was still limitated to the Balkans and if some of the pathologies possibly connected with ochratoxicosis are occuring somewhere else in the world. The international symposium on human ochratoxicosis and associated pathologies in Africa and developing countries held in Bordeaux, 4-6 July 1993 intends to answer at least partially these difficult questions.

For several reasons of which enumeration is not needed here, the attentive study of human ochratoxicosis in developing countries may help to predict the outcome of this problem in other parts of the world, i.e. in industrialized countries and to set up in due time preventive measures.

The scientific committee and the organizers thank the EEC Commission XII, INSERM, UNEP and Aquitaine local government for their financial support. Thanks also to the generous donors listed on an attached sheet.

The Editors

Préface

L'ochratoxine A (OTA) est une mycotoxine produite essentiellement par des moisissures des genres *Aspergillus* et par *Penicillium verrucosum*. Elle a été découverte en Afrique du Sud par van der Merwe et coll. (1965) qui en ont déterminé la structure.
Son mécanisme d'action moléculaire a été établi par Röschenthaler et coll. pour les procaryotes, Creppy et Dirheimer pour les eucaryotes.
Parmi les mycotoxines auxquelles l'homme est exposé, l'OTA a pris, depuis les années 70, une place de plus en plus importante, car elle est impliquée dans la néphropathie endémique des Balkans (NEB). Dans ces régions d'Europe Centrale (Bulgarie, Roumanie, Yougoslavie), l'OTA a été retrouvée comme contaminant des aliments (céréales, volailles et porc) puis identifiée dans le sang des personnes souffrant de tubulonéphrite interstitielle évolutive, avec une forte fréquence des tumeurs associées siégeant au niveau du tractus urinaire (Krogh et coll., Petkova-Bocharova et coll.). Sa cancérogénicité ne fait plus de doute chez les rongeurs (NTP 1989), et elle est considérée par le Centre International de Recherche sur le Cancer (CIRC) comme un cancérigène potentiel chez l'homme (classée 2B).
Au cours de ces dix dernières années, l'ochratoxicose humaine a été identifiée dans des pays d'Europe de l'Ouest, en Amérique du Nord, en Afrique du Nord et en Asie, comme un facteur potentiel de risques, car l'OTA a été retrouvée dans le sang de personnes vivant en Allemagne, en Scandinavie, en France, au Canada, en Tunisie, en Algérie, en Inde et au Japon. En plus de sa néphrotoxicité chez toutes les espèces animales étudiées, y compris chez l'homme et sa cancérogénicité, l'OTA est, tout au moins chez l'animal, immunosuppressive, tératogène et génotoxique. Elle perturbe sans doute de nombreux systèmes enzymatiques et métaboliques.
Devant cette nouvelle situation, les spécialistes mondiaux de l'ochratoxine A réunis au CIRC à Lyon en 1991 se sont posé la question de savoir si la NEB est encore limitée aux Balkans et si l'une ou l'autre des pathologies qui peuvent lui être imputées existent ailleurs que dans les Balkans.
C'est pour tenter de répondre à ces questions que le symposium international sur l'«ochratoxicose humaine et pathologies associées en Afrique et dans les pays en voie de développement» se tient à Bordeaux du 4 au 6 juillet 1993.
Pour de nombreuses raisons qu'il n'est pas nécessaire d'énumérer ici, l'étude attentive de l'ochratoxicose humaine et des pathologies éventuellement associées dans les pays sous-développés peut permettre de prévoir l'avenir du problème dans d'autres régions du globe, notamment dans les pays industrialisés, et de rechercher, quand il est encore temps, les moyens de prévention.
Le comité scientifique et les organisateurs tiennent à remercier la Commission XII de la CEE, l'INSERM, l'UNEP et la Région Aquitaine pour leur soutien financier.
Que tous les généreux donateurs dont la liste est jointe en annexe trouvent ici l'expression de notre gratitude.

Les Éditeurs

Scientific Committee
Comité Scientifique

G. Dirheimer, CNRS, Université Louis Pasteur de Strasbourg (France)
P. Bach Polytechnic of East London (Royaume-Uni)
H. Bacha, Université de Monastir (Tunisie)
H. Barstch, IARC, Lyon (France)
M. Castegnaro, IARC, Lyon (France)
E.E. Creppy, Université Bordeaux II (France)
T. Kuiper-Goodman, Health and Welfare, Ottawa (Canada)
B. Hald, Royal Veterinary Agricultural University (Danemark)
R.P. Plestina, W.H.O., Geneva (Suisse)
H.P. van Egmond, National Institute of Public Health (Pays-Bas)

Organizing Committee
Comité d'Organisation

Creppy Eric Edmond, Université Bordeaux II (France)
Castegnaro Marcel, IARC, Lyon (France)
Andrieux Monique, Université Bordeaux II (France)
Betbeder Anne Marie, Université Bordeaux II (France)
Bonini Michelle, Université Bordeaux II (France)
Sanchez Denise, Université Bordeaux II (France)
Bacha Hassen, Université de Monastir (Tunisie)
Guilcher Jocelyne, Université Bordeaux II (France)

Acknowledgements
Remerciements

We would like to express our gratitude and special appreciation to :

ICS, Instrumentation Consommable Service, Moulin de Caupian, 33160 Saint-Médard-en-Jalles, France
Cofralab, Z.I. du Haut Vigneau, 37, rue de la Source, 33170 Gradignan, France
Palmer Research, Parc Cadéra-Sud, 30, avenue Ariane, 33700 Mérignac, France
Biovalori, J. Tastet S.A., 40380 Cassen, France

List of participants

Liste des participants

Achour A., Service de Néphrologie et Hémodialyse, C.H.U., Monastir, Tunisie

Andrieux M., Laboratoire de Toxicologie, Université Bordeaux II, 3 ter, place de la Victoire, 33076 Bordeaux Cedex, France
Tel: (33) 56 91 84 07
Fax: (33) 56 91 14 16

Aparicio M., C.H.R. Le Tripode, Place Amélie Raba-Léon, 33076 Bordeaux Cedex, France

Bach P., School of Science, Polytechnic of East London, Romford Road, London E 15 4LZ, Royaume-Uni
Tel: 44 81 849 3496
Fax: 44 81 519 37 40

Bacha H., Laboratoire de Biochimie et Toxicologie Moléculaire, Faculté de Chirurgie dentaire, Monastir, Tunisie
Tel: 216/361 000
Fax: 216/ 361 150

Badria Farid A., Pharmacology Department, Faculty of Pharmacy Mansoura University, Mansoura, Egypte

Bartsch H., International Agency for Research on Cancer, 150, cours Albert Thomas, 69372 Lyon Cedex 08, France
Tel: (33) 72 73 84 85
Fax: (33) 72 73 85 75

Baudrimont I.
Laboratoire de Toxicologie, Université Bordeaux II, 3 ter, place de la Victoire, 33076 Bordeaux Cedex, France

Benabadji M., CHU, Thenia, Boumerdes, Algérie

Berbiche O., El Watan, Alger, Algérie

Betbeder A.M., Laboratoire de Toxicologie, Université Bordeaux II, 3 ter, place de la Victoire, 33076 Bordeaux Cedex, France
Tel: (33) 56 91 84 07
Fax: (33) 56 91 14 16

Bose M., Industrial Toxicology Research Centre, Mahatma Gandhi Marg, P.O. Box n 80, Lucknow 226001, UP, Inde

Boudra H., Laboratoire de Pharmacologie-Toxicologie, INRA, 180, chemin de Tournefeuille, 31931 Toulouse Cedex, France

Burgat V., Laboratoire de Pharmacie et Toxicologie, Ecole Vétérinaire de Toulouse, 26, Chemin des Capelles, 31076 Toulouse, France

Castegnaro M., International Agency for Research on Cancer, 150, cours Albert Thomas, 69372 Lyon Cedex, France
Tel: (33) 72 73 84 85
Fax: (33) 72 73 85 75

Chernozemsky I.N., National Oncology Centre, 6, Plovdivosko Pole street, 1156 Sofia, Bulgarie
Tel: 35 92 71 23 294
Fax : 35 92 72 06 51

Creppy E.E., Laboratoire de Toxicologie, Université Bordeaux II, 3 ter, place de la Victoire, 33076 Bordeaux Cedex, France
Tel: (33) 56 91 84 07
Fax: (33) 56 91 14 16

Dano-Djedje S., Laboratoire de Chimie Analytique et Toxicologie, Faculté de Pharmacie, Université d'Abidjan, Abidjan, Côte-d'Ivoire

Dirheimer G., Institut de Biologie Moléculaire et Cellulaire, CNRS, 15, rue René Descartes, 67084 Strasbourg, France
Tel: 33 /88 41 70 55
Fax: 33/ 88 61 06 80

El-May M., Département de Médecine Interne, Université de Monastir, Rue Fattouma Bourguiba, 5019 Monastir, Tunisie

Fink-Gremmels J., Faculty of Veterinary Medicine, Utrecht University, Yalelaan 2, P.O. Box 80.176, 3508 TD Utrecht, Pays-Bas
Tel: 31 30 53 54 53
Fax: 31 30 53 50 77

Fremy J., Ministère de l'Agriculture et du Développement Rural, 78, rue de Varenne, 75007 Paris, France

Fuchs R., Department of Toxicology, Institute for Medical Research and Occupational Health, University of Zagreb, Ksaverska cesta 2, PO Box 201, 41000 Zagreb, Croatie

Galtier P., Laboratoire de Pharmacologie-Toxicologie, INRA, 180, chemin de Tournefeuille, 31931 Toulouse Cedex, France
Tel: 61 28 50 28
Fax: 61 28 53 10

Gharbi A., Laboratoire de Toxicologie, Université Bordeaux II, 3 ter, place de la Victoire, 33076 Bordeaux Cedex, France
Tel: 56 91 84 07
Fax 56 91 14 16

Gugnani H.C., Department of Microbiology, University of Nigeria, Nsukka, Nigeria

Hadlock R.M., Institute of Veterinary Meat-and Food Hygien., Justus-Liebig-University of Giessen, Frankfurter Strasse 92, Giessen, Allemagne
Tel: 49 641 702 49 75
Fax 49 641 702 74 08

Hald B., Department of Veterinary Microbiology, Royal Veterinary and Agricultural University, 13 Bülowsvej, 1870 Frederiksberg C, Danemark

Hammami M., Faculté de Médecine, Laboratoire de Biochimie, Monastir, Tunisie
Fax: 216 36 07 37

Hult K., Department of Biochemistry, Royal Institute of Technology, 100 44 Stockholm, Suède

Jonzyn F.E., Liverpool School of Tropical Medicine, Pembroke Place, Liverpool L3 5QA, Royaume-Uni
Tel: 051 708 93 93
Fax 051 708 87 33

Kane A., Laboratoire des Mycotoxines, Institut de Technologie Alimentaire, B.P. 2765, Dakar, Sénégal

Khalef A., UGTA, Maison du Peuple, Place du 1er mai, Alger, Algérie

Kuiper-Goodman T., Toxicological Evaluation Division, Bureau of Chemical Safety, Food Directorate, Health Protection Branch, Health and Welfare Canada Tunney's Pasture, Ottawa, K1A OL2, Canada
Tel: 613 957 16 76
Fax: 613 941 26 32

Le Bars J., Laboratoire de Pharmacologie-Toxicologie, INRA, 180, chemin de Tournefeuille, 31931 Toulouse Cedex, France
Tel: 61 28 50 28
Fax: 61 28 53 10

Maaroufi K., Laboratoire de Biochimie et Toxicologie Moléculaire, Faculté de Chirurgie dentaire, Monastir, Tunisie
Tel: 216/361 000
Fax: 216/ 361 150

Mantle P.G., Biochemistry Department, Imperial College of Science, Technology and Medicine, London SW7 2AY, Royaume-Uni
Tel: 071 589 51 11
Fax: 071 225 09 60

Maxwell S.M., Liverpool School of Tropical Medicine, Pembroke Place, Liverpool L3 5QA, Royaume-Uni
Tel: 051 708 93 93
Fax: 051 708 87 33

Miraglia M., Istituto Superiore de Sanita, Laboratorio Alimenti, viale Regina Elena 299, 00161 Roma, Italia
Tel: 06 4990 377
Fax: 06 4451 767

Murn M., Department of Preclinical Research, Laboratory of Toxicology, KRKA Pharmaceutical, 45, C Herojev, 68000 Novo Mesto, Slovénie
Tel: 38 68 23 233
Fax: 38 68 21 537

Netter K.J., Institut für Pharmakologie, Philips Universität Marburg, Karl von Frisch Strasse, 3550 Marburg, Allemagne

Olsen J.H., Danish Cancer Registry, Danish Cancer Society, Rosenvaengets Hovedvej 35, Box 839, 2100 Copenhagen, Danemark

Petzinger E., Institut für Pharmacologie und Toxikologie, Fachbereich Veterinärmedizin, Justus-Liebig Universität Giessen, Frankfurter Strasse 107, D6300 Giessen , Allemagne
Tel: 49/641 702 49 50
Fax: 49/ 641/702 73 90

Pfohl-Leszkowicz A., Institut de Biologie Moléculaire et Cellulaire, CNRS, 15, rue René Descartes, 67084 Strasbourg, France
Tel: (33) 88 41 70 63
Fax: (33) 88 61 06 80

Plestina R., International Programme on Chemical Safety, World Health Organization, 1211 Geneva 27, Suisse
Tel: (41) 22 791 35 92
Fax: (41) 22 791 07 46

Scudamore K.A., Central Science Laboratory, London road, Slough, Berkshire SL3 7HJ, Royaume-Uni
Tel: (44) 0753 53 46 26
Fax: (44) 0753 82 40 58

Speijers G.J.A., Department of Toxicology, National Institute of Public Health and Environmental Protection, Postbox 1, 3720 BA Bilthoven, Pays-Bas
Tel: (030) 74 91 11
Fax: (030) 74 29 71

Steyn P.S., The CSIR Fellow, P.O. Box 395, Pretoria 0001, République d'Afrique du Sud
Tel: (012) 841 43 82
Fax: (012) 841 41 74

Stormer F.C., Department of Toxicology, National Institut of Public Health, Postuttak Oslo 1, Oslo, Norvège

Ueno Y., Department of Toxicology and Microbial Chemistry, Faculty of Pharmaceutical Sciences, Science University of Tokyo, Ichigaya, Tokyo 162, Japon
Tel: 03 32 60 67 25
Fax: 03 32 68 30 45

Zellou Aziza, Unité de Mycologie, Département de Parasitologie, Institut National d'Hygiène, 27, avenue Ibn Batouta, Rabat, Maroc
Tel: 719 02
Telex: 319 97 INH Rabat

Zidane C., CHU Panet, Hussein Dey, Alger, Algérie

Contents
Sommaire

V Foreword
VII Préface
XI List of participants
Liste des participants

I. GENERAL VIEW ON OCHRATOXIN A AND ASSOCIATION WITH OTHER MYCOTOXINS
I. GÉNÉRALITÉS SUR L'OCHRATOXINE A ET AUTRES MYCOTOXINES ASSOCIÉES

3 **P. S. Steyn**
Mycotoxins of human health concern
Les mycotoxines qui affectent la santé de l'homme

33 **P. H. Bach and M. McLean**
Research priorities for assessing the risk of multiple mycotoxin exposure to domestic animals and man: what we know and what we need to know !
Recherches prioritaires pour l'évaluation du risque dû à l'exposition de l'homme ou de l'animal domestique aux mycotoxines multiples : ce que nous savons et ce qu'il nous faut savoir

43 **P. H. Bach**
The modulation of mycotoxin exposure in domestic animals and man: can we affect what we can't control ?
Modulation de l'exposition de l'homme et de l'animal domestique aux mycotoxines : pouvons-nous agir sur des paramètres que nous ne pouvons contrôler ?

51 **P. S. Steyn**
Ochratoxin A : its chemistry, conformation and biosynthesis
Ochratoxine A : chimie, conformation et biosynthèse

59 **P. Galtier, G. Larrieu et M. Alvinerie**
Influence de l'ochratoxine A sur le devenir des xénobiotiques
Influence of ochratoxin A on the outcome of xenobiotics

67 **J. Fink-Gremmels, M. Blom and F. Woutersen van Nijnanten**
In vitro investigations on ochratoxin A metabolism
Etude du métabolisme de l'ochratoxine A in vitro

75 **P. Deberghes, G. Deffieux, A. Gharbi, A.M. Betbeder, F. Boisard, R. Blanc, J.F. Delaby et E.E. Creppy**
Détoxification de l'ochratoxine A par des moyens physiques, chimiques et enzymatiques
Detoxification of ochratoxin A by physical, chemical and enzymatic processes

II. OCCURENCE OF OCHRATOXIN A IN FOODSTUFFS AND HUMAN BLOOD
II. PRÉSENCE DE L'OCHRATOXINE A DANS LES ALIMENTS ET LE SANG HUMAIN

85 **G.J.A. Speijers and H.P. van Egmond**
Worldwide ochratoxin A levels in food and feeds
Taux de contamination des aliments par l'ochratoxine à travers le monde

101 **Y. Bouraïma, L. Ayi-Fanou, I. Kora, J. Setondji, A. Sanni et E.E. Creppy**
Mise en évidence de la contamination des céréales par les aflatoxines et l'ochratoxine A au Bénin
Food contamination by aflatoxin and ochratoxin A in Benin

111 **H. Bacha, K. Maaroufi, A. Achour, M. Hammami, F. Ellouz, et E.E. Creppy**
Ochratoxines et ochratoxicoses humaines en Tunisie
Ochratoxins and human ochratoxicoses in Tunisia

123 **A. Khalef, C. Zidane, A. Charef, A. Gharbi, M. Tadjerouna, A.M. Betbeder et E.E. Creppy**
Ochratoxicose humaine en Algérie
Human ochratoxicosis in Algeria

129 **M. Miraglia, C. Brera, S. Corneli and R. De Dominicis**
Ochratoxin A in Italy : status of knowledge and perspectives
Ochratoxine A en Italie : point de la situation et perspectives

141 **R.M. Hadlock**
Human ochratoxicosis in Germany updating 1993
Ochratoxicose humaine en Allemagne : le point de la situation en 1993

147 **E.E. Creppy, M. Castegnaro, Y. Grosse, J. Mériaux, C. Manier, P. Moncharmont, C. Waller et coll.**
Etude de l'ochratoxicose humaine dans trois régions de France : Alsace, Aquitaine et Rhône-Alpes
Study of human ochratoxicosis in three French regions : Alsace, Aquitaine and Rhône-Alpes

159 **O. Kawamura, S. Maki, S. Sato and Y. Ueno**
Ochratoxin A in livestock and human sera in Japan quantified by a sensitive ELISA
Ochratoxine A dans les aliments et le sang humain au Japon quantifiée par une méthode ELISA sensible

167 **T. Kuiper-Goodman, K. Ominski, R.R. Marquardt, S. Malcolm, E. McMullen, G.A. Lombaert and T. Morton**
Estimating human exposure to ochratoxin A in Canada
Estimation de l'exposition humaine à l'ochratoxine A au Canada

III. OCHRATOXICOSIS AND PATHOLOGIES IN HUMAN AND ANIMAL
III. OCHRATOXICOSE ET PATHOLOGIES CHEZ L'HOMME ET L'ANIMAL

177 **A. Pfohl-Leszkowicz, Y. Grosse, A. Kane, A. Gharbi, I. Baudrimont, S. Obrecht, E.E. Creppy and G. Dirheimer**
Is the oxidative pathway implicated in the genotoxicity of ochratoxin A ?
La voie oxydative est-elle impliquée dans la génotoxicité de l'ochratoxine A ?

189 **I. Baudrimont, A.M. Betbeder, A. Gharbi, A. Pfohl-Leszkowicz, G. Dirheimer et E.E. Creppy**
Influence de la superoxyde dismutase associée à la catalase sur la néphrotoxicité induite par l' ochratoxine A chez le rat
Influence of superoxide dismutase and catalase on the OTA-induced nephrotoxicity in rat

199 **A. Pfohl-Leszkowicz, Y. Grosse, S. Obrecht, A. Kane, M. Castegnaro, E.E. Creppy and G. Dirheimer**
Preponderance of DNA-adducts in kidney after ochratoxin A exposure
Prépondérance des adduits à l'ADN dans le rein après exposition à l'ochratoxine A

209 **J.H. Olsen, B. Hald, I. Thorup and B. Carstensen**
Distribution in Denmark of porcine nephropathy and chronic disorders of the urinary tract in humans
Distribution de la néphropathie porcine et des pathologies chroniques du tractus urinaire chez l'homme au Danemark

217 **P.H. Bach**
Markers for mycotoxin nephrotoxicity in domestic animals and man. Why are there no selective or specific ways of assessing the lesion ?
Marqueurs des néphropathies mycotoxiniques chez l'homme et les animaux domestiques : raisons de l'absence de méthode sélective et spécifique d'évaluation des lésions

227 **A. Achour, M. El-May, H. Bacha, M. Hamammi, K. Maaroufi et E.E. Creppy**
Néphropathies interstitielles chroniques. Approches cliniques et étiologiques : ochratoxine A
Chronic interstitial nephropathy. Clinical and etiological approaches : ochratoxin A

235 **A. Khalef, M. Benabadji, T. Rayan et F. Haddoumi**
Présence d'ochratoxine A et néphropathies en Algérie
Presence of ochratoxin A and nephropathies in Algeria

239 **Author index**
Index des auteurs

I. General view on ochratoxin A and association with other mycotoxins

I. Généralités sur l'ochratoxine A et autres mycotoxines associées

Human ochratoxicosis and its pathologies. Eds E.E. Creppy, M. Castegnaro, G. Dirheimer. Colloque INSERM/ John Libbey Eurotext Ltd. © 1993, Vol. 231, pp. 3-31.

Mycotoxins of human health concern

Pieter S. Steyn

CSIR, P.O. Box 395, Pretoria, 0001 South Africa

ABSTRACT Mycotoxins are produced by five genera of fungi: *Aspergillus*, *Penicillium*, *Fusarium*, *Alternaria* and *Claviceps*. The occurrence and associated human health risks of the aflatoxins, ergotoxins, citreoviridin, trichothecenes, ochratoxins, tremorgenic mycotoxins and the fumonisins are described.

INTRODUCTION

The famous mycotoxicologist Forgacs stated 30 years ago that *of the innumerable diseases that affect man and domestic animals, the mycotoxicoses are perhaps the most unfamiliar and least investigated, presumably due to lack of proper techniques and interested scientists. The situation has prompted many to regard this elusive group of diseases as hypothetical when, in fact, there is abundant evidence to suspect mycotoxicoses as causal factors in diseases of unknown etiology.*

Mycotoxicoses were referred to in the early 1960's as the most neglected diseases, although during World War II, many people died in Russia owing to alimentary toxic aleukia. The world was rudely awakened to the importance of mycotoxins by the discovery of the Turkey X disease during 1960 in the United Kingdom. Turkey X disease was soon linked to the contamination of groundnuts with the ubiquitous fungus, *Aspergillus flavus*, and its propensity to elaborate the aflatoxins, a group of structurally related hepatocarcinogens. The aflatoxins contaminate a wide variety of food commodities and have also been implicated in the etiology of human primary liver cancer in countries in Africa and South East Asia. Since this seminal finding there was an unabated growth in interest in mycotoxins as evidenced by the numerous international conferences and symposia, monographs, reviews and research papers which were devoted to this fascinating topic.

Mycotoxins are a structurally diverse group of mostly small molecular weight compounds which are produced by the secondary metabolism of filamentous fungi and are toxic to mammals, fish and poultry. The human ingestion of mycotoxins is due to the consumption of the mycotoxins in plant-based foods and the residues and metabolites in animal-derived food, such as milk, cheese and meat.

Mycotoxins have assumed world-wide importance owing to the global distribution of toxinogenic fungi, thereby putting crops and animals at risk; and by their contamination of harvested crops destined for internal markets and export to international markets. This accounts for the long standing involvement of IUPAC in this crucially important area of science.

Concerns about the effects of mycotoxins to human and animal health and their structural complexity attracted much attention from organic chemists, and several sterling achievements have been recorded since the epoch-making discovery of the aflatoxins. The toxins of most of the food-borne fungi have subsequently been identified. The advent of sophisticated chromatographic techniques, high-resolution mass spectroscopy, high-field nuclear magnetic resonance spectroscopy and single crystal X-ray crystallography contributed to these successes (Cole, 1986) (Steyn, 1989). The discovery of the mycotoxins stimulated outstanding contributions on the biosynthesis of secondary fungal metabolites (Steyn, 1980), their biochemical mechanism of action (Wogan, 1992b) as well as pathological effects (Uraguchi and Yamazaki, 1978) and immunobased analysis (particularly ELISA) (Morgan, 1989) of mycotoxins.

The impact of the potent carcinogenic effect of the aflatoxins on the mycotoxin scene worldwide justifies a non-chronological sequence in this paper.

HEALTH RISKS OF MYCOTOXINS

Mycotoxins have been recognized as an important class of compounds in the human food chain. Since humans would normally avoid foods visually contaminated by moulds, human health problems resulting from exposure to acutely toxic levels of mycotoxins are relatively rare. The greatest health concern about mycotoxins is the cancer risks based on long-term, low level exposures to toxins. In nature most of the energy-rich commodities such as cereal grains, oil seeds, tree nuts and dehydrated fruits are susceptible to fungus contamination and mycotoxin formation under certain environmental conditions such as high humidity and high water activity (Bullerman *et al* 1984).

Under laboratory conditions (control of temperature, moisture and nutrients) at least three hundred mycotoxins have been produced by pure cultures of fungi and chemically characterized. Fortunately, only about 20 are known to occur in foodstuffs at significant levels and frequency to be of food safety concern (Hsieh, 1992). These toxins are produced by five genera of fungi : *Aspergillus*, *Penicillium*, *Fusarium*, *Alternaria* and *Claviceps*.

The mycotoxins produced by these fungi are :

- *Aspergillus* toxins
  Aflatoxin B_1, G_1, M_1, sterigmatocystin, cyclopiazonic acid.

- *Penicillium* toxins
  Patulin, ochratoxin A, citrinin, penitrem A, and cyclopiazonic acid.

- *Fusarium* toxins
  Deoxynivalenol, nivalenol, zearalenone, T-2 toxin, diacetoxyscirpenol, fumonisins and moniliformin.

- *Alternaria* toxins
  Tenuazonic acid, alternariol, alternariol methyl ether.

- *Claviceps* toxins
  Ergot alkaloids.

Mycotoxins induce powerful biological effects. Some are carcinogenic, mutagenic, teratogenic, estrogenic, hemorrhagic, immunotoxic, nephrotoxic, hepatotoxic and neurotoxic, whereas others display antitumour, cytotoxic and antimicrobial properties. Some of these toxic effects are consistent with the characteristic symptoms seen in a number of human diseases associated with mycotoxins. Other effects may be an indication of mammalian diseases still to be discovered. For example, the ochratoxins (Van der Merwe *et al* 1965) were discovered many years prior to the recognition of their role in Danish porcine nephropathy (Kuiper-Goodman and Scott, 1989), and possible role in Balkan endemic nephropathy (Krogh, 1978) (Pohland *et al* 1992).

Health concerns regarding mycotoxins depend on the amount of mycotoxin consumed, the toxicity of the compound, the body weight and physical condition of the individual, the presence of other mycotoxins, and other dietary factors (Kuiper-Goodman, 1991a). The evidence that a certain form of cancer is caused by a mycotoxin, or a human disease is a mycotoxicosis, requires all of the following five criteria being satisfied (Hsieh, 1990b) :

(1) Occurrence of the mycotoxin in food supplies.
(2) Human exposure to the mycotoxin.
(3) Correlation between exposure and incidence.
(4) Reproducibility of characteristic symptoms in experimental animals.
(5) Similar mode of action in humans and animal models.

Of these criteria, the most important evidence indicating that a mycotoxin is involved in the etiology of a form of human cancer is the correlation between the exposure to the mycotoxin in question and the incidence of the disease. The *sufficient evidence* of a chemical agent to be considered as a human carcinogen as specified by IARC is *a positive relationship between exposure to the agent and the occurrence of the cancer in question, in which chance, bias, and confounding could be ruled out with reasonable confidence.*

Documented human mycotoxicoses such as ergotism in Europe, alimentary toxic aleukia in Russia, acute aflatoxicoses in South East Asia, and human primary liver cancer in Africa and South East Asia, confirm the global nature of the problem. The role of ochratoxin A in endemic nephropathy in the Balkans is currently seriously questioned.

Kuiper-Goodman (1991a) mentioned that the most difficult and controversial aspect of the risk assessment, especially for carcinogenicity data, is the extrapolation of toxicological data from animals to humans, using safety factors or other methods, to arrive at an estimate of safe intake. Pre- and cocarcinogens surely may play an important role; the quantification of their effects is a daunting task. In the extrapolation of animal toxicity data to humans, it is important to consider species differences in disposition, such as differences in absorption, as well as differences in binding to plasma and tissue constituents. These latter species differences may affect the mobility of the mycotoxin between different body compartments (organs), and the ultimate target tissue. Species differences in biotransformation and in plasma and tissue half-life are also important.

Kuiper-Goodman (1991a) concluded that overall risk extrapolation from animal data to humans involves :

(1) extrapolation from high doses within the experimental range of animal experiments to low doses, usually outside the experimental range, to which humans might be exposed;

(2) extrapolation from test species to humans; and

(3) extrapolation to the most sensitive subgroup of humans.

Although aflatoxin B_1 (AFB_1) is a potent carcinogenic and genotoxic agent, its potency in humans appears to be about 10-fold lower than in the Fischer rat, based on biotransformation data and on epidemiologic data from several other ecological studies. In the case of AFB_1, other biological factors, such as hepatitis B infection, may also play a role in human tumour development (Kuiper-Goodman, 1991a).

It is evident that it is currently not possible to state with certainty the degree of harm that would occur from the intake of low levels of mycotoxins (Kuiper-Goodman, 1991a). Mycotoxins as a whole have been suspected as a significant class of naturally occurring carcinogens in food (Ames *et al*, 1987) (Gold *et al*, 1992). Since humans normally avoid heavily mouldy foods, human health problems resulting from exposure to acute levels of mycotoxins are relatively rare.

From the complexity seen in the identification of AFB_1 as a causative agent of human PLC, one can appreciate the degree of uncertainty associated with the cause-effect relationship between mycotoxins and human diseases. Therefore, one may recommend that (a) for regulatory agencies, evidence for the involvement of mycotoxins in human diseases needs very careful substantiation to justify massive investment in their regulatory actions; and (b) for an individual household, mouldy foods and poorly controlled fermented foods should be avoided as much as possible (Hsieh, 1990b). In industrialised countries most food is processed in factories. Visual recognition of fungal contamination is therefore difficult.

AFLATOXINS

Aflatoxin is the generic name for the structurally related group of hepatocarcinogenic compounds produced by the moulds *Aspergillus flavus* and *Aspergillus parasiticus* and for their animal metabolites. The fungal-produced compounds are frequent contaminants of food commodities such as corn (maize), peanuts and cotton seed, and other energy-rich foodstuffs. Serious aflatoxin contamination of harvested crops has been reported from all continents. While contamination is popularly associated with poor agricultural practices (frequently involving drought stress, insect damage and mechanical damage (Bullerman *et al*, 1984), modern intensive farming systems at times promote the worst known contamination episodes. Human exposure to AFB_1 is difficult to avoid, owing to its widespread occurrence, especially in parts of the world where the temperature and humidity support the growth of the fungi.

In the early 1960's, experiments with laboratory rats showed that aflatoxin was a potent hepatocarcinogen. This finding led to extensive studies on the carcinogenic properties of the aflatoxins, and much information was gained on their mechanism of action, their occurrence in foods and feeds, and their putative importance as risk factors for PLC in humans (Wogan, 1992a). Ingestion of naturally contaminated feeds, or *per os.* administration of AFB_1 induces PLC in many species of experimental animals, including fish (rainbow trout, sockeye salmon, and guppy), bird (duck and chickens), rodents (rats, mice and tree shrew), a carnivore (ferret) and subhuman primates (rhesus, cynomolgus, African green and squirrel monkeys). The liver is the primary target organ in the different animal species, however, tumours of other organs have also been observed in the aflatoxin-treated species (Wogan, 1992a). It is of importance to note that a wide variation in sensitivity existed between species and within species.

AFB_1 potency in induction of liver tumours also differed widely in subhuman primates (Adamson, 1989). In case of the cynomolgus monkey, tumours in extrahepatic tissues (including adenocarcinomas of the pancreas, adenomas of the biliary and pancreatic ducts, hemangiocarcinoma of the liver, and osteosarcomas) occurred at higher frequency than liver tumours. AFB_1, therefore, may also be a risk factor for cancers other than PLC in humans (Wogan, 1992a).

Aflatoxin and primary liver cancer

According to Bruce (1990) a review of epidemiology literature revealed that only studies conducted in Africa and Asia included adequate data to permit quantitative assessment of the dose-response relationship between aflatoxin exposure and liver cancer rates. Van Rensburg (1985 and 1986) referred to the difficulties encountered with obtaining an accurate assessment of aflatoxin consumption levels; e.g. the amounts of aflatoxin found in the food sampled from the rural grain store or supermarket can increase precipitously or be much reduced prior to consumption. Van Rensburg (1985) then advocated that the only alternative to laborious table-food surveys are actual levels in human tissues and excretions. Researchers studied the presence of aflatoxin metabolites in human urine, human livers and in human blood serum (Van Rensburg, 1985).

The epidemiological studies of Van Rensburg *et al* (1985) established a strong statistical association between HCC incidence and aflatoxin consumption in several populations in Africa and Asia. Original work was done by Van Rensburg *et al* (1985) in Mozambique and the Transkei region; results from Kenya, Swaziland and Thailand were summarized. The studies in Africa were derived from aflatoxin exposures based upon an extensive sampling of prepared meals. Under-registration of cancer cases did pose a problem in the studies of Van Rensburg.

Linsell and Peers (1977) described cancer incidence data from Swaziland and Kenya as well as aflatoxin exposure based upon the analysis of food; the quality of cancer data was again suspect.

Peers *et al* (1987) extended the study in Swaziland and included newer data as well as information on the presence of HBV markers in blood of blood-bank donors. Peers *et al* (1987) claimed that their analysis showed aflatoxin exposure to be a more important determinant of the variation in liver cancer incidence than HBV exposure. Shank *et al* (1972a and 1972b) described efforts to compare aflatoxin and liver cancer rates in two selected regions of Thailand.

Stoloff (1983) compared the mortality ratio for liver cancer (ratio of liver cancer deaths to all deaths) for rural white males between the Southeast area of the United States and the "North and West". The mortality ratio for this group is about 10% higher in the Southeast than in the North and West. Stoloff (1983) estimated the aflatoxin exposure for these groups of people (rural white males) over their life spans using data on aflatoxin contamination of corn and peanut products. He found that the group for the Southeast showed excess of liver cancer, and concluded that the excess liver cancer found in the group was *far from the manyfold difference that would have been anticipated from experiments with rats and from prior epidemiological studies in Africa and Asia*. In a more recent study Stoloff (1989) reported that most of the epidemiological studies in the 1980's of aflatoxin and PLC were either in the United States, where HBV-infected groups could be excluded from the study, or when in areas of chronic HBV infection, attempts were made to include that factor. The study of United States populations showed no difference in mortality rates from PLC that could be attributed to aflatoxin exposure. The studies of populations with endemic HBV infection produced no convincing evidence to support a primary role for aflatoxin in the induction of human PLC, although an accessory role to HBV infection for aflatoxin could not be ruled out.

In an extensive study of Yeh *et al* (1989) a cohort of 7 917 Chinese men, age 25 to 64, and judged to be free of liver cancer, were followed for 3 to 4 years to determine the number of deaths due to liver cancer. In this study four areas of different HCC incidences were chosen to compare correlations between the disease and HBV infection or aflatoxin exposure as a risk factor. While a very strong linear dose-response relationship was observed between HCC incidence and aflatoxin exposure, no correlation was found between the disease and HBV infection.

In a recent study, Ross *et al* (1992) employed assays for urinary aflatoxin B_1, its metabolites AFP_1 and AFM_1, and DNA-adducts (AFB_1-N^7-Gua) to assess the relation between aflatoxin exposure and liver cancer. The study involved 18 244 middle-aged men in Shangai, People's Republic of China. Subjects with liver cancer were more likely than were controls to have detectable concentrations of any of the aflatoxin metabolites, particularly aflatoxin P_1. Ross *et al* (1992) concluded that their study provided the most direct evidence for an etiological role of aflatoxin in hepatocellular carcinogenesis, in fact that 50% of liver cancer cases in Shangai are related to aflatoxin exposure. Supporting results were obtained by Groopman's group in establishing the urinary aflatoxin - DNA adducts in human urine in Gambia, West Africa (Groopman *et al* 1992a) and in Guangxi autonomous region, People's Republic of China (Groopman *et al* 1992b).

Although IARC (1987) decided that aflatoxin is a human carcinogen (the classification was confirmed by re-evaluation in 1992), this decision is still questioned (Hsieh, 1990) on (i) the high uncertainties in the dose and response databases; (ii) wide species difference in sensitivity to aflatoxin carcenogencity and (iii) the strong evidence that hepatitis B virus (HBV) is the causative agent of HCC. Aflatoxin is also not found in the diets in regions of Alaska where HCC incidences are high. From the foregoing evidence it is clear that there is no unanimity on the role of aflatoxin in HCC.

The need to limit aflatoxin exposure is based on two major concerns :

(1) The adverse effects of aflatoxin contaminated commodities on human and animal health.

(2) The presence of aflatoxin residues or toxic metabolites in animal tissues used as human foods.

There is a general consensus that human and animal exposure to carcinogens should be at the lowest practical level. Furthermore, there is ample epidemiological evidence that humans are not immune to acute aflatoxicosis, with reported major poisonings in India where 74 people died and Kenya with 12 deaths (Park, 1992a) (Park and Stoloff, 1989).

Aflatoxin control

Mycotoxin control programmes, i.e. risk management, include establishment of regulatory limits or guidelines, monitoring programmes for mycotoxins in susceptible products, and decontamination procedures and/or diversion of contaminated products to lower risk uses (Park, 1992a). Physical removal of discoloured, damaged or inadequately developed kernels is the technique mostly used in the peanut industry. The approach taken with corn or cottonseed involves chemical inactivation of the toxin. Ammonia or ammonia-related compounds appear to have the most practical application for the decontamination of aflatoxin in most agricultural commodities (Park, 1992a). The severity of aflatoxin contamination in selected agricultural commodities worldwide led to the approval and application of ammonia decontamination (Park, 1992a and 1992b). The current aflatoxin levels established by the US Food and Drug Administration are shown in Table 1 (Park, 1992a). Excellent methods are currently available for the determination of aflatoxins in contaminated products (Park *et al*, 1990) (Dell *et al*, 1990).

Stoloff *et al* (1991) evaluated the rationales for the establishment of limits and regulations for mycotoxins by assessing the situation in 50 countries. Internationally, over 50 countries have enacted or proposed regulations for control of aflatoxins in food and feed; 15 of these countries also have regulations for permitted levels of mycotoxins other than aflatoxin, e.g. patulin, ochratoxin A,

ergotamine, ergotixine, and ergoxine groups. Ergotamine, ergostine, ergocristine and ergocornine are among the most pharmacologically active peptides and are the main alkaloids of *C. purpurea* (Lacey 1991) (Stoll and Hoffmann, 1943).

CITREOVIRIDIN

A disease known as cardiac beri-beri or *shoshin kakke* occurred in Japan and Asian countries for over three centuries; it was investigated from the standpoint of being either an infection, an avitaminosis, or an intoxication. The disease was characterized by convulsions, paralysis, and respiratory arrest (Richard 1990).

In 1891 it was shown that an ethanol extract of naturally contaminated rice caused neurotoxic effects in mice, frogs and rabbits. Subsequently a toxinogenic fungus *Penicillium citreoviride* was isolated from *yellow rice*, and Sakabe *et al* (1964) characterized citreoviridin. Nagel *et al* (1972) and Steyn *et al* (1982) studied the biosynthesis of citreoviridin employing radio-active and stable isotopes, respectively. Ueno and Ueno (1972) established the causal relationship of citreoviridin to cardiac beri-beri. At present, cardiac beri-beri is clearly not an important disease, owing to improved inspection of rice quality and the consumption of vitamin-supplemented food. However, citreoviridin still remains as a potential causal agent owing to its recognized occurrence in corn and other food and feedstuffs (Richard 1990), and its production by a number of *Penicillium* and *Eupenicillium* species (Franck and Gehrken, 1980) (Pitt 1979). Wicklow *et al* (1988) investigated corn infected with *Eupenicillium ochrasalmoneum* and detected citreoviridin in bulk samples from five of eight Georgia fields in amounts varying from 19 to 2 790 ug/kg.

Nishie and co-workers (1988) studied the toxicity of citreoviridin in mice and rabbits. The major effects of the mycotoxin were catalepsy, hypothermia and hypotension; the targets of the toxin were the respiratory and cardiovascular systems. Ueno (1971) showed that, in dosed rats, the liver contained the greatest concentrations of citreoviridin.

TRICHOTHECENES

The trichothecenes comprise a group of closely related chemical compounds designated *sesquiterpenoids*. All the naturally occurring toxins contain an olefinic bond at C-9, 10 and an epoxy group at C-12,13; the latter characterizes them as 12, 13-epoxytrichothecenes. Dangerous levels of trichothecene mycotoxins can occur in mouldy grains, cereals and agricultural products (ApSimon *et al*, 1990).

The trichothecenes are produced by various species of *Fusarium*, *Trichoderma*, *Cephalosporium*, *Verticimonisporium* and *Stachybotrys*.

The genus *Fusarium* contains important mycotoxin-producing species (Marasas *et al* 1985). Toxinogenic *fusaria* have been implicated in human health diseases such as alimentary toxic aleukia (ATA) (Yagen *et al* 1977), Kashin-Beck disease, akakabi-byo (scabby grain intoxication) and oesophageal cancer, as well as in a number of animal diseases such as skin toxicity, bone marrow damage, haemorrhagic and oestrogenic syndromes (zearalenone), and equine leukoencephalomalacia (fumonisins). The trichothecenes gained some notoriety in the 1980s when it was alleged that these toxins were employed as components of the chemical warfare agent Yellow Rain (Rosen and Rosen, 1982) and (Haig, 1982).

ATA : Over the period 1942 to 1947 more than 10% of the population in the Orenburg district, close to Siberia, were fatally affected by consuming overwintered millet wheat and barley (Ueno (1980), Wannemacher *et al* (1991)). Yagen *et al* (1977) reported that a disease whose symptoms are similar to those of ATA were in fact reported in Russia from the beginning of the 19th century. ATA has become associated with the consumption of food made from grain which remained unharvested under snow and became mouldy from contamination with a variety of micro-organisms. Symptoms of ATA include vomiting, diarrhoea, skin inflammation, leukopenia, multiple haemorrhage, necrotic angina, sepsis, and exhaustion of bone marrow.

Ueno *et al* (1972) demonstrated that T-2 toxin was the likely agent in ATA. Yagen *et al* (1977) also concluded from studies employing cats as experimental animals that the trichothecene mycotoxin, T-2 toxin, was responsible for an ATA-like syndrome in cats.

The trichothecenes have been involved in a number of intoxications. Therefore, a large body of literature has been developed on the toxicology, pathology, and potential therapies for acute trichothecene toxicosis in a variety of laboratory species (Wannemacher *et al* 1991).

Wannemacher *et al* (1991) excellently reviewed the acute toxicity of trichothecene mycotoxins in rats and mice, as well as the histological findings in mice after acute exposure to T-2 toxin. T-2 toxin is the most potent of the simple trichothecenes that have been found to date (Wannemacher *et al* 1991), and leads to necrosis in the skin of mice 6 hours after exposure, and to inflammation within 12 hours.

The intense international interest in the analysis and natural occurrence of mycotoxins produced by *Fusarium* species, in particular the trichothecenes, testifies to the importance of this group of toxins to human and animal health (ApSimon *et al* 1990). Of the simple trichothecenes, the ones most frequently detected in agricultural commodities are T-2 toxin, diacetoxyscirpenol (DAS), nivalenol, and 4-deoxynivalenol (DON). (Tanaka *et al* 1985, 1986, 1988, 1990) (Lee *et al* 1986) (Gilbert *et al* 1992a and 1992b) (Ueno *et al* 1986) (Hagler *et al* 1984) (Abrahamson *et al* 1987) (Blaney *et al* 1987) (Faifer *et al* 1990) (Osborne and Willis, 1984) (Lauren *et al* 1991) (Shotwell *et al* 1985) (Perkowski *et al* 1990).

Rabie *et al* (1986) reported the first T-2 toxin producing *Fusarium* species, viz *F. acuminatum*, from Africa in their investigations of oats and barley. The finding that large amounts of T-2 toxin were produced at relatively high temperatures (25°C) is significant, owing to the optimum temperature previously reported for the production of T-2 toxin as comparatively low (8 to 14°C). In this study, *F acuminatum* MRC 3936 produced T-2 toxin (2 600 mg/kg corn) at 25°C.

Sapa-Associated Press (February 19, 1993) reported that poisoned wheat has killed 24 people and is threatening thousands more in war-torn southern Tajikistan. "The victims died after eating bread made from the tainted wheat, said members of the International Committee of the Red Cross in the neighbouring Central Asian state of Uzbekistan. Most died during the past two months from internal bleeding and liver complications. About 1 600 people remained in hospitals. A further 2 000 victims were believed to be unable to obtain any medical help. - **Sapa-AP**". It is possible that the '*tainted wheat*' referred to mould-spoiled wheat, and that humans were dying from trichothecene-related mycotoxicoses.

OCHRATOXINS

The ochratoxins were the first group of naturally occurring mycotoxins discovered after the aflatoxins (Van der Merwe *et al*, 1965). The ochratoxins formed the substance of many reviews, [Steyn (1984), Harwig *et al* (1983), Kuiper-Goodman and Scott (1989), Pohland *et al* (1992), as well as the thesis of Holmberg (1992)].

The ochratoxins comprise a polyketide-derived dihydroisocoumarin moiety linked via its 12-carboxy group by an amide bond to L-β-phenylalanine. Ochratoxin A and its ethyl ester are the most toxic compounds. Ochratoxin B, the dechloro derivative of ochratoxin A, is essentially non-toxic.

The ochratoxins were at first discovered as metabolites of *Aspergillus ochraceus* (Van der Merwe *et al*, 1965). Since then, ochratoxin A has been found to be produced by a number of both *Aspergillus* and *Penicillium* species, as shown in Table 2.

Aspergillus species are associated with ochratoxin A production in tropical areas, whereas ochratoxin producing *Penicillium* species thrive and can produce ochratoxin A in a colder climate with temperatures as low as 5°C. For several years *Penicillium viridicatum* was regarded as the main ochratoxin A producer. Ochratoxin A became regarded as a very important mycotoxin since it plays a major role in the nephropathy occurring in swine (mycotoxic porcine nephropathy also referred to as Danish porcine nephropathy) and poultry.

TABLE 2 : SPECIES OF ASPERGILLUS AND *PENICILLIUM* WHICH HAVE BEEN REPORTED TO PRODUCE OCHRATOXIN A (Holmberg 1992)

ASPERGILLUS	*PENICILLIUM*
*A.ochraceus (A.alutaceus) group:*1)	*P.verrucosum (P.viridicatum)* 2)
A.ochraceus (A.alutaceus)	
A.alliaceus	Other Penicillia:
A.elegans	*P.chrysogenum*
A.fresenii	*P.commune*
A.melleus	*P.cyclopium*
A.ostianus	*P.expansum*
A.petrakii	*P.palitans*
A.sclerotiorum	*P.purpurescens*
A.sulphureus	*P.nordicum*
	P.variabile
	P.verruculosum

1) *A.ochraceus* is also referred to as *A.alutaceus*.
2) Ochratoxin A producing groups (II and III) of *P.viridicatum* are now referred to as *P.verrucosum* (Pitt, 1987).

Samson *et al* (1976) proposed that *P. viridicatum* should belong to a single species of *P. verrucosum var. verrucosum*. Among the common grain, food and feed-related species in subgenus *Penicillium*, only *P. verrucosum* is now considered to produce ochratoxin A (Pitt 1987). Excellent methods are available for the analysis of ochratoxin A (Sharman *et al* 1992) (Nesheim *et al* 1992).

The occurrence of ochratoxin A in several plant and animal products has been extensively reviewed (Kuiper-Goodman and Scott, 1989). Ochratoxin A contamination is typically associated with grain stored in the temperate climate of Europe and North America. There are big variations in ochratoxin A content of the grains from year to year, caused by differences in weather conditions. Most positive samples are found in Denmark and Sweden, and the highest values measured among rye and bran products (Olsen *et al* 1991).

The kidneys are the organs most susceptible to ochratoxin A; it can cause both acute and chronic kidney lesions. Ochratoxin A principally acts on the first part of the proximal tubules in the kidney and induces a defect on the anion transport mechanism on the brush border of the proximal convoluted tubular cells and basolateral membranes (Endou *et al*, 1986). In addition, the immunotoxic and teratogenic effects of ochratoxins are well-established.

Ochratoxin A is also reputed to be a carcinogen. Administration of ochratoxin A over a two-year period in doses of 3 500 to 6 000 ug/kg body weight/day to mice and 70 ug/kg body weight/day to rats, proved it possible to provoke tumours, primarily kidney tumours (Olsen *et al*, 1991).

The primary toxic effect of ochratoxin A is inhibition of protein synthesis (Röschenthaler *et al*, 1984; Kuiper-Goodman and Scott, 1989). The effect is specific and involves a competitive inhibition of phenylalanine $tRNA^{Phe}$ synthetase which is necessary for fundamental amino-acylation steps in protein synthesis. In mice and cell cultures it has been shown that the competitive inhibition and toxicity can be reduced by administration of phenylalanine (Creppy *et al* 1980). It is furthermore of importance to note that bovine carboxypeptidase hydrolyses ochratoxin B almost a hundred-fold faster than ochratoxin A; this characteristic is utilized in the determination of ochratoxin B in the presence of ochratoxin A (Hult *et al* 1979).

Ochratoxin A and human health

Ochratoxin A has clearly been shown to be a toxic substance, with nephrotoxic, immunosuppressive, teratogenic and carcinogenic effects in many species.

Evidence has emerged that humans are being exposed to ochratoxin A (Krogh, 1978) (Olsen *et al* 1991), in that measurements taken on blood serum collected from six countries have shown quantities up to 40 ng ochratoxin A/ml serum (Kuiper-Goodman and Scott, 1989). Big fluctuations were observed in the serum levels, indicating the local, seasonal and annual fluctuations in the ochratoxin A content in food products. The highest value measured in blood serum is 1 800 ng/ml.

In a recently completed Swedish study it was found that in the region where the highest values were measured, 30% positive samples were found with a mean plasma concentration of 0,06 ng/ml (Breitholtz *et al* 1991). The highest plasma concentration measured on Gotland corresponds to an intake of 9,6 ng ochratoxin A/kg body weight/day (Hagelberg *et al* 1989). Kuiper-Goodman (1991b) reported on recent studies in Manitoba, Canada that indicated the presence of low levels of ochratoxin A in the human serum of about 40% of persons sampled. The occurrence of ochratoxin A in human serum can be attributed to the probable long half-life of ochratoxin A.

Balkan Endemic Nephropathy (BEN) is a serious, chronic kidney disease, which is found in high frequency in certain areas of the Balkan countries, such as the Vratza region of Bulgaria which is a hyperendemic area (rural agricultural communities which have traditionally produced and stored much of their own food). BEN seems to require residence of at least 10 to 15 years in an endemic village, and most people are not diagnosed clinically until into middle age. No clear picture of the early histopathology of the disease has been described.

Some findings indicate that ochratoxin A may be a factor in the etiology of Balkan endemic nephropathy, a disease in which there appears to be a highly significant relationship with tumours of the urinary tract (particularly tumours of the renal pelvis and ureters). This evidence includes (1) surveys of foodstuffs in Yugoslavia and Bulgaria which indicate a higher level of ochratoxin A contamination in endemic areas than in non-endemic areas; (2) a striking similarity between renal pathology associated with Balkan endemic nephropathy and renal pathology in field cases of porcine nephropathy induced by ochratoxin A; and (3) the finding of higher levels of ochratoxin A in blood of patients with urinary system tumours than in blood of healthy persons from the same area (Bulgaria). However, data on human blood levels of ochratoxin A should not be interpreted as proof of the causality of human cancer by ochratoxin A, given the normally accepted long latency period between exposure and the appearance of cancer in humans (Pohland *et al* 1992).

In an extensive study Cvetnic and Pepeljnjak (1990) assessed the ochratoxinogenicity of *Aspergillus ochraceus* strains from nephropathic and nonephropathic areas in Yugoslavia. In this study 855 samples of stored grains and dried meat were collected from households in nephropathic and non-nephropathic areas in Croatia in Yugoslavia; 10% of the samples were contaminated with *Aspergillus ochraceus*. 37% ochratoxin A producers were found when 70 *A.ochraceus* strains were analysed. Cvetnic and Pepeljnjak (1990) found no significant difference in the toxinogenicity of *A.ochraceus* strains between the nephropathic and non-nephropathic areas.

Several researchers consider the nephrotoxicity of *Penicillium aurantiogriseum* as a possible factor in the aetiology of BEN (Yeulet *et al* 1988; Mantle *et al* 1991a and 1991b). *P.aurantiogriseum* does not produce citrinin or ochratoxin A, however, it causes histopathological changes in proximal tubules at the renal cortico-medullary junction of rats. The effect was seen as possibly offering an experimental model for the human disease. The phenomenon is more prominent than that induced by ochratoxin A over a similar period, raising the question of the putative role of ochratoxin in the aetiology of chronic human renal disease. The toxic factor was shown to be a water-soluble amphoteric fraction of the fungus (Yeulet *et al* 1988).

Mantle *et al* (1991a and b) concluded that, although it is not yet possible to quantify the *P.aurantiogriseum* nephrotoxin that induces karyomegaly, and thereby to compare its specific activity to that of ochratoxin A, it is possible that *P.aurantiogriseum* is much more potent than ochratoxin A in inducing karyomegaly, and is much more target-specific.

In a recent study Mantle and McHugh (1992) investigated the *Penicillium* spp. and *Aspergillus* spp. associated with 83 samples of foods from Bulgarian households, most of which had a history of BEN. This study strengthened the previous findings of Mantle *et al* (1991b) on the importance of *P. aurantiogriseum* in the aetiology of BEN, namely that *P. aurantiogriseum* was the most common of species of the *Penicillium* genus, and that all the *P. aurantiogriseum* isolates tested were extremely nephrotoxic in the rat. *P. verrucosum*, a frequent producer of ochratoxin A, was rarely isolated, and furthermore all its isolates failed to produce ochratoxin A in pure culture. Several isolates of *P. citrinum* produced abundant quantities of citrinin, however, these had no effect on rat kidneys. In conclusion, Mantle and McHugh (1992) stated that the natural occurrence of ochratoxin A in food in the region was ascribed mainly to fungi which do not fit the current concept of ochratoxin producing *penicillia*.

Mantle and McHugh (1992) expressed the view that the low general natural incidence both of ochratoxinogenic fungi and also of ochratoxin A in foods of Bulgarian households do not support the popular epidemiological role of ochratoxin A in BEN.

The conformation of the ochratoxins

The conformation of the ochratoxins may play a dominant role in their interaction with enzymes, and this may have a bearing on the rates of hydrolysis and in turn on the difference in toxicities. The amide carbonyl could compete with the lactone carbonyl for hydrogen-bonding involving the phenolic hydroxy-proton, should the amide carbonyl and phenolic hydroxy groups be *syn* oriented.

The conformations of ochratoxin A and B in the solid state were determined by X-ray crystallography : both substances crystallize in the space group $P2_1$ (Bredenkamp *et al* 1989).

Only very minor differences were observed for the torsional angles of the side-chain, and no significant differences in bond lengths and angles were observed for these two toxins. In both cases the amide carbonyl and phenolic hydroxy groups are *anti* with respect to each other and only the phenolic hydroxy group is involved in hydrogen-bonding with the lactone carbonyl group. The interoxygen distances are 2,542 Å and 2,562 Å for ochratoxins A and B, respectively. The amide carbonyls of both toxins are involved mainly in hydrogen-bonding with the hydrogen atom of the acid group.

Ochratoxin B was submitted to the MM2 force-field calculation of Allinger by using the X-ray co-ordinates as inputs to investigate the possible low energy conformations of the side chain. The N-13 -- C-14 -- C-15 -- C-16 torsional angle was rotated 360° in steps of 10°, and the energy was minimized after each step. A global energy minimum was found at -60° compared to the solid state value of -63,4°.

Infrared spectroscopic studies of ochratoxins A and B in solution and solid state also point to a β form conformation, i.e. the proton of the phenolic hydroxy-group is involved in hydrogen-bonding with the lactone carbonyl group. This observation was confirmed by the observed three-bond coupling constant for the hydroxy-group proton to C-7 and C-9 in ochratoxins A and B (Bredenkamp *et al* 1989).

TREMORGENIC MYCOTOXINS

An inspection of the various mycotoxin structural types reveals nitrogen to be a fairly common feature; it is frequently derived from amino-acids, albeit occasionally in modified form. In the case of the tetracyclines and viridicatumtoxin, the amide nitrogen is likely derived from an ammonia source (De Jesus *et al*, 1982). The fumonisins, a novel group of aliphatic mycotoxins isolated from *Fusarium moniliforme* Sheldon, contain an amino group most likely derived from ammonia (Bezuidenhout *et al*, 1988).

Amino-acids form the backbone of the toxin rhizonin A, an unique cyclic heptapeptide from *Rhizopus microsporus* van Tieghem (Steyn *et al*, 1983). The molecule contains three *N*-methyl amino acids as well as two pairs of like amino acids with opposite α-carbon stereochemistry, i.e. *L*-valine and *D*-valine, as well as *N*-methyl-3-(3-furyl)-*L*-alanine and *N*-methyl-3-(3-furyl)-*D*-alanine. The liquid and solid conformation of rhizonin was investigated by Potgieter *et al* (1989).

Phompsin A, the main mycotoxin from cultures of *Phomopsis leptostromiformis* and the cause of lupinosis, is a linear hexapeptide containing 3-hydroxy-L-isoleucine, 3,4-didehydrovaline, *N*-methyl-3-(3-chloro-4,5-dihydroxyphenyl)serine, *E*-2,3-didehydroaspartic acid, *E*-2,3-didehydroisoleucine, and 3,4-didehydro-L-proline (Culvenor *et al*, 1983). An X-ray crystallographic study of phomopsin A confirmed the amino acid sequence and showed that the linear hexapeptide is modified by an ether bridge in place of the 5-hydroxy group of the *N*-methyl-3-(3-chloro-4,5-dihydroxyphenyl)serine and the hydroxy group of the 3-hydroxyisoleucine units (Culvenor *et al*, 1989).

The cyclopiazonic acids, metabolites from *Penicillium cyclopium* Westling, were isolated and characterized by Holzapfel and Wilkins (1971). The bio-origin of α-cyclopiazonic acid via L-tryptophan, L-cycloacetylacetyl tryptophan and β-cyclopiazonic acid was established by De Jesus *et al* (1981b), and Steyn *et al* (1975). In a related tetramic acid, tenuazonic acid, the nitrogen group is derived from isoleucine.

Tryptophan is, in fact, a common constituent of many secondary metabolites, e.g. α-cyclopiazonic acid, as well as of several affecting the central nervous system, such as the ergot alkaloids and the tremorgenic toxins, *viz.* fumitremorgens A and B, verruculogen, and TR2. In the structurally related substances, the brevianamides and austamides, tryptophan and proline contribute the dioxopiperazine part of the molecules.

L-Tryptophan and L-histidine are the precursors of the dioxopiperazines oxaline and roquefortine, metabolites of *Penicillium oxalicum* and *Penicillium roqueforti*, respectively. The bioformation of oxaline, via roquefortine and involving several complex molecular rearrangements, was studied by Steyn and Vleggaar (1983).

Two tremor-producing tetrapeptide metabolites, tryptoquivaline and tryptoquivalone, were isolated from *Aspergillus clavatus*. Tryptophan is again a building block of the tryptoquivalines in addition to anthranillic acid, valine, and methylalanine.

The penitrems, the most important group of tremorgenic mycotoxins, were discovered in 1968, however, their structural elucidation remained a great challenge owing to the lack of adequate quantities of the toxins and their extreme acid lability. In their structural investigations, De Jesus *et al* (1981a), (1983), (the findings on the penitrems were obtained from these references) extensively utilized n.m.r. spectroscopic techniques, linked to biosynthetic studies employing stable isotopes. Steyn and Vleggaar (1985) reviewed their findings on the structures of penitrems A-F.

P. crustosum (Sol-7) was grown in stationary culture at 25°C in Czapek medium enriched with 2% yeast extract. The mixture of penitrems was purified by chromatography on silica gel using benzene-acetone (85:15, v/v) to yield six penitrems : penitrem A-F. The similarity in the molecular formulae (mass spectroscopy), indicates that the penitrems share the same basic structure and differ from each other only in the nature of substituents at certain carbon atoms, e.g. the molecular formulae of penitrems A and E indicate that chlorine is replaced in the latter by hydrogen.

Penicillium janthinellum Biourge was studied owing to its potential role in ryegrass staggers. De Jesus *et al* (1984) isolated three tremorgens, janthitrems E, F, and G, as colourless amorphous solids from a modified Czapek medium by chromatography on silica gel. The molecular composition of janthitrems E, F, and G (mass spectrometry) is $C_{37}H_{49}NO_6$, $C_{39}H_{51}NO_7$ and $C_{39}H_{51}NO_6$, respectively. The similarity in the molecular formulae and the biological action of the janthitrems and the penitrems strongly indicated closely related structural features. Absorptions at λ_{max}228 (ϵ17 700), 258sh (ϵ27 300), 265 (ϵ30 000), and 330 nm (ϵ17 000) in the u.v. spectrum of the janthitrems, e.g. janthitrem E suggested the presence of a 2,3-disubstituted indole nucleus with a double bond in conjugation with the indole.

Lolitrems B and C were isolated by Gallagher *et al* (1984) from perennial ryegrass (*Lolium perenne L.*) infected with an endophytic fungus, viz. an *Acremonium* species. In New Zealand this type of ryegrass is associated with a neurotoxic syndrome affecting sheep, cattle and horses. The molecular formula of lolitrem B, $C_{43}H_{45}NO_7$ indicated the presence of an additional C_5 unit compared with the penitrems and janthitrems.

A brief survey of the structural properties of the fungal tremorgens, viz. penitrems, janthitrems, lolitrems, aflatrem, paxilline, paspaline, paspalicine, paspalinine and paspalitrems A and B, reveal their close biogenetic relationship. In the case of aflatrem and paspalitrems A and B, a C_5 unit is attached to the paspaline type structure (Steyn and Vleggaar, 1985).

FUMONISINS

Fusarium moniliforme Sheldon, a common contaminant of corn throughout the world (Booth 1971), has been implicated in various animal and human diseases (Marasas *et al* 1984, 1988a, 1988b). Strains of the fungus have been shown to be highly toxic and carcinogenic (Gelderblom *et al* 1988).

Contamination of corn with *F. moniliforme* has been associated with human oesophageal cancer in the Transkei part of Southern Africa (Marasas *et al* 1988a), and in China (Yang 1980), as well as with field outbreaks of equine leukoencephalomalacia (ELEM) in many countries such as Egypt, South Africa and the United States of America (Marasas *et al* 1988b) and pulmonary edema in swine (Ross *et al*, 1990). ELEM is a fatal neurological disease of horses, characterized by liquefactive necrosis of the white matter of the brain. ELEM has been experimentally induced in horses by either supplementing their diets with *F. moniliforme*-contaminated corn or by the oral administration of fumonisin B_1(FB_1), a toxin produced by *F. moniliforme* (Kellerman *et al* 1990). Poisoning of horses by FB_1 is unusual in that both the brain and liver lesions are caused by the same toxin; current evidence shows that small doses of the toxin (culture material) over long periods culminate in ELEM, whereas high doses over short periods result in hepatosis (Kellerman *et al* 1990). Porter *et al* (1990) observed in rats fed corn contaminated with *F. moniliforme*, increases in 5-hydroxyindoleacetic acid (5-HIAA, major metabolite of serotonin, 5-HT) and of the 5-HIAA/5-HT ratios. The results suggest a fumonisin-induced dysfunction in either 5-HT metabolism or 5-HIAA elimination in rat brains. Wang *et al* (1991) described the inhibition of sphingolipid biosynthesis by fumonisins and the implications for diseases associated with *F. moniliforme*. The sphingolipids and fumonisins are structurally closely related.

The health importance of *F. moniliforme* necessitated concerted efforts to isolate and characterize the toxins involved, as well as the development of analytical methodology to assess the level of contamination of natural foods and feeds by the toxin(s). Earlier studies led to the isolation of moniliformin and the fusarins (Gelderblom *et al* 1984), the latter a group of potent mutagens from *F. moniliforme*. Results obtained on the fumonisins will be discussed in greater detail since corn comprises the staple food of the people in several developing countries, particularly in Southern Africa. Thiel *et al* (1991) thus investigated the ability of a number of *Fusarium* species to produce the fumonisins, and showed that, in addition to *F. moniliforme*, that *F. proliferatum* and *F. nygamai* also produce fumonisins in culture (see Table 3).

TABLE 3 : FUMONISIN PRODUCING *FUSARIUM* ISOLATES

	No positive	Fumonisins (ug g^{-1})	
		FB_1	FB_2
SECTION LISEOLA			
F. moniliforme	(7/7)	85 - 7100	0 - 3000
F. proliferatum	(4/4)	20 - 870	65 - 450
F. subglutinans	(0/1)	0	0
F. anthophilum	(0/1)	0	0
NEW SPECIES			
F. nygamai	(1/2)	605	530
F. napiforme	(0/1)	0	0

Numbers tested in brackets (Thiel *et al* 1991)

Wet sterilized maize meal was inoculated with cultures of *F. moniliforme* (MRC 826) to produce mouldered material toxic to rats and ducklings. The cultures were extracted with aqueous methanol to give an extract which was active in a rat liver histopathological bio-assay; this bio-assay was used in the subsequent fractionation of the extract. Fractionation of the extract using macroreticular polystyrene (XAD-2), Sephadex LH-20, and reversed phase silica gel chromatography, resulted in the isolation of a mixture of fumonisins. Subsequent purification involved treatment of the fumonisins with excess of diazomethane to yield colourless oils of the tetramethyl esters (Bezuidenhout *et al* 1988). Cawood *et al* (1991) reported a quantitative approach to the isolation of the fumonisin mycotoxins.

The structural analysis of the fumonisins involved application of liquid secondary ion mass spectrometry of tetramethylfumonisin A_1 to give a protonated molecular ion at $^m/_z820$ $(M+H)^+$. The interpretation of data from ^{13}C n.m.r spectra of tetramethylfumonisin A_1 led to a suggested empirical formula of $C_{40}H_{69}NO_{16}$ for tetramethylfumonisin A_1; acetylation of this substance gave a triacetate $C_{46}N_{75}NO_{19}$ (M^+945). Basic hydrolysis of this compound gave a neutral nitrogen-containing substance and propane-1,2,3-tricarboxylic acid.

FB_1 and FB_2 were isolated by using a cancer promotion bio-assay (Gelderblom *et al*, 1988), and were identified by detailed application and analysis of M.S., ^1H and ^{13}Cn.m.r. data. The C-10 hydroxy group of FB_1 is replaced by a hydrogen atom in FB_2 (analogous to fumonisins A_1 and A_2)

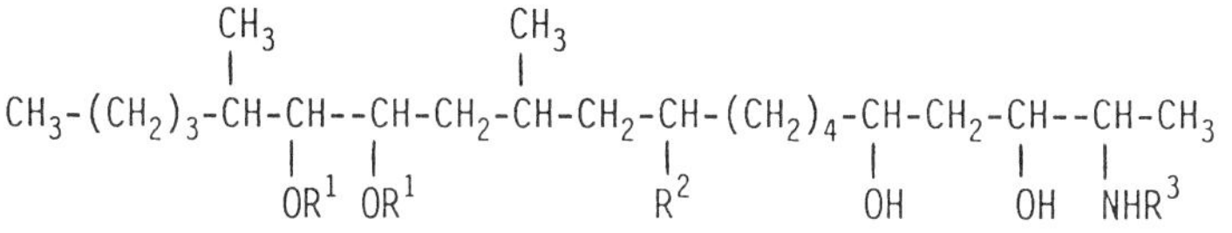

Fumonisin B_1 : $R^1 = COCH_2CH(CO_2H)CH_2CO_2H$, $R^2 = OH$, $R^3 = H$
Fumonisin B_2 : $R^1 = COCH_2CH(CO_2H)CH_2CO_2H$, $R^2 = R^3 = H$

Chemical structures of fumonisins B_1 and B_2

The analytical method of Shephard *et al* (1990) for the determination of FB_1 and FB_2 was selected as the method of choice by the Food Chemistry Commission of the International Union of Pure and Applied Chemistry (IUPAC) for an interlaboratory collaborative study. Several fungi have the ability to produce mycotoxins under laboratory and natural conditions. However, a toxin only assumes real importance once it is detected in nature at such levels as to be of importance in the etiology of human and animal diseases. Several research groups recently found both fumonisins B_1 and B_2 in alarmingly high levels in animal feeds. Plattner *et al* (1990) also described a method for the detection of fumonisins in corn samples associated with field cases of ELEM.

TABLE 4 : MYCOTOXINS IN MOLDY CORN FROM THE TRANSKEI (1985)

Mycotoxin (ug g^1)	Low OC area	High OC area
Moniliformin	3,5	0,8
Zearalenone	1,2	0,4
Nivalenol	4,6	1,8
Deoxynivalenol	2,9	0,3
FB_1	6,5	23,9
FB_2	2,5	7,6

Means of 12 samples/area Sydenham *et al* (1990b)

The high oesophageal cancer rate of the people living in Transkei stimulated many investigations. Van Rensburg (1985) implicated several factors, including dietary deficiencies and or exposure to environmental carcinogens, in the aetiology of this disease. Sydenham *et al* (1990a) found FB_1 as a natural contaminant of corn and Sydenham *et al* (1990b) also investigated the natural occurrence of some *Fusarium* mycotoxins in corn from low and high oesophageal cancer (OC) prevalence areas in

Transkei. Moldy and healthy corn samples were collected from the different cancer rate areas and screened mycologically. The moldy corn samples were analyzed for the presence of several *Fusarium* mycotoxins, including FB_1 and FB_2. The healthy corn samples were screened for the presence of FB_1 and FB_2. No T_2-toxin or diacetoxyscirpenol could be detected in the samples. High concentrations of the other toxins were recorded in moldy corn samples. The data indicate that significantly higher levels of FB_1 and FB_2 were present in both the healthy and the moldy corn samples from the high oesophageal cancer rate area than in corresponding samples from low-rate areas (Sydenham *et al* 1990b) (Tables 4 and 5). Sydenham *et al* (1990b) concluded that the possible role of the fumonisins in the etiology of oesophageal cancer merits further investigation.

TABLE 5 : FUMONISINS IN HEALTHY CORN FROM THE TRANSKEI (1985)

Mycotoxin (ug g^1)	Low OC area	High OC area
FB^1	0,06	1,6
FB_2	<0,05	0,5

Means of 12 samples/area Sydenham *et al* (1990b)

Park *et al* (1992) investigated the mutagenic potential of fumonisin-contaminated corn following the ammonia decontamination procedure.

CONCLUSION

The past ten years witnessed several accomplishments in the structure elucidation of mycotoxins which are reputedly involved in mycotoxicoses among humans and animals, e.g. phomopsin A (Culvenor *et al* 1989); the tremorgenic mycotoxins such as the penitrems, janthitrems and lolitrems (Steyn and Vleggaar 1985), and the fumonisins (Bezuidenhout *et al* 1988). The real importance of mycotoxins as environmental contaminants has been established by sophisticated analytical methodology, based on physical methods (Plattner *et al* 1989) (Roch *et al* 1992) and immunoassays (Morgan, 1989) (Park *et al* 1989).

The identification of the toxins elaborated by *Diplodia maydis* and the study of their natural occurrence is currently regarded as a major challenge. Diplodiatoxin was earlier identified as a toxic substance produced by this mould (Steyn *et al* 1972), however, it represents only a fraction of the toxigenicity of *Diplodia maydis*.

Infestation of corn (maize) grain and cobs by *Diplodia maydis* is highly prevalent in Africa and other continents. Corn infected with this fungus is toxigenic to cattle, sheep, goats and poultry. Diplodiosis in sheep and cattle is characterized by ataxia, paresis, salivation and constipation (Kellerman *et al* 1988).

The potential danger to humans of corn contaminated with *Diplodia maydis* was increased by the findings of Fincham *et al* (1991) that addition of *Diplodia maydis*-infected corn (pure culture) to the food of omnivorous primates led to demyelation of nerves, atrophy, degeneration and necrosis of muscle, and hepatitis. Fincham *et al* (1991) reported that these findings may be medically significant since neuromuscular syndromes of unknown cause are prevalent among Africans; Africans are frequently dependent on a cereal diet based on corn (maize).

Kellerman *et al* (1991) observed that *Diplodia maydis* cob-rot led to perinatal losses in flocks and herds that had been exposed to diplodiosis. The affected lambs and calves were either stillborn or died soon after birth. Histopathological examination revealed a consistent, prominent *status spongiosis* of the white matter of their brains. Kellerman *et al* (1991) reported the same findings in dosing trials when dams were exposed to cultures of *Diplodia maydis*. The experiments also clearly demonstrated that the foetuses were much more susceptible to diplodiosis than the adults.

From a perusal of the mycotoxin literature it is clearly evident that the assessment of the health risk to man associated with the consumption of mycotoxin-contaminated food is still a vexing problem. The situation is exacerbated by the great need for food in many drought-stricken developing countries - there can be no human dignity under such conditions of famine and hunger. Schaeffer and Hamilton (1991) stated that no levels of mycotoxins have been demonstrated to be safe. Safe levels of mycotoxins depend on *legal requirements, sensitivity of the analytical methods, presumed safety, and estimates of what level can be afforded economically and politically* (Schaeffer and Hamilton (1991)).

At present researchers are very dependent on results obtained on experimental animals. However, the quantitative extrapolation of findings obtained by testing carcinogens on rodents to humans, particularly at low levels, was regarded by Ames *et al* (1987) as guesswork with no way of validating it. Ames *et al* (1987) based their statements on lack of knowledge in the following major areas :

(1) The basic mechanisms of carcinogenicity.
(2) The relation of cancer, aging, and life-span.
(3) The timing and order of the steps in the carcinogenic process that are being accelerated.
(4) Species differences in metabolism and pharmacokinetics.
(5) Human heterogeneity - for example, pigmentation affects susceptibility to skin cancer from ultraviolet light.

These sources of uncertainty are so numerous, and so substantial, that only empirical data will resolve them (Ames *et al* 1987).

The greatest challenge to plant (cereal) breeders should be the development of cultivars resistant to infection by toxin, e.g. aflatoxin-producing fungi, and to insects. Plant biotechnologists have at present a variety of techniques available to challenge the foregoing problems, as well as to breed cereals and groundnuts which are resistant to pre-/postharvest mycotoxin contamination.

REFERENCES

Abrahamson, D., Clear, R.M., Nowick, T.W. (1987): *Fusarium* species and trichothecene mycotoxins in suspect samples of 1985 Manitoba wheat. *Can. J. Plant. Sci.* 67, 611-619.

Adamson, R.H. (1989): Induction of hepatocellular carcinoma in nonhuman primates by chemical carcinogens. *Cancer Detect Prev.* 14, 215-220.

Ames, B.N., Magaw, R., Gold, L.S. (1987): Ranking possible carcinogenic hazards. *Science* 236, 271-280.

ApSimon, J.W., Blackwell, B.A., Blais, L., Fielder, D.A., Greenhalgh, R., Kasitu, G., Miller, J.D., Savard, M. (1990): Mycotoxins from *Fusarium* species: detection, determination and variety. *Pure Appl. Chem.* 62, 339-1346.

Bezuidenhout, S.C., Gelderblom, W.C.A., Gorst-Allman, C.P., Horak, R.M., Marasas, W.F.O., Spiteller, G., Vleggaar, R. (1988): Structure elucidation of the fumonisins, mycotoxins and *Fusarium moniliforme*. *J. Chem. Soc., Chem. Commun.* 743-745.

Blaney, B.J., Moore, C.J., Tyler, A.L. (1987): The mycotoxins - 4-deoxynivalenol, zearalenone and aflatoxin - in weather-damaged wheat harvested 1983-1985 in South-eastern Queensland. *Aust. J. Agric. Res.* 38, 993-1000.

Booth, C. (1971): *The genus Fusarium*, Commonwealth Mycological Institute: Kew, Surrey, England.

Bredenkamp, M.W., Dillen, J.L.M., Van Rooyen, P.H., Steyn, P.S. (1989): Crystal structures and conformational analysis of ochratoxin A and B : Probing the chemical structure causing toxicity. *J. Chem. Soc. Perkin Trans. II*, 1835-1839.

Breitholtz, A., Olsen, M., Dahlbäck, A., Hult, K. (1991): Plasma ochratoxin A levels in three Swedish populations surveyed using an ion-pair HPLC technique. *Food Addit. Contam.* 8, 183-192.

Bressac, B., Kew, M., Wands, J., Ozturk, M. (1991): Selective G to T mutations of p53 gene in hepatocellular carcinoma from southern Africa. *Nature* 350, 429-431.

Bruce, R.D. (1990): Risk assessment for aflatoxin II. Implications of human epidemiology data. *Risk Analysis* 10(4), 561-569.

Bullerman, L.B.N., Schroeder, L.L., Park, K-Y. (1984): Formation and control of mycotoxins in food. *J Food Protection* 47(8), 637-646.

Busby, W.F. Jr, Wogan, G.N. (1984): Aflatoxins. In *Chemical Carcinogens* Second Edition Vol 2, ACS Monograph 182, Searle, C.E. (ed.): Washington, DC, pp 945 - 1136.

Castegnaro M., Bartsch, H., Chernozemsky, I. (1987): Endemic nephropathy and urinary tract tumours in the Balkans. *Cancer Res.* 47, 3608-3609.

Cawood, M.E., Gelderblom, W.C.A., Vleggaar, R., Behrend, Y., Thiel, P.G., Marasas, W.F.O (1991): Isolation of the fumonisin mycotoxins: A quantitative approach. *J Agric Food Chem* 39, 1958-1962.

Cole, R.J. (1986): Modern methods in the analysis and structural elucidation of mycotoxins. New York: Academic Press.

Cole, R.J., Cotty, P.J. (1990): Biocontrol of aflatoxin production by using biocompetitive agents. In perspective on aflatoxin in field crops and animal food products in the United States, A symposium. *ARS-83*, pp. 62-66.

Cole, R.J., Dorner, J.W. (1992): Aflatoxin management during peanut production and processing: Current and future strategies. Personal communication.

Creppy, E.E., Schlegel, M., Röschentaler, R., Dirheimer, G. (1980): Phenylalanine prevents acute poisoning by ochratoxin A in mice. *Toxicol. Lett.*, 6, 77-80.

Cuero, R.G., Duffus, E., Osuji, G., Pettit, R. (1991): Aflatoxin control in prehavest maize: effects of chitosan and two microbial agents. *J. Agr. Science, Cambridge, 117*, 165-169.

Culvenor, C.C.J., Cockrum, P.A., Edgar, J.A., Frahn, J.L., Gorst-Allman, C.P., Jones, A.J., Marasas, W.F.O, Murray, K.E., Smith, L.W., Steyn, P.S., Vleggaar, R., Wessels, P.L. (1983): Structure elucidation of phompsin A, a novel cyclic hexapeptide mycotoxin produced by *Phomopsis leptostromifornis*. *J. Chem. Soc., Chem. Commun.*: 1259-1262.

Culvenor, C.C.J., Edgar, J.A., Mackay, M.F., Gorst-Allman, C.P., Marasas, W.F.O., Steyn, P.S., Vleggaar, R., Wessels, P.L. (1989): Structure elucidation and absolute configuration of phomopsin A, a hexapeptide mycotoxin produced by *Phomopsis leptostromiformis*. *Tetrahedron* 45, 2351-2372.

Cvetnic, Z., Pepeljnjak, S. (1990): Ochratoxinogenicity of *Aspergillus ochraceus* strains from nephropathic and non-nephropathic areas in Yugoslavia. *Mycopathologia* 110, 93-99.

De Jesus, A.E., Steyn, P.S., Van Heerden, F.R., Vleggaar, R., Wessels, P.L., Hull, W.E. (1981a): Structure and biosynthesis of the penitrems A-F, six novel tremorgenic mycotoxins from *Penicillium crustosum*. *J. Chem. Soc., Chem. Commun.*, 289-291.

De Jesus, A.E., Steyn, P.S., Vleggaar, R., Kirby, G.W., Varley, M.J., Ferreira, N.P. (1981b): Biosynthesis of a-cyclopiazonic acid. Steric course of proton removal during the cyclisation of ß-cyclopiazonic acid in *Penicillium griseofulvum*. *J. Chem. Soc., Perkin Trans. I*, 3292-3294.

De Jesus, A.E., Hull, W.E., Steyn, P.S., Van Heerden, F.R., Vleggaar, R. (1982): Biosynthesis of viridicatumtoxin, a mycotoxin from *Penicillium expansum*. *J. Chem. Soc., Chem. Commun.*, 902-904.

De Jesus, A.E., Steyn, P.S., Van Heerden, F.R., Vleggaar, R., Wessels, P.L., Hull, W.E. (1983): Tremorgenic mycotoxins from *Penicillium crustosum*. Isolation of penitrems A-F and the structure elucidation and absolute configuration of penitrem A_1. *J. Chem. Soc., Perkin Trans. I*, 1847-1856.

De Jesus, A.E., Steyn, P.S., Van Heerden, F.R., Vleggaar, R. (1984): Structure elucidation of the janthitrems, novel tremorgenic mycotoxins from *Penicillium janthinellum*. *J. Chem. Soc., Perkin Trans I*, 697-701.

Dell, M.P.K., Haswell, S.J., Roch, O.G., Coker, R.D., Medlock, V.F.P., Tomlins, K. (1990): Analytical methodology for the determination of aflatoxins in peanut butter: Comparison of High-performance thin-layer chromatographic, enzyme-linked immunosorbent assay and high-performance liquid chromatographic methods. *Analyst*, 115, 1435-1439.

Dorner, J.W., Cole, R.J., Sanders, T.H., Blankenship, P.D. (1989): Interrelationship of kernel water activity, soil temperature, maturity and phytoalexin production in preharvest aflatoxin contamination of drought-stressed peanuts. *Mycopathologia*, 105, 117-128.

Endou, H., Koseki, C., Yamada, H., Obara, T. (1986): Evaluation of nephrotoxicity using isolated nephron segments. *Dev. Toxicol. Environ. Sci.*, 14, 207-216.

Faifer, G.C., De Miguel, M.S., Godoy, H.M. (1990): Patterns of mycotoxin production by *Fusarium graminearum* isolated from Argentine wheat. *Mycopathologia*, 109, 165-170.

Fincham, J.E., Hewlett, R., De Graaf, A.S., Taljaard, J.J.F, Steytler, J.G., Rabie, C.J., Seier, J.V., Venter, F.S., Woodroof, C.W., Wynchank, S. (1991): Mycotoxic peripheral myelinopathy, myopathy, and hepatitis caused by *Diplodia maydis* on vervet monkeys. *J. Med. Primatol.* 20, 240-250.

Franck, B., Gehrken, H.P. (1980): Citreoviridins from *Aspergillus terreus*. *Chem. Int. Ed. Engl.* 19, 461-462.

Gallagher, R.T., Hawkes, A.D., Steyn, P.S., Vleggaar, R. (1984): Tremorgenic neurotoxins from perennial ryegrass causing ryegrass staggers disorders of livestock: Structure elucidation of lolitrem B. *J. Chem. Soc., Chem. Commun.*, 614-616.

Gelderblom, W.C.A, Marasas, W.F.O, Steyn, P.S., Thiel, P.G., Van der Merwe, K.J., Van Rooyen, P.H., Vleggaar, R., Wessels, P.L. (1984): Structure elucidation of fusarin

C, a mutagen produced by *Fusarium moniliforme*. *J. Chem. Soc., Chem. Commun.*, 122-124.

Gelderblom, W.C.A, Thiel, P.G., Marasas, W.F.O, Van der Merwe, K.J. (1984): Natural occurrence of Fusarin C, a mutagen produced by *Fusarium moniliforme*, in corn. *J. Agric. Food Chem.*, 32, 1064-1067.

Gelderblom, W.C.A., Jaskiewicz, K., Marasas, W.F.O., Thiel, P.G., Horak, R.M., Vleggaar, R., Kriek, N.P.J (1988): Fumonisins - novel mycotoxins with cancer-promoting activity produced by *Fusarium moniliforme*. *Appl. Environ. Microbiol.*, 54, 1806-1811.

Gilbert, J., Boenke, A., Wagstaffe, P.J. (1992a): Deoxynivalenol in wheat and maize flour reference materials. 1. An intercomparison of methods. *Food Additives and Contaminants*, 9(1), 71-81.

Gilbert, J., Sharman, M., Patel, S., Boenke, A., Wagstaffe, P.J. (1992b): Deoxynivalenol in wheat and maize flour reference materials. 2. Preparation and certification. *Food Additives and Contaminants*, 9(2), 119-135.

Gold, L.S., Slone, T.H., Stern, B.R., Manley, N.B., Ames, B.N. (1992): Rodent carcinogens: setting priorities. *Science*, 258, 261-265.

Groopman, J.D., Hall, A.J., Whittle, H. (1992a): Molecular dosimetry of aflatoxin-N7-guanine in human urine obtained in Gambia, West Africa. *Cancer Epidemiol Biomarkers Prevention*, 1, 221-227.

Groopman, J.D., Jiaqi, Z., Donahue, P.R., Pikul, A., Lisheng, Z., Jun-shi, C. (1992b): Molecular dositometry of urinary aflatoxin - DNA adducts in people living in Guangxi Region, People's Republic of China. *Cancer Research 52*, 45-52.

Hagelberg, S., Fuchs, R., Hult, K. (1989): Toxicokinetics of ochratoxin A in several species and its plasma-binding properties. *J. Appl. Toxicol.*, 9, 91-96.

Hagler Jr, W.M., Tyczkowska, K., Hamilton, P.B. (1984): Simultaneous occurrence of deoxynivalenol, zearalenone, and aflatoxin in 1982 scabby wheat from the Midwestern United States. *Appl. Environ. Microbiol.*, 47 (1), 151-154.

Haig, A.M. (1982): Chemical warfare in southeast Asia and Afghanistan. *US State Department Special Report 1.*

Harwig, J., Kuiper-Goodman, T., Scott, P.M. (1983): Microbial food toxicants: Ochratoxins In.: handbook of food borne diseases of biological origin. M. Recheigl (ed.). Boca Raton: CRC Press, pp. 193-238.

Holmberg, T. (1992): Ph.D Thesis. Ochratoxin in cereal grain and its potential effects on animal health. Swedish University of Agricultural Science, Uppsala, Sweden.

Holzapfel, C.W., Wilkins D.C. (1971): On the biosynthesis of the cyclopiazonic acids. *Phytochemistry*, 10, 351.

Hsieh, D.P.H. (1990a): Aflatoxin and human liver carcinogenesis. *Proceedings of the Sino-US Bi-National Conference on Toxicology*, 69-78.

Hsieh, D.P.H (1990b): Health risks posed by mycotoxins in foods. *Korean J. Toxicol.* 6(2), 159-166.

Hsieh, D.P.H (1992): An overview of mycotoxin risks. (Presentation at the Symposium on naturally occurring substances in traditional and biotechnology-derived foods: Their potential toxic and antitoxic effects, Irvine, California (March 1992).

Hsu, I.C., Metcalf, R.A., Sun, T., Welsh, J.A., Wang, N.J., Harris, C.C. (1991): Mutational hotspot in the p53 gene in human hepatocellular carcinomas. *Nature* 350, 427-428.

Hult, K., Hökby, E., Hägglund, U., Gatenbeck, S., Rutqvist, L., Sellyey, G. (1979): Ochratoxin A in pig blood: Method of analysis and use as a tool for feed studies. *Appl. Environ. Microbiol.*, 38, 772-776.

IARC (International Agency for Research on Cancer (1987): IARC Monographs on the Evaluation of Caranogenic Risks to Humans. Supplement 7. Overall Evaluation of Carcinogenicity: An Updating of IARC Monographs Volumes 1 to 42. IARC, Lyon, France.

Kellerman, T.S., Coetzer, J.A.W., Naudé, T.W. (1988): Plant poisonings and mycotoxicoses of livestock in southern Africa. Cape Town: Oxford University Press. p. 47.

Kellerman, T.S., Marasas, W.F.O, Thiel, P.G., Gelderblom, W.C.A, Cawood, M., Coetzer, J.A.W (1990): Leukoencephalomalacia in two horses induced by oral dosing of fumonisin B1. *Onderstepoort J. Vet. Res.*, 57, 269-275.

Kellerman, T.S., Prozesky, L., Schultz, R.A., Rabie, C.J., Van Ark, H., Maartens, B.P., Lübben, A. (1991): Perinatal mortality in lambs of ewes exposed to cultures of *Diplomia maydis (= Stenocarpella maydis)* during gestation. *Onderstepoort J. Vet. Res.*, 58, 297-308.

King, B. (1979): Outbreak of ergotism in Wollo, Ethiopia. *The Lancet*, 1411.

Krishnamachari, K.A.V.R., Bhat, R.V. (1976): Poisoning by ergot bajra (pearl millet) in man. *Indian J. Med. Res.*, 64, 1624.

Krogh, P., (1978): Causal associations of mycotoxic nephropathy. *Acta Pathol. Microbiol. Scand. Sect.* A supl. 269, 1-28.

Kuiper-Goodman T., Scott, P.M. (1989): Review: Risk assessment of the mycotoxin ochratoxin A. *Biomed. Environ. Sci.*, 2, 179-248.

Kuiper-Goodman, T. (1991a): Risk assessment to humans of mycotoxins in animal-derived food products. *Vet. Human. Toxicol.*, 33(4), 325-333.

Kuiper-Goodman, T. (1991b): Risk assessment of ochratoxin A residues in food. In: Mycotoxins, endemic nephropathy and urinary tract tumours. Castegnaro M., Plestina, R., Dirheimer, G., Chernozemsky, I.N., Bartsch, H. (eds.): Lyon: IARC, pp. 307-320.

Lacey, J. (1991): Natural occurrence of mycotoxins in growing and conserved forage crops. In: Mycotoxins and Animal Foods. Smith, J.E., Henderson, R.S. (eds.): Boston: CRC Press, pp. 363-397.

Lauren, D.R., Agnew, M.P., Smith, W.A., Sayer, S.T. (1991): A survey of the natural occurrence of *Fusarium* mycotoxins in cereals grown in New Zealand in 1986-1989. *Food Additives and Contaminants*, 8(5), 599-605.

Lee, U-S., Jang, H-S., Tanaka, T., Hasegawa, A., Oh, Y-J., Cho, C-M., Sugiura, Y., Ueno, Y. (1986): Further survey on the *Fusarium* mycotoxins in Korean cereals. *Food Additives and Contaminants*, 3(3), 253-261.

Linsell, C.A., Peers, F.G. (1977): Aflatoxin and liver cell cancer. *Trans. Royal Soc. of Tropical Medicine and Hygiene, 71*, 471-473.

Mantle, P.G., McHugh, K.M., Adatia, R., Gray, T., Turner, D.R. (1991a): Persistent karyomegaly caused by *Penicillium* nephrotoxins in the rat. *Proc. R. Soc. Lond. B.*, 246:251-259.

Mantle, P.G., McHugh, K.M., Adatia, R., Heaton, J.M., Gray, T., Turner, D.R. (1991b): *Penicillium aurantiogriseum*-induced, persistent renal histopathological changes in rats, an experimental model for Balkan endemic nephropathy competitive with ochratoxin A In: Mycotoxins, Endemic Nephropathy and Urinary Tract Tumours. In Castegnaro, M., Plestina, R., Dirheimer, G., Chernozemsky, I.N., Barstch, H. (eds.): Lyon: ARC, pp. 119-127.

Mantle, P.G., McHugh, K.M. (1992): Nephrotoxic fungi in food from nephropathy households in Bulgaria. *Mycol. Res.*: Accepted for publication.

Marasas, W.F.O., Kriek, N.P.J., Fincham, J.E., Van Rensburg, S.J. (1984): Primary liver cancer and esophageal basal cell hyperplasia in rats caused by *Fusarium moniliforme*. *Int. J. Cancer*, 34, 383-387.

Marasas, W.F.O., Nelson, P.E., Tousson, T.A. (1985): Taxonomy of toxigenic *fusaria* in *Trichothecenes* and other mycotoxins. Lacey, J. (ed.): New York: John Wiley and Sons, pp. 3-14.

Marasas, W.F.O., Jaskiewicz, K., Venter, F.S., Van Schalkwyk, D.J. (1988a): *Fusarium moniliforme* contamination of maize in esophageal cancer areas in Transkei. *S. Afr. Med. J.*, 74, 110-114.

Marasas, W.F.O., Kellerman, T.S., Gelderblom, W.C.A., Coetzer, J.A.W., Thiel, P.G., Van der Lugt, J.J. (1988b): Leukoencephalomalacia in a horse induced by fumonisin B_1, isolated from *Fusarium moniliforme*. *Onderstepoort J. Vet. Res.*, 55, 197-203.

McMahon, G., Davis, E.F., Huber, L.J., Kim, Y., Wogan, G.N. (1990): Characterization of c-K-*ras* and N-*ras* oncogenes in aflatoxin B_1-induced rat liver tumors. *Proc. Natl. Acad. Sci. USA*, 87, 1104-1108.

Morgan, M.R.A. (1989): Mycotoxin immunoassays (with special reference to ELISAs). *Tetrahedron*, 45, 2237-2249.

Nagel, D.W., Steyn, P.S., Ferreira, N.P. (1972): The biosynthesis of citreoviridin. *Phytochemistry*, 11, 3215-3218.

Nesheim, S., Stack, M.E., Trucksess, M.W., Eppley, R.M. (1992): Rapid solvent-efficient method for liquid chromatographic determination of ochratoxin A in corn, barley and kidney: Collaborative study. *J. of AOAC International*, 75 (3), 481-487.

Nishie, K., Cole, R.J., Dorner, J.W. (1988): Toxicity of citreoviridin. *Res. Comm. Chem. Pathol. Pharmacol.*, 59, 31-52.

Olsen, M., Thorup, I., Knudsen, I., Larsen, J-J., Hald, B., Olsen, J. (1991): Health evaluation of ochratoxin A in food products. The Nordic Working Group on Food Toxicology and Rish Evaluation. Nordic Council of Ministers, Copenhagen.

Osborne, B.G., Willis, K.H. (1984): Studies into the occurrence of some trichothecene mycotoxins in UK home-grown wheat and in imported wheat. *J. Sci. Food Agric.*, 35, 579-583.

Park, D.L. (1992a): Aflatoxin control - managing risks associated with an unavoidable natural toxicant in food and feed. Presented at the 1992 IFT meeting, June 1992, New Orleans, LA.

Park, D.L. (1992b): Perspectives on mycotoxin decontamination procedures. *Microb. Hyg. All.* 4 (9), 21-27.

Park, D.L., Stoloff, L. (1989): Aflatoxin control - How a regulatory agency managed Risk from an unavoidable natural toxicant in food and feed. *Regulatory Toxicology and Pharmacology*, 9, 109-130.

Park, D.L., Miller, B.M., Hart, L.P., Yang, G., McVey, J., Page, S.W., Pestka, J., Brown, L.H. (1989): Mycotoxins : Enzyme-linked immunosorbent assay for screening aflatoxin B_1 in cottonseed products and mixed feed: Collaborative study. *J. Assoc. Off. Anal. Chem.*, 72 (2), 326-332.

Park, D.L., Nesheim S., Trucksess, M.W., Stack, M.E., Newell, R.F. (1990): Liquid chromatographic method for the determination of aflatoxins B_1, B_2, G_1, and G_2 in corn and peanut products: Collaborative study. *J. Assoc. Off. Anal. Chem.*, 73 (2), 260-266.

Park, D.L., Rua Jr, S.M., Mirocha, C.J., Abd-Alla, E-S.A.M. and Weng, C.Y. (1992): Mutagenic potentials of fumonisin contaminated corn following ammonia decontamination procedure. *Mycopathologia*, 117, 105-108.

Peers, F., Bosch, X., Kaldor, J., Linsell, A., Pluijmen, M. (1987): Aflatoxin exposure, hepatitis B virus infection and liver cancer in Swaziland. *Int. J. Cancer.*, 39, 545-553.

Perkowski, J., Plattner, R.D., Golinski, P., Vesonder, R.F., Chelkowski, J. (1990): Natural occurrence of deoxynivalenol, 3-acetyl-deoxynivalenol, 15-acetyl-deoxynivalenol, nivalenol, 4,7-dideoxynivalenol, and zearalenone in Polish wheat. *Mycotoxin Research*, 6, 7-12.

Pitt, J.I. (1979): The genus *Penicillium* and its telemorphic states *Eupenicillium* and *Talaromyces*. New York: Academic Press, pp. 219-276.

Pitt, J.I. (1987): *Penicillium viridicatum*, *Penicillium verrucosum* and production of ochratoxin A. *Appl. Environ. Microbiol.*, 53, 266-269.

Plattner, R.D., Beremand, M.N., Powell, R.G. (1989): Analysis of trichothecene mycotoxins by mass spectrometry and tandem mass spectrometry. *Tetrahedron*, 45, 2251-2262.

Plattner, R.D., Norred, W.P., Bacon, C.W., Voss, K.A., Peterson, R., Shackelford, D.D., Weisleder D. (1990): A method of detection of fumonisins in corn samples associated with field cases of equine leukoencephalomalacia. *Mycologia*, 82 (6), 698-702.

Pohland, A.E., Nesheim, S., Friedman, L. (1992): Ochratoxin A : A Review (Technical Report). *Pure and Appl. Chem.*, 64 (7), 1029-1046.

Porter, J.K., Voss, K.A., Bacon, C.W., Norred, W.P. (1990): Effects of *Fusarium moniliforne* and corn associated with equine leukoencephalomalacia on rat neurotransmitters and metabolites. *Proc. Soc. Exp. Biol. Med.*, 194, 265-269.

Potgieter, M., Steyn, P.S., Van Heerden, F.R., Van Rooyen, P.H., Wessels, P.L. (1989): Conformational analysis of the cyclic peptide rhizonin A in solution and in crystalline state. *Tetrahedron*, 45, 2337-2350.

Rabie, C.J., Sydenham, E.W., Thiel, P.G., Lübben, A., Marasas, W.F.O. (1986): T-2 Toxin production by *Fusarium acuminatum* isolated from oats and barley. *Appl. Environ. Microbiol.*, 52 (3), 594-596.

Richard, J.L. (1990): Additional mycotoxins of potential importance to human and animal health. *Vet. Hum. Toxicol.*, 32, 63-70.

Roch, O.G., Blunden, G., Coker, R.D., Nawaz, S. (1992): The development and validation of a solid phase extraction/HPLC method for the determination of aflatoxins in groundnut meal. *Chromotagraphia*, 33 (5-6), 208-212.

Röschenthentaler, R., Creppy, E.E., Dirheimer, G. (1984): Ochratoxin A on the mode of action of a ubiquitous mycotoxin. *J. Toxicol. Toxin. Rev.*, 1, 53-86.

Rosen, R.T., Rosen, J.D. (1982): Presence of four *Fusarium* mycotoxins and synthetic material in *yellow rain* : Evidence for the use of chemical weapons in Laos. *Biomed. Mass. Spectrom.*, 9, 443-450.

Ross, P.F., Nelson, P.E., Richard, J.L., Osweiler, G.D., Rice, L.G., Plattner, R.D., Wilson, T.M. (1990): Production of fumonisins by *Fusarium moniliforme* and *Fusarium proliferatum* isolates associated with equine leukoencephalomalacia and a pulmonary edema syndrome in swine. *Appl. Environ. Microbiol.*, 56, 3225-3226.

Ross, R.K., Yuan, J-M., Yu, M.C., Wogan, G.N., Qian, G-S., Tu, J-T., Groopman, J.D., Gao, Y-T., Henderson, B.E.(1992): Urinary aflatoxin biomarkers and risk of hepatocellular carcinoma. *The Lancet*, 339, 943-946.

Sakabe, N., Goto, T., Hirata, Y. (1964): Structure of citreoviridin, a toxic compound produced by *P. citreoviride* molded on rice. *Tetrahedron Lett.*, 27, 1825-1830.

Samson, R.A., Stolk, A.C., Hadlok, R. (1976): Revision of the subsection of *Fasciculata* of *Penicillium* and some allied species. *Stud. Mycol.*, 11, 47.

Schaeffer, J.L., Hamilton, P.B. (1991): Interactions of mycotoxins with feed ingredients. Do safe levels exist. In *Mycotoxins and Animal Foods*. Smith, J.E., Henderson R.S. (eds.): Boston: CRC Press Inc., pp. 827-843.

Shank, R.C., Gordon, J.E., Wogan, G.N. (1972a): Dietary aflatoxins and human liver cancer. III Field survey of rural Thai families for ingested aflatoxins. *Food and Cosmetics Toxicology*, 10, 71-84.

Shank, R.C., Bhamarapravati, N., Gordon, J.E., Wogan, G.N. (1972b): Dietary aflatoxins and human liver cancer. IV Incidence of primary liver cancer in two municipal populations of Thailand. *Food and Cosmetics Toxicology*, 10, 171-179.

Sharman, M., MacDonald, S., Gilbert, J. (1992): Short Communication. Automated liquid chromatographic determination of ochratoxin A in cereals and animal products using immunoaffinity column clean-up. *J. Chrom.*, 603, 285-289.

Shephard, G.S., Sydenham, E.W., Thiel, P.G., Gelderblom, W.C.A (1990): Quantitative determination of fumonisins B_1 and B_2 by high-performance liquid chromatography with fluorescence detection. *J. Liquid Chromat.*, 13, 2077-2087.

Shotwell, O.L., Bennett, G.A., Stubblefield, R.D., Shannon, G.M., Kwolek, W.F., Plattner, R.D. (1985): Deoxynivalenol in hard red winter wheat: relationship between toxin levels and factors that could be used in grading. *J. Assoc. Off. Anal. Chem.*, 68 (5), 954-955.

Steyn, P.S. (1980): The biosynthesis of mycotoxins: A study in secondary metabolism. New York: Academic Press.

Steyn, P.S. (1984): Ochratoxins and related dihydroisocoumarins. In *Mycotoxins - production, isolation, separation and purification.* Betina, V. (ed.): Amsterdam: Elsevier Sci. Publ., pp. 183-216.

Steyn, P.S. (1989): Tetrahedron Symposia - in Print. Nr. 37. *Mycotoxins.*

Steyn, P.S., Vleggaar, R. (1983): Roquefortine and intermediate in the biosynthesis of oxaline in cultures of *Penicillium oxalicum. J. Chem. Soc., Chem. Commun.*, 560-561.

Steyn, P.S., Vleggaar, R. (1985): Tremorgenic mycotoxins. *Fortschr. Chem. Org. Naturst.*, 48, 1-80 (and references cited therein).

Steyn, P.S., Wessels, P.L., Holzapfel, C.W., Potgieter, D.J.J., Louw, W.K.A. (1972): The isolation and structure of a toxic metabolite from *Diplodia maydis* (Berk.) Sacc. *Tetrahedron*, 28, 4775-4778.

Steyn, P.S., Vleggaar, R., Ferreira, N.P., Kirby, G.W., Varley, M.J. (1975): The steric course of proton removal during the cyclization of ß-cyclopiazonic acid in *Penicillium cyclopium. J. Chem. Soc., Chem. Commun.*, 465-466.

Steyn, P.S., Vleggaar, R., Wessels, P.L., Woudenberg, M. (1982): Biosynthesis of citreoviridin. A carbon-13 NMR study. *J. Chem. Soc., Perkin Trans. I*, 2175-2178.

Steyn, P.S., Tuinman, A.A., Van Heerden, F.R., Van Rooyen, P.H., Wessels, P.L., Rabie, C.J. (1983): The isolation, structure and absolute configuration of the mycotoxin, rhizonin A, a novel cyclic heptapeptide containing *N*-methyl-3-(3-furyl) alanine, produced by *Rhizopus microporus. J. Chem. Soc., Chem. Commun.*, 47-49.

Stoll, A. (1952): Recent investigations on ergot alkaloids. *Fortsch. Chem. Org. Naturstoffe*, 9, 114-174.

Stoll, A., Hofmann, A. (1943): Die Alkaloide der Ergotoxingruppe: Ergocristin, Ergokryptin und *Ergocornin. Helv. Chim. Acta.*, 26, 1570-1601.

Stoloff, L. (1983): Aflatoxin as a cause of primary liver-cell cancer in the United States: A probability study. *Nutrition and Cancer* 5, 165-185.

Stoloff, L. (1989): Aflatoxin is not a probable human carcinogen: The published evidence is sufficient. *Regulatory Toxicology and Pharmacology*, 10 (3), 272-283.

Stoloff, L. Van Egmond, H.P., Park, D.L. (1991): Rationales for the establishment of limits and regulations for mycotoxins. *Food Additives and Contaminants*, 8(2), 213-222.

Sydenham, E.W., Gelderblom, W.C.A, Thiel, P.G., Marasas, W.F.O. (1990a): Evidence for the natural occurrence of fumonisin B_1, a mycotoxin produced by *Fusarium moniliforme* in corn. *J. Agric. Food Chem.*, 38, 285-290.

Sydenham, E.W., Thiel, P.G., Marasas, W.F.O., Shephard, G.S., Van Schalkwyk, D.J., Koch, K.R. (1990b): Natural occurrence of some *Fusarium* mycotoxins in corn from low and high esophageal cancer prevalence areas of the Transkei, Southern Africa. *J. Agric. Food. Chem.*, 38, 1900-1903.

Tanaka, T., Hasegawa, A., Matsuki, Y., Ueno, Y. (1985): A survey of the occurrence of nivalenol, deoxynivalenol and zearalenone in foodstuffs and health foods in Japan. *Food Additives and Contaminants*, 2(4), 259-265.

Tanaka, T., Hasegawa, A., Yamamoto, S., Matsuki, Y., Ueno, Y. (1986): Residues of *Fusarium* mycotoxins, nivalenol, deoxynivalenol and zearalenone, in wheat and processed food after milling and baking. *J. Food. Hyg. Soc. Japan*, 27 (6), 653-655.

Tanaka, T., Hasegawa, A., Yamamoto, S., Lee, U-S., Sugiura, Y., Ueno, Y. (1988): Worldwide contamination of cereals by the *Fusarium* mycotoxins nivalenol, deoxynivalenol, and zearalenone. 1. Survey of 19 countries. *J. Agric. Food Chem.*, 36, 979-983.

Tanaka, T., Yamamoto, S., Hasegawa, A., Aoki, N., Besling, J.R., Sugiura, Y., Ueno, Y. (1990): A survey or the natural occurrence of *Fusarium* mycotoxins, deoxynivalenol, nivalenol and zearalenone, in cereals harvested in the Netherlands. *Mycopathologia*, 110, 19-22.

Thiel, P.G., Marasas, W.F.O, Sydenham, E.W., Shephard, G.S., Gelderblom, W.C.A., Nieuwenhuis, J.J. (1991): Survey of fumonisin production by *Fusarium* species. *Appl. Environ. Microbiol.*, 57, 1089-1093.

Ueno, Y. (1971): Production of citreoviridin, a neurotoxic mycotoxin of *Penicillium citreoviride* Biourge. In Symposium on mycotoxins in Human Health. Purchase, I.F.H. (ed.) Pretoria: McMillan, pp. 115-132.

Ueno, Y. (1980): Trichothecene mycotoxins: mycology, chemistry and toxicology. *Adv. Nutr. Res.*, 3, 301.

Ueno, Y., Ueno, I. (1972): Isolation and acute toxicity of citreoviridin, a neurotoxic mycotoxin of *Penicillium citreoviride* Biourge. *Jap. J. Exp. Med.*, 42, 91-105.

Ueno, Y., Sato, N., Ishii, K., Saiha, K., Enomoto, M. (1972): Toxicological approaches to the metabolites of the *Fusaria V.* Neosolaneol, T-2 toxin and butenolide, toxic metabolites of *Fusarium sporotrichioides* NRRL 3510 and *Fusarium poae* 3287. *Jpn. J. Exp. Med.*, 42, 461.

Ueno, Y., Lee, U-S., Tanaka, T., Hasegawa, A., Matsuki, Y. (1986): Examination of Chinese and USSR cereals for the *Fusarium* mycotoxins, nivalenol, deoxynivalenol and zearalenone. *Toxicon.*, 24 (6), 618-621.

Uraguchi, K., Yamazaki, M. (ed.) (1978): Toxicology - Biochemistry and Pathology of Mycotoxins. New York: Halsted Press.

Van der Merwe, K.J., Steyn, P.S., Fourie, L. (1965): Mycotoxins. Part II. The constitution of ochratoxins A, B and C, metabolites of *Aspergillus ochraceus* Wilh. *J. Chem. Soc.*, 7083-7088.

Van Egmond, H.P. (1989). Current situation on regulations on mycotoxins. Overview of tolerances and status of standard methods of sampling and analysis. *Food Additives and Contaminants 6*, 139-188.

Van Rensburg, S.J. (1985): Recent studies on the etiology of esophageal cancer. *S. Afr. Cancer Bull.*, 29, 22-31.

Van Rensburg, S.J. (1986): Role of mycotoxins in endemic liver cancer and oesophageal cancer. In Steyn, P.S., Vleggaar, R. (eds.): *Mycotoxins and Phycotoxins*. Amsterdam: Elsevier Science Publishers, pp. 483-494.

Van Rensburg, S.J., Altenkirk, B. (1974): *Claviceps purpurea* - ergotism. In Mycotoxins. Purchase, I.F.H. (ed.): Amsterdam: Elsevier, p. 69.

Van Rensburg, S.J., Cook-Mozaffari, P., Van Schalkwyk, D.J., Van der Walt, J.J., Vincent, T.J., Purchase, I.F. (1985): Hepatocellular carcinoma and dietary aflatoxin in Mozambique and Transkei. *Brit. Cander.*, 51, 713-726.

Wang, E., Norred, W.P., Bacon, C.W., Riley, R.T., Merril, A.H. (1991): Inhibition of sphingolipid biosynthesis by fumonisins. *J. Biol. Chem.* 266(2), 14486-14490.

Wannemacher, R.W., Bunner, D.L., Neufeld, H.A. (1991): Toxicity of trichothecenes and other related mycotoxins in laboratory animals. In *Mycotoxins and Aminal Foods*. Smith, J.E., Henderson, R.S. (eds.): Boston: CRC Press, pp. 499-552.

Wicklow, D.T., Stubblefield, R.D., Horn, B.W., Shotwell, O.L. (1988): Citreoviridin levels in *Eupenicillium ochrasalmoneum* infested maize kernels at harvest. *Appl. Environ. Microbiol.*, 54, 1096-1098.

Wogan, G.N. (1992a): Aflatoxins as risk factors for hepatocellular carcinoma in humans. *Cancer Research (Suppl.)*, 52, 2114s-2118s.

Wogan, G.N. (1992b): Aflatoxin carcinogenesis : interspecies potency differences and relevance for human risk assessment. In *Relevance of animal studies to the evaluation of human cancer risk*. New York: Wiley-Liss, pp. 123-137.

Yagen, B., Joffe, A.Z., Horn, P., Mor, N., Lutsky, I.I. (1977): Toxins from a strain involved in ATA: In *Mycotoxins in human and animal health*. Rodricks, J.V., Hesseltine, C.W., Mehlman, M.A. (eds.): Ill., USA: Pathotox Publishers, pp. 327-336.

Yang, C.S. (1980): Research on esophageal cancer in China : a review. *Cancer Res.*, 40, 2633-2644.

Yeh, F-S., Yu, M.C., Mo, C-C., Luo, S., Tong, M.J., Henderson, B.J. (1989): Hepatitis B virus, aflatoxins, and hepatocellular carcinoma in southern Guangxi, China. *Cancer. Res.*, 49, 2506-2509.

Yeulet, S.E., Mantle, P.G., Rudge, M.S., Greig, J.B. (1988): Nephrotoxicity of *Penicillium aurantiogriseum*, a possible factor in the actiology of Balkan endemic nephropathy. *Mycopathologia*, 102, 21-30.

Résumé

Les mycotoxines sont produites par cinq genres de moisissures : Aspergillus, Penicillium, Fusarium, Alternaria et Claviceps. Le présent article décrit les risques pour la santé liés à la présence des mycotoxines suivantes et de leur association : aflatoxines, ergotoxines, citréoviridine, trichothécènes, **ochratoxines**, mycotoxines trémorgéniques et les fumonisines.

Human ochratoxicosis and its pathologies. Eds E.E. Creppy, M. Castegnaro, G. Dirheimer. Colloque INSERM/ John Libbey Eurotext Ltd. © 1993, Vol. 231, pp. 33-41.

Research priorities for assessing the risk of multiple mycotoxin exposure to domestic animals and man: what we know and what we need to know !

Peter H. Bach[(1)*] and Michelle McLean[(2)]

[(1)]*Drug Development and Chemical Safety Research Unit, Faculty of Science, University of East London, Romford Road, London E15 4LZ, England and* [(2)]*Department of Physiology, Faculty of Medicine, University of Natal, P.O. Box 17039, Congella 4013, Durban, Natal, Republic of South Africa*

**Corresponding author*

Introduction

The ubiquity of mycotoxins and their well established toxicity (WHO, 1979, 1990) is a focus of ongoing concern. The presence of one mycotoxin in feed often signifies the presence of other fungal toxins. There is an increasing awareness that domestic animals are exposed to multiple mycotoxins under field conditions, often each at a level that would be considered unacceptably high, and examples of mycotoxicosis have been reported. There is virtually no information on the effects of multichemical exposure in man and limited data on the effects of multiple chemical administered in animals. The more common mycotoxin combinations have been documented and quantitative ranges have been defined (Chelkowski *et al.*, 1987; Ciegler, 1972; Hayes *et al.* 1977; Huff *et al.*, 1986, 1988a,b; Jelinek *et al.*, 1989; Jonsyn, 1988; Krogh *et al.*, 1973; Lillehoj and Ciegler, 1975; Pathre and Mirocha, 1977; Scott *et al.*, 1972; Scott, 1978; Soares and Rodriguez-Amaya, 1989; Ueno, 1987; Watson, 1985).

A rational basis for experimental design

Exposure to multiple chemicals represents a major toxicological dilemma as man is already almost always exposed to more than one substance, but most risk assessment is based on the investigation of single chemicals. Most of the available data on multiple chemical exposure has been derived from polypharmacy, where the combination of drugs, the dosage of each and a pharmacological and toxicological profile in animals and man is well defined. More recently, the toxicological consequences of exposure to toxic mixtures such as light petroleum have been addressed (Mehlmann *et al.*, 1984).

Mycotoxins present a unique problem and challenge - current knowledge .

The relatively limited data on multiple mycotoxin dosing (listed in Table 1) in poultry, rats, mice, guinea pigs, calves, piglets and the dog using appropriate combinations of ochratoxin A (OA), aflatoxin (AF), citrinin, deoxynivalenol, T-2 toxin, penicillic acid and rubratoxin have been reviewed by McLean and Bach (1993). Interactions between AFB1 and oxytetracycline (Blaude *et al.*, 1990; Blaude *et al.*, 1992) raises the possibility of mycotoxins interacting with other chemicals and also therapeutic agents. The mechanisms that underlie these interactions are not understood, and at present there is no rational basis to predict them.

In vivo investigations

Species and sex-related differences in the metabolism and effects of single mycotoxins (WHO, 1979, 1990, Micco *et al.*, 1987) are well documented. The investigation of multiple fungal toxin exposure is logistically difficult and compounded by the number of possible combinations, and their qualitative and quantitative seasonal variation. The statistically significant changes listed in Table 1 do not have a clear toxicological interpretation, and variability between studies with respect to a diverse modes of toxin administration, sometimes with no regard of realistic exposure levels or routes, regimen, species and strain, taken together with limited experimental information about dose-related effects and inter-species and inter-strain differences (Ruff *et al.*, 1990, 1992b; Huff *et al.*, 1992) makes these data difficult to compare and interpret. The validity of extrapolating data from poultry and rats to other domestic animals and man is further complicated by the use of animals at a range of ages, both sexes and applying criteria for defining adverse interactions, most often morbidity, organ weight, routine urine and serum biochemistry etc., rather than assessing target organ morphology and function.

Recovery from multiple mycotoxin exposure

Thus far there is very little data on recovery from single or multiple mycotoxins feeding. Rati *et al.*, (1991) showed that 24 weeks after AFB1 and OA exposure there was no obvious difference between the clinical biochemistry or pathology reflecting liver and renal function. Thus there appears to be no compromised recovery after the two toxins had been given.

Biomarkers and molecular dosimetry

Molecular dosimetry shows a relationship between urinary AF-N7-guanine and serum AF-albumin adducts and carcinogenesis in rodents (Groopman *et al.*, 1992a). More importantly, AF serum albumin adducts and AFB-N7-guanine in urine correlated well with exposure of humans to contaminated feeds (Autrup *et al.*, 1991; Groopman *et al.*, 1992b) and subjects with liver cancer were more likely to have detectable AF metabolites (Ross *et al.*, 1992; Groopman *et al.*, 1992c). These data highlight the value of biomonitoring and the need to develop biomarkers as the ways for assessing other mycotoxins.

In vitro investigations

Increasingly *in vitro* methods are being applied to the investigation of the toxicity and mechanism of mycotoxin and their interactions. The genotoxic potential of a range of mycotoxins have been assessed *in vitro* (Krivobok *et al.*, 1987; Mayura *et al.*, 1989) but studies involving two or more chemicals does not appear to have been reported. The differential cytotoxicity of aqueous and organic extracts from *Fusarium moniliforme* in rat primary hepatocytes shows that other toxins in addition to fusarin C and fumonisin B1, were produced (Norred *et al.*, 1991). The neutral red viability assay using human hepatoma HepG2 cell line ranked cytotoxicity as T-2 > HT-2 > T-2 triol > T-2 tetraol. Rotter *et al.*, (1991) used the chick embryotoxicity screening test to examine possible interactions between deoxynivalenol, 15-acetyldeoxynivalenol and HT-2 toxin and showed the combined toxicity of any two to be additive.

In vitro methods could speed investigations of multichemical exposure, but there is little published data on toxicodynamic interactions. Riley and Showker (1991) have shown that the cytotoxic effects (lipid peroxidation, abrupt calcium influx, extensive blebbing, and total LDH release) of patulin in LLC-PK_1 cells were prevented by butylated hydroxytoluene, deferoxamine and cyclopiazonic acid, but depletion of nonprotein sulphydryls, increased $^{86}Rb^+$ efflux and dome collapse were not prevented. This is the first published example of cyclopiazonic acid protecting against the cytotoxicity of patulin, and highlights the possiblity that some interactions may be advantageous. There is a synergistic interaction between OA and citrinin *in vitro* with regard to the inhibition of protein, RNA and DNA synthesis (Creppy *et al.*, 1980).

The molecular biology of interactions

A mechanistic understanding can provide part of the rational basis for risk assessment. The carcinogenesis of AFB1 is understood in term of reactive intermediates (Kodama *et al.*, 1990; Baertschi *et al.*, 1989), but this is not the case for other types of AF-induced target cell injury nor are there accepted mechanisms to explain the toxicity of other mycotoxins. There is an increasing range of the *in vitro* effects of mycotoxins on cells being described, some of which will help facilitate a molecular understanding of how these compounds affect cells. For example, Siraj *et al.* (1981) studied the effects of OA and citrinin in neonatal rats and showed that renal oligomycin-sensitive Mg^{2+}-ATPase was not potentiated by combined treatment, renal Na^+-K^+-ATPase activity was

Table 1. Summary of the most well described clinical and toxicological manifestations caused by combinations of mycotoxins.

Ochratoxin A and aflatoxin B_1 *

Increase mortality, decrease body weight in poultry, with a delayed recovery and enhanced bruising, and increased liver weights and enhanced kidney injury.

Brownie and Brownie, 1988; Campbell *et al.*, 1983; Harvey *et al.*, 1989; Huff and Doerr, 1981; Huff *et al.*, 1983; 1984; Micco *et al.*, 1988; Patterson *et al.*, 1981; Pier *et al.*, 1980; Rati *et al.*, 1981; Tapia and Seawright, 1985

Ochratoxin A and citrinin

Enhanced renal tumourigenesis in mice, but not in rats, and had few other adverse effects. This is also the case in pigs.

Brown *et al.*, 1986; Creppy *et al.*, 1980a; Glahn *et al.*, 1988; Kanisawa 1983a,b, 1984; Manning *et al.*, 1985; Mayura *et al.*, 1984; Sansing *et al.* 1976; Siraj *et al.*, 1981; Sandor *et al.*, 1991.

Ochratoxin A and deoxynivalenol

Decreased body weight and increased relative major organ weights suggesting an additive interaction.

Kubena *et al.*, 1988

Ochratoxin A and penicillic acid

Depressed body weight, and increased rodent mortality and renal proximal tubular necrosis synergistically.

Kubena *et al.* 1984; Sansing *et al.*, 1976; Shepherd *et al.*, 1981

Ochratoxin A and T-2 toxin

Decreased body weights and increased organs weights additive or synergistically, and was additively teratogenic on day 8 and synergistically so on day 10, with a possible shift of target organ toxicity.

Hood *et al.*, 1978; Kubena *et al.*, 1989

Aflatoxin and cyclopiazonic acid

Increase relative organ weights, serum glutamic oxalacetic transaminase and blood urea nitrogen, and decreased serum albumin and phosphorus, and the relative weight of the bursa of Fabricius additively in broiler chickens.

Smith *et al.*, 1992.

Aflatoxin and kojic acid

interacted antagonistically and increased mean corpuscular haemoglobin and haemoglobin concentration in broiler chickens.

Giroir *et al.*, 1991.

Aflatoxin B_1 and deoxynivalenol

Caused a significant decrease in body weight and increase in major organ weights.

Huff *et al.*, 1986.

Aflatoxin B_1 and T-2 toxin

Interact synergistically to decrease body and increase major organ weights, affected haematological parameters additively and impaired the immune system. Mortality was significantly increased in mice and guinea pigs, but without histopathological changes and less than additive in pigs.

Huff *et al.*, 1988b; Lindenfelser *et al.*, 1974; Pier *et al.*, 1986; Harvey *et al.*, 1990a,b

Aflatoxin B_1 and Rubratoxin B

Additive or synergistic lethality and weight changes, but without enhancing the carcinogenic potential of aflatoxin.

Hayes *et al.*, 1977; Wogan *et al.*, 1971.

* For information on the species, strain, sex, concentration and dosing routes, etc see McLean and Bach (1993) or original papers.

decreased synergistically. Whereas cytochrome P-450, NADPH-dependent dehydrogenase and NADPH-cytochrome c reductase activities were enhanced after each toxin, they were drastically reduced after concomitant exposure. Using both chicken and mouse tissue *in vitro* and *in vivo*, penicillic acid inhibits pancreatic carboxypeptidase A activity (Parker *et al.*, 1982). In the gastrointestinal tract, this enzyme is responsible for converting OA to its non-toxic metabolite, OA-alpha. Parker *et al.* (1982) suggest that the synergistic interaction between OA and penicillic acid is explained by an increased OA burden reaching the kidneys, following penicillic acid inhibition of carboxypeptidase A activity. This mechanistic explanation warrants further investigation.

Research priorities

Future research needs to have a strong multidisciplinary basis, and as such, can only be undertaken as collaborative national or international efforts. While each mycotoxin combination should be investigated *in vivo*, using doses and routes of administration that reflect the conditions encountered under field conditions, in reality, the permutations are too vast. There is a need for a rational approach to undertaking risk assessment in multiple mycotoxin exposure. In order to address the question of the roles of multiple mycotoxin exposure in animal and human health priorities must be established to rationalize how best to undertake a realistic risk assessment.

Such priorities include:-

1. There is still a need to document the levels and types of co-contaminating mycotoxins occurring in commodities and animal feeds in order to more clearly define the mycotoxin combinations and concentrations to which animals (and man) are likely to be exposed; and how these vary seasonally under field and storage conditions in different regions. In order to achieve this, there is a need for appropriate robust, rapid and sensitive analytical techniques (ELISA-based) in the field and data validation with the more sophisticated GLC, HPLC and MS. There is also a need for such data to be archived and analyzed by national agencies.
2. The highly variable and unpredictable exposure to mycotoxins demands that more effort is taken to establish biomarkers from which to document molecular dosimetry for animals and man for a range of high risk fungal toxins. It will then be possible to undertake investigations in animals that reflect human exposure.
3. Studying tissues from animals exposed to multiple mycotoxins in order to document a broader spectrum of histopathological changes. This will give a better indication of target cell type that should be used *in vitro* to screen for interactions and provide a mechanistic understanding, rather than the application of toxicologically non-selective methods (weight change, etc.) that provide little hard data for risk assessment.
4. While OA, AF and citrinin absorption, distribution, metabolism and excretion have been well investigated in a range of animal species, more kinetic data is needed for other mycotoxins. There is also a need to probe how an additional mycotoxin (or more that one) affects the pharmacokinetic parameters of toxins, especially their bioavailability, their organ distribution, half-lives and metabolites.
5. While much of the published toxicology has used experiments designed to produce lesions there has been insufficient effort to establish the "no observed effect" levels. Similarly, the dynamics of injury and recovery have rarely been documented. Such data is essential for the risk assessment process.
6. There is a strong rationale to investigate the potential for mycotoxin interactions using *in vitro* systems, ranging from simple assessment of changes in plasma binding characteristics such as displacement of one mycotoxin by another, to more extensive teratogenicity screening using cultured embryos and investigate cytotoxicity in target cell types (both animals and human) exposed to a range of mycotoxin mixtures.
7. Data relating to the full spectrum of pathophysiological changes and biochemical mechanisms of toxicological action of each mycotoxin, from which rational strategies for risk assessment can be built up, need to be accumulated and related to species differences.

8. Very little is known about how humans handle single mycotoxins and there is no data on the pharmacokinetics or dynamics of multiple mycotoxins in humans. These data need to be obtained in those regions of the world where human exposure is common. This can only be achieved by a concerted and collaborative scientific effort involving appropriate experimental design and the use of sophisticated analysis.

It is difficult to envisage how any shift to a new target organ can be predicted from our current understandings, but it is likely that a composite of information on pathology and toxicokinetics based on investigation from points 1-8 will give a clearer picture of such changes.

Future prospects for risk assessment in animals and man

The estimation of a "no (adverse) observed effect level" or maximum tolerated levels of contamination permissible for specific food or feeds, which would not endanger animal well-being (Park and Pohland, 1986), is a pre-requisite for making any rational risk assessment.

Conclusion

The available data do not provide an adequate basis for risk assessment of multiple mycotoxin exposure to domestic animals or man, but do highlight examples of toxicological interactions. In order to assess the health significance there is a need to consolidate data on exposure levels in different parts of the world, study the metabolic handling of each mycotoxin in a range of species, both on its own and in the presence of interacting mycotoxins. Information is needed on the mechanistic basis of the adverse effects of each mycotoxin. Once such data has been assembled, they can be used to establish a hierarchy of risk, and thence a series of priorities of how to prevent the presence of multiple mycotoxins in food or limit the adverse health effects they cause.

Acknowledgements

Jess Clark and Sarita Mohur helped prepare the manuscript. Dr. A.T. Evans and Stephen Brant provided useful critical comment. The authors' research was supported by the International Agency for Research in Cancer, European Commission, Humane Research Trust, British Council and (MMcL) University of Natal.

References

Autrup, J.L., Schmidt, J., Seremet, T., and Autrup, H. (1991): Determination of exposure to aflatoxins among Danish workers in animal-feed production through the analysis of aflatoxin B1 adducts to serum albumin. *Scand. J. Work. Environ. Health,* **17**: 436-40.

Babich, H. and Borenfreund, E. (1991): Cytotoxicity of T-2 toxin and its metabolites determined with the neutral red cell viability assay. *Appl. Environ. Microbiol.*, **57**: 2101-3

Baertschi, S.W., Raney, K.D., Shimada, T., Harris, T.M., and Guengerich., F.P. (1989): Comparison of rates of enzymatic oxidation of aflatoxin B1, aflatoxin G1, and sterigmatocystin and activities of the epoxides in forming guanyl-N7 adducts and inducing different genetic responses. *Chem. Res. Toxicol.*, **2**: 114-2.

Blaude, M.N., Goethals, F.M., Ansay, M.A., and Roberfroid., M.B. (1990): Interaction between aflatoxin B1 and oxytetracycline in isolated rat hepatocytes. *Cell Biol. Toxicol.*, **6**: 339-51.

Blaude, M.N., Goethals, F.M., Ansay, M.A., and Roberfroid, M.B. (1992): Synergism between aflatoxin B1 and oxytetracycline on fatty acid esterification in isolated rat hepatocytes. *Toxicol. Lett.*, **61**: 159-66.

Brown, T.P., Manning R.O., Fletcher O.J., and Wyatt R.D. (1986): The individual and combined effects of citrinin and ochratoxin A on renal ultrastructure in layer chicks. *Avian Dis.*, **30**: 191-198.

Brownie, C.F., and Brownie, C. (1988): Preliminary study on serum enzyme changes in Long Evans rats given parenteral ochratoxin A, aflatoxin B_1 and their combination. *Vet. Human Toxicol.*, **30**: 211-214.

Campbell, M.L., May J.D., Huff W.E., and Doerr J.A. (1983): Evaluation of immunity of young broiler chickens during simultaneous aflatoxicosis and ochratoxicosis. *Poult. Sci.*. **62**: 2138-2144.

Chelkowski, J., Golinski, P., and Wiewiorowska, M. (1987): Mycotoxins in cereal grain. Part 12. Contamination with ochratoxin A and penicillic acid as indicator of improper storage of cereal grain. *Nahrung*, **31**: 81-84.

Ciegler, A. (1972): Bioproduction of ochratoxin and penicillic acid by members of the *Aspergillus ochraceus* group. *Canad. J. Microbiol.*, **18**: 631-636.

Creppy, E.E., Lorkowski, G., Beck, G., Rschenthaler, R., and Dirheimer, G. (1980): Combined action of citrinin and ochratoxin A on hepatoma tissue culture cells. *Toxicol. Lett.*, **5**: 375-380.

Giroir, L.E., Huff, W.E., Kubena, L.F., Harvey, R.B., Elissalde, M.H., Witzel, D.A., Yersin, A.G., and Ivie, G.W. (1991): The individual and combined toxicity of kojic acid and aflatoxin in broiler chickens. *Poult. Sci.*, **70**: 1351-6.

Glahn, R.P., Wideman, R.F., Evangelisti, J.W., and Huff, W.E. (1988): Effects of ochratoxin A alone and in combination with citrinin on kidney function of single comb white Leghorn pullets. *Poult. Sci.*, **67**: 1034-1042.

Groopman, J.D., Roebuck, B.D., and Kensler, T.W. (1992a): Molecular dosimetry of aflatoxin DNA adducts in humans and experimental rat models. *Prog. Clin. Biol. Res.*, **374**: 139-55.

Groopman., J.D., Zhu, J.Q., Donahue, P.R., Pikul, A., Zhang, L.S., Chen, J.S., and Wogan, G.N. (1992b): Molecular dosimetry of urinary aflatoxin-DNA adducts in people living in Guangxi Autonomous Region, People's Republic of China. *Cancer. Res.,* **52**: 45-52.

Groopman, J.D., DeMatos, P., Egner, P.A., Love-Hunt, A., and Kensler, T.W. (1992c): Molecular dosimetry of urinary aflatoxin-N7-guanine and serum aflatoxin-albumin adducts predicts chemoprotection by 1,2-dithiole-3-thione in rats. *Carcinogenesis*. **13**: 101-6.

Harvey, R.B., Huff, W.E., Kubena, L.F., and Phillips, T.D. (1989): Evaluation of diets cocontaminated with aflatoxin and ochratoxin fed to growing pigs. *Am. J. Vet. Res.*, **50**: 1400-1405.

Harvey, R.B., Kubena, L.F., Corrier, D.E., and Huff, W.E., Rottinghaus, G.E. (1990a): Cutaneous ulceration and necrosis in pigs fed aflatoxin- and T-2 toxin-contaminated diets. *J. Vet. Diagn. Invest.*, **2**: 227-9.

Harvey, R.B., Kubena, L.F., Huff, W.E., Corrier, D.E., Rottinghaus, G.E., and Phillips, T.D. (1990b): Effects of treatment of growing swine with aflatoxin and T-2 toxin. *Am. J. Vet. Res.*, **51**: 1688-93.

Hayes, A.W., Cain, J.A., and Moore, B.G. (1977): Effect of aflatoxin B1, ochratoxin A and rubratoxin B on infant rats. *Food Cosmet. Toxicol.*, **15**: 23-27.

Hood, R.D., Kuczuk, M.H., and Szczech, G.M. (1978): Effects in mice of simultaneous prenatal exposure to ochratoxin A and T-2 toxin. *Teratology*, **17**: 25-30.

Huff, W.E., and Doerr, J.A. (1981): Synergism between aflatoxin and ochratoxin A in broiler chickens. *Poult. Sci.*, **60**: 550-555.

Huff, W.E., Doerr, J.A., Wabeck, C.J., Chaloupka, G.W., May, J.D., and Merkley, J.W. (1983): Indiviual and combined effects on aflatoxin and ochratoxin A on bruising in broiler chickens. *Poult. Sci.*, **62**: 1764-1771.

Huff, W.E., Doerr, J.A., Wabeck, C.J., Chaloupka, G.W., May, J.D., and Merkley, J.W. (1984): The individual and combined effects of aflatoxin and ochratoxin A on various processing parameters of broiler chickens. *Poult. Sci.*, **63**: 2153-2161.

Huff, W.E., Kubena, L.F., Harvey, R.B., Hagler, W.M., Swanson, S.P., Phillips, T.D., and Creger, C.R. (1986): Individual and combined effects of aflatoxin and deoxynivalenol (DON, Vomitoxin) in broiler chickens. *Poult. Sci.*, **65**: 1291-1298.

Huff, W.E., Kubena, L.F., Harvey, R.B., and Doerr, J.A. (1988a): Mycotoxin interactions in poultry and swine. *J. Animal Sci.*, **66**: 2351-2355.

Huff, W.E., Harvey, R.B., Kubena, L.F. and Rottinghaus, G.E. (1988b): Toxic synergism between aflatoxin and T-2 toxin in broiler chickens. *Poult. Sci.*, **67**: 1418-1423.

Huff, W.E., Ruff, M.D., and Chute, M.B. (1992): Characterization of the toxicity of the mycotoxins aflatoxin, ochratoxin, and T-2 toxin in game birds. II. Ringneck pheasant. *Avian-Dis*. **36**: 30-3.

Jelinek, C.F., Pohland, A.E., and Wood, G.E. (1989): Worldwide occurrence of mycotoxins in foods and feeds--an update. *J.A.O.A.C.*, **72**: 223-230.

Jonsyn, F.E. (1988): Seedborne fungi of sesame (Sesamum indicum L) in Sierra Leone and their potential aflatoxin/mycotoxin production. *Mycopathologia*, **104**: 123-127.

Kanisawa, M. (1984): Synergistic effect of citrinin on hepatorenal carcinogenesis of ochratoxin A in mice. *Devel. Food Sci.*, **7**: 245-254.

Kodama, M., Inoue, F., and Akao, M. (1990): Enzymatic and non-enzymatic formation of free radicals from aflatoxin B1. *Free Radic. Res. Commun.*, **10**: 137-42.

Krivobok, S., Olivier, P., Marzin, D.R., Seigle-Murandi, F., and Steiman, R. (1987): Study of the genotoxic potential of 17 mycotoxins with the SOS Chromotest. *Mutagenesis*, **2**: 433-439.

Krogh, P., Hald, B., and Pederson, E.J. (1973): Occurrence of ochratoxin A and citrinin in cereals associated with mycotoxic porcine nephropathy. *Acta Pathol. Microbiol. Scand. B*, **81**: 689-695.

Kubena, L.F., Phillips, T.D., Witzel, D.A., and Heidelbaugh, N.D. (1984): Toxicity of ochratoxin A and penicillic acid to chicks. *Bull. Environ. Contam. Toxicol.*, **32**: 717-723.

Kubena, L.F., Huff, W.E., Harvey, R.B., Corrier, D.E., Phillips, T.D., and Creger, C.R. (1988): Influence of ochratoxin A and deoxynivalenol on growing broiler chicks. *Poult. Sci.*, **67**: 253-260.

Kubena, L.F., Harvey, R.B., Huff, W.E., Corrier, D.E., Phillips, T.D., and Rottinghaus, G.E. (1989): Influence of ochratoxin A and T-2 toxin singly and in combination on broiler chickens. *Poult. Sci.*, **68**: 867-872.

Lillehoj, E.B., and Ciegler, A. (1975): Mycotoxin Synergism. In *Microbiology*, Edited by D. Schlessinger. pp. 344-358. American Society for Microbiology, Washington DC.

Lindenfelser, L.A., Lillehoj, E.B., and Burmeister, H.R. (1974): Aflatoxin and trichothecene toxins: Skin tumour induction and synergistic acute toxicity in white mice. *J.N.C.I.*, **52**: 113-116.

Manning, R.O., Brown, T.P., Wyatt, R.D., and Fletcher, O.J. (1985): The individual and combined effects of citrinin and ochratoxin A in broiler chicks. *Avian Dis.*, **29**: 986-997.

Mayura, K., Edwards, J.F., Maull, E.A., and Phillips, T.D. (1989): The effects of ochratoxin A on postimplantation rat embryos in culture. *Arch. Environ. Contam. Toxicol.*, **18**: 411-415.

Mayura, K., Parker, R., Berndt, W.O., and Phillips, T.D. (1984): Effect of simultaneous prenatal exposure to ochratoxin A and citrinin in the rat. *J. Toxicol. Environ. Health*, **13**: 553-561.

McLean, M. and Bach, P.H. (1993) Multiple mycotoxin exposure: Risk to domestic animals and man. *Food Chem. Toxicol.*, **Submitted**

Mehlmann, M.A., Hemstreet, C.P., Thorpe, J.J. and Weaver, N.K., (1984). [Editors] Renal Effects of Petroleum Hydrocarbons. Princeton Scientific Publishers, Inc., Princeton, NY.

Micco, C., Miraglia, M., Onori, R., Ioppolo, A., and Mantovani, A. (1987): Long-term administration of low doses of mycotoxins in poultry. 1. Residues of ochratoxin A in broilers and laying hens. *Poult. Sci.*, **66**: 47-50.

Micco, C., Miraglia, M., Benelli, L., Onori, R., Ioppolo, A., and Mantovani, A.L. (1988): Long term administration of low doses of mycotoxins in poultry. 2. Residues of ochratoxin A and aflatoxin in broilers and laying hens after combined administration of ochratoxin A and aflatoxin B1. *Food Add. Contam.*, **5**: 309-314.

Micco, C., Grossi, M., Miraglia, M., and Brera, C. (1989): A study of the contamination by ochratoxin A of green and roasted coffee beans. *Food Add. Contam.*, **6**: 333-339.

Norred, W.P., Bacon, C.W., Plattner, R.D. and Vesonder, R.F. (1991): Differential cytotoxicity and mycotoxin content among isolates of Fusarium moniliforme. *Mycopathologia*, **115**: 37-43

Park, D.L., and Pohland, A.E. (1986): A rationale for the control of aflatoxin in animal feeds. In *Mycotoxins and Phycotoxins.*, Edited by P.S. Steyn, and R. Vleggaar. pp. 473-481. Elsevier Science Publishers, Amsterdam, The Netherlands.

Parker, R.W., Phillips, T.D., Kubena, L.F., Russell, L.H., and Heidelbaugh, N.D. (1982): Inhibition of pancreatic carboxypeptidase A: A possible mechanism of interaction between penicillic acid and ochratoxin A. *J. Environ. Sci. Health*, **B17**, 77-91.

Pathre, S.V., and Mirocha, C.J. (1977): Assay methods for trichothecenes and review of their natural occurrence. In *Mycotoxins in Human and Animal Health*, Edited by J.V. Rodricks, C.W. Hesseltine and M.A. Mehlman. pp. 229-253 Pathotox Publishers, Park Forest South, Illinois.

Patterson, D.S.P., Roberts, B.A., Shreeve, B.J., Wrathall, A.E., and Gitter, M. (1977): Aflatoxin, ochratoxin, and zearelonone in animal feedstuffs: Some clinical and experimental observations. *Ann. Nutrit. aliment.*, **31**: 643-650.

Patterson, D.S.P., Schreeve, B.J., Roberts, B.A., Berrett, S., Brush, P.J., Glancy, E.M., and Krogh, P. (1981): Effect on calves of barley naturally contaminated with ochratoxin A and groundnut meal contaminated with low concentrations of aflatoxin B1. *Res. Vet. Sci.*, **31**: 213-218.

Pier, A.C., Richard, J.L., and Cysewski, S.J. (1980): Implications of mycotoxins in animal diseases. *J. Am. Vet. Med. Assoc.*, **176**: 719-724.

Pier, A.C., Varman, M.J., Dahlgren, R.R., Belden, E.L., and Maki, L.R. (1986) Aflatoxic suppression of cell mediated response and interaction with T-2 toxin. In *Mycotoxins and Phycotoxins.*, Edited by P.S. Steyn and R. Vleggaar pp. 423-434. Elsevier Science Publishers, Amsterdam.

Rati, E.R., Basappa, S.C., Sreenivasa Murthy V., Ramesh, H.P., and Ramesh, B.S. (1981): The synergistic effect of aflatoxin B1 and ochratoxin A in rats. *J. Food Sci. Technol.,* **18**: 176-179.

Rati, E.R., Shantha, T., and Ramesh, H.P. (1991): Effect of long term feeding and withdrawal of aflatoxin B1 and ochratoxin A on kidney cell transformation in albino rats. *Indian. J. Exp. Biol.*, **29**: 813-7.

Riley, R.T. and Showker, J.L. (1991): The mechanism of patulin's cytotoxicity and the antioxidant activity of indole tetramic acids. *Toxicol. Appl. Pharmacol.*, **109**: 108-26

Ross, R.K., Yuan, J.M., Yu, M.C., Wogan, G.N., Qian, G.S., Tu, J.T., Groopman, J.D., Gao, Y.T, and Henderson, B.E. (1992): Urinary aflatoxin biomarkers and risk of hepatocellular carcinoma. *Lancet*. **339**: 943-6.

Rotter, B.A., Thompson, B.K., Prelusky, D.B. and Trenholm, H.L. (1991): Evaluation of potential interactions involving trichothecene mycotoxins using the chick embryotoxicity bioassay. *Arch. Environ. Contam. Toxicol.*, **21**: 621-4

Ruff, M.D., Huff, W.E., and Wilkins, G,C. (1990): Characterization of the toxicity of the mycotoxins aflatoxin, ochratoxin, and T-2 toxin in game birds. I. Chukar partridge. *Avian Dis.*, **34**: 717-20.

Ruff, M.D., Huff, W.E., and Wilkins, G.C. (1992): Characterization of the toxicity of the mycotoxins aflatoxin, ochratoxin, and T-2 toxin in game birds. III. Bobwhite and Japanese quail. *Avian Dis.*, **36**: 34-9.

Sandor, G., Busch, A., Watzke, H., Reek, J., and Vanyi, A. (1991): Subacute toxicity testing of ochratoxin A and citrinin in swine. *Acta Vet. Hung.*, **39**: 149-60.

Sansing, G.A., Lillehoj, J.B., Detroy, R.N., and Miller, M.A. (1976): Synergistic toxic effects of citrinin, ochratoxin A and penicillic acid in mice. *Toxicon*, **14**: 213-220.

Scott P.M. (1978): Mycotoxins in feeds and ingredients and their origin. *J. Food Protect.*, **41**: 385-398.

Scott, P.M., van Walbeek W., Kennedy, B., and Anyeti, D. (1972): Mycotoxins (ochratoxin A, citrinin, and sterigmatocystin) and toxigenic fungi in grains and other agricultural products. *J. Agric. Food Chem.*, **20**: 1103-1109.

Shepherd, E.C., Phillips, T.D., Joiner, G.N., Kubena, L.F., and Heidelbaugh, N.D. (1981): Ochratoxin A and penicillic acid interaction in mice. *J. Environ. Sci. Health*, **B16**: 557-573.

Siraj, M.Y., Phillips, T.D., and Hayes, A.W. (1981): Effects of the mycotoxins citrinin and ochratoxin A on hepatic mixed-function oxidase and adenosinetriphosphatase in neonatal rats. *J. Toxicol. Environ. Health*, **8**: 131-140.

Smith, E.E., Kubena, L.F., Braithwaite, C.E., Harvey, R.B., Phillips, T.D., and Reine, A.H. (1992): Toxicological evaluation of aflatoxin and cyclopiazonic acid in broiler chickens. *Poult. Sci.* **71**: 1136-44.

Soares, L.M., and Rodriguez-Amaya, D.B. (1989): Survey of aflatoxins, ochratoxin A, zearalenone, and sterigmatocystin in some Brazilian foods by using multi-toxin thin-layer chromatographic method. *J.A.O.A.C.*, **72**: 22-26.

Tapia, M.O., and Seawright, A.A. (1985): Experimental combined aflatoxin B_1 and ochratoxin A intoxication in pigs. *Aust. Vet. J.*, **62**: 33-37.

Ueno Y. (1987): Trichothecenes in food. In *Mycotoxins in Food.*, Edited by P. Krogh pp 123-147. Academic Press, Cambridge.

Watson, D.H. (1985): Toxic fungal metabolites in foods. *CRC Critical Reviews in Food Science and Nutrition*, **22**: 177-198.

WHO (1979): World Health Organisation -IPCS Environmental Health Criteria Document 11, Mycotoxins. WHO/IPCS, Geneve.

WHO (1990): World Health Organisation -IPCS Environmental Health Criteria Document 105, Selected Mycotoxins: Ochratoxin, Trichothecenes, Ergot. WHO/IPCS, Geneve.

Wogan, G.N., Edwards, G.S., and Newberne, P.M. (1971): Acute and chronic toxicology of rubratoxin B. *Toxicol. Appl. Pharmacol.*, **19**: 712-720.

Résumé

A l'heure actuelle, nous savons que les effets biologiques des mycotoxines peuvent être atténués de diverses manières permettant de protéger contre les mycotoxicoses.

Ces différentes méthodes sont compatibles avec la récolte, le stockage, la formulation et la consommation des aliments. Toutes ces stratégies peuvent être spécifiquement étudiées pour les adapter à différentes situations afin de faciliter la protection de l'animal ou de l'homme exposés aux mycotoxines.

Human ochratoxicosis and its pathologies. Eds E.E. Creppy, M. Castegnaro, G. Dirheimer. Colloque INSERM/ John Libbey Eurotext Ltd. © 1993, Vol. 231, pp. 43-49.

The modulation of mycotoxin exposure in domestic animals and man: can we affect what we can't control ?

Peter H. Bach

Drug Development and Chemical Safety Research Unit, Faculty of Science, University of East London, Romford Road, London E15 4LZ, United Kingdom

Introduction

Mycotoxins are ubiquitous in a broad range of commodities and feeds (Chelkowski *et al.*, 1987; Ciegler, 1972; Hayes *et al.* 1977; Huff *et al.* 1986, 1988a,b; Jelinek *et al.*, 1989; Jonsyn, 1988; Krogh *et al.*, 1973; Lillehoj and Ciegler, 1975; Pathre and Mirocha, 1977; Scott *et al.*, 1972; Scott, 1978; Soares and Rodriguez-Amaya, 1989; Ueno, 1987; Watson, 1985) and both animals and man are therefore exposed to levels that are unacceptably high. Risk assessments for aflatoxin (Bruce, 1990: Gorelick, 1990; Bosch and Peers, 1991; Hendrickse, 1991; Robens and Richard, 1992; Hoseyni, 1992), ochratoxin A (OA - Castegnaro *et al.*, 1990; Kuiper-Goodman, 1990; Marquardt *et al.*, 1990; Plestina *et al.*, 1990; Ewald *et al.*, 1991), citrinin (Castegnaro *et al.*, 1990) and zearalenone (Ewald *et al.*, 1991) have been published. These have stimulated the establishment of guideline regulations for mycotoxins limits (Stoloff *et al.*, 1991), although these levels cannot be applied in most countries due to prevailing amounts of mycotoxins and the lack of any strategy on how to reduce them!

Furthermore, the economic importance of commodities to developing nations may be such that even if feeds are heavily contaminated with fungi and mycotoxins there is not necessarily the legislative pressures to destroy them especially because they may be "blended down" with non-contaminated produce. Even if feeds are free of mycotoxins *per se*, sufficient fungal spores may be present such that they reach the market contaminated after transport and storage.

Commodities will continue to be contaminated by one or more mycotoxin producing fungi or thus mycotoxicoses are an ongoing problem. As the world demand for food increases mycotoxin contamination and mycotoxicoses will have to be assessed and controlled. It is difficult to predict mycotoxin production under real agricultural conditions from *in vitro* studies with surface disinfected wheat (Muller and Boley, 1991; see Bach and McLean, 1993; McLean, and Bach 1993), or indeed the range of mycotoxins that will be present since seed moisture content and available seed nutrients determines the spectrum of fungi that might prevail at any one time. This makes the levels of exposure highly variable and unpredictable, and necessitates the continuous monitoring of mycotoxin contamination important for risk assessment, but does not help address the key questions of risk management when heavily contaminated batches of food are identified. Depending on the extent of mycotoxin exposure there are a growing series of solutions that need to be considered in relation to the level of contamination, the types of mycotoxins present and the cost-benefit of the strategy.

Modulating exposure

There are may potential ways of modulating the degree of exposure to mycotoxins in order to affect the body burden in animals and man.

Preventing crop contamination and fungal growth

Preharvest contamination is currently unavoidable, despite efforts to identify and breed genetically resistant plant strains (Widstrom and Zuber, 1983). In addition, insect damage in the field provides opportunistic fungi like *Aspergillus flavus* a portal of entry. Furthermore, there is increasing evidence that seed to seed transmission of fungi could account for some of the contamination observed in preharvest crops (Mycock and Berjak, 1990). Prevention of post-harvest contamination should be easier to control, provided suitable storage conditions, including low relative humidities and prevention of insect infestation are available to curb the activities of many of the toxin-producing fungi. Propionic acid, used as a food preservatives, inhibited aflatoxin production on betel nut for up to 4 weeks. This shows the potential to safeguard the consumers from mycotoxins (Raisuddin and Misra, 1991). However, in reality, most food or feed commodities are unavoidably contaminated with mycotoxins and rarely stored optimally. Clearly, more focused research is needed to document the most acceptable approach to controlling the levels of mycotoxin contamination in foods and feeds between harvesting and consumption.

Decontamination.

There is conflicting data on decontamination. Parboiling steeping (precooking/soaking) of rice grains and bran removal had no effect on aflatoxin levels in Sri Lankan rice and increased the susceptibility to subsequent fungal contamination (Bandara *et al.*, 1991). Roasting coffee samples has been reported to destroy 48-100% of up to 15 mg/kg OA (Micco *et al.*, 1989), but Tsubouchi *et al.*, (1987) showed only a 0-12% reduction. Increasing gamma-irradiation decreased greatly or inhibited production of tenuazonic acid by *Alternaria alternata* in tomato paste and juice (Aziz *et al.*, 1991) and irradiation of mycotoxins solutions decomposed up to 50% OA, whereas heating had no effect (Kostecki *et al.*, 1991).

Ammoniation effectively detoxifies aflatoxins in corn and cottonseed, but recent data (Norred *et al.*, 1991) shows this process did not alter toxicity despite the 30-45% reduced fumonisin B1 levels. Aflatoxin-polluted peanut cakes (up to 3,500 ppb total aflatoxin) have been detoxified with monomethylamine - $Ca(OH)_2$ (Giddey *et al.*, 1992), a promising approach based on cost, efficiency and toxicological safety of the product in rats. Other methods of reducing the feed burden of mycotoxins have been reviewed (WHO, 1979, 1990).

Limiting or preventing mycotoxin effects in exposed species

Some types of spoiled food may be refused by animals. There is one report (Harvey *et al.*, 1987) that swine refused to consume grain sorghum contaminated with 5 mg/kg OA, but the occurrence of residues in porcine kidneys all around the world (Rousseau and Van Peteghem, 1989) suggests that this is at best an extreme example of food spoilage. The actual diet of the livestock may be more important. For example, OA is converted to the non-toxic alpha metabolite much faster in the rumen of hay-fed than grain-fed sheep (Xiao *et al.*, 1991a) due to faster hydrolysis arising from different ruminal microbes (Xiao *et al.*, 1991b).

This opens a number of interesting possibilities such as inoculating susceptible herds of animals with genetically manipulated micro-organisms in order to directly influenced the rate of hydrolysis of OA and so reduce toxicity. This has the added advantage of being preventive in areas where mycotoxin exposure is endemic and also a rapid way of protecting animals exposed to contaminated feeds. In addition, the ability to introduce a number of recently isolated genes (eg., different cytochrome P-450 isoforms, and conjugation enzymes, etc.) makes it feasible to greatly affect the gastric handling of the mycotoxin to reduce or prevent the absorption of toxins. Naturally, genetically engineered bacteria could also be important in protecting humans from OA in high risk areas.

Modulating excretion

Harmane, a beta-carboline formed during pyrolysis of tryptophan, caused a limited increase in the excretion of aflatoxin B1 in faeces but not in urine, without altering the growth process in rats (Billaud, 1991). Bicarbonate pretreatment to alter urinary pH followed by OA decreased the frequency of histological lesions in kidneys, liver, lung, and heart and also decreased acute toxicity (LD_{50}) by 20% in rats (Yong *et al.*, 1987). There is convincing evidence to show probenecid, which inhibits active proximal tubule uptake at the anion transport system (Tune, 1975), reduces citrinin nephrotoxicity (Berndt and Hayes, 1982). The importance of the organic anion transport system to OA nephrotoxicity is, however, controversial. OA competes with p-aminohippurate and phenolsulphophthalein transport in pig kidney cortex slices (Friis *et al.*, 1988) in a dose-dependent manner and rat renal brush border and basolateral membrane vesicles (Sokol *et al.*, 1988) but does not affect organic cation N'-methylnicotinamide. Probenecid has been reported to be protective and prevent the enzymuria caused by a single oral dose of OA (Sokol *et al.*, 1988), but Stein *et al.*, (1985) showed decreased urine osmolality, Na^+ and K^+ concentrations, body weight and increased OA concentrations in the kidneys, suggesting an increased nephrotoxicity. Other transport systems may also be involved, as OA toxicity is closely related to its aromatic amino acid moiety, the significance of which is shown by phenylalanine protecting against OA toxicity (Creppy *et al.*, 1980). This suggests an aromatic amino acid transport system upon which OA enters the cells. Therefore aromatic amino acids may be a suitable competitive inhibitor that prevents the injurious effects of OA (see below).

Modulating gastrointestinal absorption

Kubena *et al.*, (1990a,b) were amongst the first to demonstrate that hydrated, sodium calcium aluminosilicate (HSCAS, 0.5% of the total diet) an anticaking agent for mixed feed, ameliorated aflatoxicosis (caused by purified aflatoxin B1) in chickens whereas activated charcoal did not. HSCAS decreases the level of aflatoxin M1 residues in the milk of lactating dairy cattle (Phillips *et al.*, 1990) and significantly reduced the amount of aflatoxin M1 in liver, kidney, and muscle tissue in swine fed an aflatoxin-contaminated diet. Aflatoxin B1 was not reduced in liver or kidney, but was decreased in muscle (Beaver *et al.*, 1990). HSCAS sequestrates aflatoxin B1 in the gastrointestinal tract and chemisorption reduces bioavailability. It has no toxicological effects on its own (Kubena *et al.*, 1991), and other studies have shown that it also provides significant protection in growing turkey poults (Kubena *et al.*, 1991), chickens and swine (Phillips *et al.*, 1990) and lambs fed aflatoxin for 42 days (Harvey *et al.*, 1991), but did not alter the toxicity of T-2 toxin (Kubena *et al.*, 1990b) in chicks nor did it effect toxicity of OA alone or in combination with aflatoxin and OA (Huff *et al.*, 1992). Interestingly both HSCAS and/or 1.0% activated charcoal reduced or essentially eliminated histopathological lesions in the livers of mink fed diets that contained aflatoxins (Bonna *et al.*, 1991), but neither had any affect on the renal lesions. Neither 1% HSCAS nor 1% and 10% acid bentonite had any effect on blood or tissue levels of pigs fed OA-contaminated feed. Activated charcoal 1% slightly decrease OA in the blood, whereas 10% reduced OA levels in both blood and tissue by 50% to 80%. Reduction of OA absorption via the dietary administration of activated charcoal (5%) was confirmed in a 16 week feeding experiment (Plank *et al.*, 1990).

Modulating cellular absorption

L-phenylalanine supplementation in the diets of young broilers fed diets containing OA reduced mortality in a dose related response (Gibson *et al.*, 1990) and favourably affected body weight, feed conversion, and relative organ weights. The effects on clinical chemistry (Bailey *et al.*, 1990) suggest an improved health status based on a significant regression slopes for uric acid, creatinine, total protein, albumin, and cholesterol. By contrast, Rogers *et al.*, (1991) showed that tryptophan supplemented to a diet containing aflatoxin increased the severity of aflatoxicosis in laying hens.

Conclusion

There is already adequate evidence that the biological effects of mycotoxins can be affected by a range of activities which provide a protection from mycotoxicosis. Many of these are compatible with feed harvesting, storage, formulation or consumption. The range of strategies outlined have the potential to be further investigated and applied to a range of different situations to facilitate the protection of animals and man in risk situations.

Acknowledgements

Dr. A.T. Evans and Stephen Brant provided useful critical comment. The authors' research was supported by the International Agency for Research in Cancer, the Cancer Research Campaign and the European Commission.

References

Aziz, N.H., Farag, S., and Hassanin, M.A. (1991): Effect of gamma irradiation and water activity on mycotoxin production of Alternaria in tomato paste and juice. *Nahrung*. **35**: 359-62.

Bach, P.H. and McLean, M. (1993) Research Priorities for Assessing the Risk of Multiple Mycotoxin Exposure to Domestic Animals and Man: What we know and what we need to know! *This volume*.

Bailey, C.A., Gibson, R.M., Kubena, L.F., Huff, W.E., and Harvey, R.B. (1990): Impact of L-phenylalanine supplementation on the performance of three-week-old broilers fed diets containing ochratoxin A. 2. Effects on hematology and clinical chemistry. *Poult. Sci.* **69**: 420-5.

Bandara, J.M., Vithanege, A.K., and Bean, G.A. (1991): Effect of parboiling and bran removal on aflatoxin levels in Sri Lankan rice. *Mycopathologia*. **115**: 31-5.

Beaver, R.W., Wilson, D.M., James, M.A., Haydon, K.D., Colvin, B.M., Sangster, L.T., Pikul, A.H., and Groopman, J.D. (1990): Distribution of aflatoxins in tissues of growing pigs fed an aflatoxin-contaminated diet amended with a high affinity aluminosilicate sorbent. *Vet. Hum. Toxicol.*, **32**: 16-8.

Berndt W.O., and Hayes A.W. (1982): The effect of probenecid on citrinin-induced nephrotoxicity. *Toxicol. Appl. Pharmacol.*, **64**: 118- 124.

Billaud, C. (1991): Effects of harmane on growth and in vivo metabolism of aflatoxin B1 in male and female rats. *Food Addit. Contam.*, **8**: 713-22.

Bonna, R.J., Aulerich, R.J., Bursian, S.J., Poppenga, R.H., Braselton, W.E., and Watson, G.L. (1991): Efficacy of hydrated sodium calcium aluminosilicate and activated charcoal in reducing the toxicity of dietary aflatoxin to mink. *Arch. Environ. Contam. Toxicol.*, **20**: 441-7.

Bosch, F.X., and Peers, F. (1991): Aflatoxins: data on human carcinogenic risk. *IARC Sci. Publ.*, 48-53.

Bruce, R.D. (1990): Risk assessment for aflatoxin: II. Implications of human epidemiology data. *Risk Anal.*, **10**: 561-9.

Castegnaro, M., Chernozemsky, I.N., Hietanen, E., and Bartsch, H. (1990): Are mycotoxins risk factors for endemic nephropathy and associated urothelial cancers? *Archiv. Geschwulst.*, **60**: 295-303.

Chelkowski, J., Golinski, P., and Wiewiorowska, M. (1987): Mycotoxins in cereal grain. Part 12. Contamination with ochratoxin A and penicillic acid as indicator of improper storage of cereal grain. *Nahrung*, **31**: 81-84.

Ciegler, A. (1972): Bioproduction of ochratoxin and penicillic acid by members of the *Aspergillus ochraceus* group. *Can. J. Microbiol.*, **18**: 631-636.

Creppy, E.E., Schlegel, M., Roschenthaler, R. and Dirheimer, G. (1980): Phenylalanine prevents acute poisoning by ochratoxin A in mice. *Toxicol. Lett.*, **6**: 77-80.

Ewald, C., Rehm, A., and Haupt, C. (1991): Mykotoxine als Risikofaktor fur das Entstehen von Krankheiten und Leistungsminderung in Schweinebestanden - eine epidemiologische Studie. *Berl. Munch. Tierarztl. Wochenschr.*, **104**: 161-6.

Friis C., Brinn R., and Hald B. (1988): Uptake of ochratoxin A by slices of pig kidney cortex. *Toxicol.*, **52**: 209-217.

Gibson, R.M., Bailey, C.A., Kubena, L.F., Huff, W.E. and Harvey, R.B. (1990): Impact of L-phenylalanine supplementation on the performance of three-week-old broilers fed diets containing ochratoxin A. 1. Effects on body weight, feed conversion, relative organ weight, and mortality. *Poult. Sci.* **69**: 414-9.

Giddey, C., Bunter, G., Larroux, R., Jemmali, M., and Rossi, J. (1992): Detoxification of aflatoxin-polluted peanut cakes with monomethylamine/Ca(OH)2: pilot industrial application, nutrition experiments, toxicity evaluation. *J. Environ. Pathol. Toxicol. Oncol.*, **11**: 60-3.

Gorelick, -N.J. (1990): Risk assessment for aflatoxin: I. Metabolism of aflatoxin B1 by different species. *Risk Anal.*, **10**: 539-59.

Harvey R.B., Kubena L.F., Lawhorn D.B., Fletcher O.J., and Phillips T.D. (1987): Feed refusal in swine fed ochratoxin-contaminated grain sorghum: evaluation of toxicity in chicks. *J. Am. Vet. Med. Ass.*, **190**: 673-675.

Harvey, R.B., Kubena, L.F., Phillips, T.D., Corrier, D.E., Elissalde, M.H., and Huff, W.E. (1991): Diminution of aflatoxin toxicity to growing lambs by dietary supplementation with hydrated sodium calcium aluminosilicate. *Am. J. Vet. Res.*, **52**: 152-6.

Hayes A.W. (1985): *Mycotoxin Teratogenicity*. CRC Press Inc., Boca Raton, Florida.

Hayes A.W., Cain J.A., and Moore B.G. (1977): Effect of aflatoxin B1, ochratoxin A and rubratoxin B on infant rats. *Food Cosmet. Toxicol.*, **15**: 23-27.

Hendrickse, R.G. (1991): Clinical implications of food contaminated by aflatoxins. *Ann. Acad. Med. Singapore.*, **20**: 84-90.

Hoseyni, M.S. (1992): Risk assessment for aflatoxin: III. Modeling the relative risk of hepatocellular carcinoma. *Risk Anal.*, **12**: 123-8.

Huff, W.E., Kubena, L.F., Harvey, R.B., Hagler W.M., Swanson S.P., Phillips, T.D., and Creger, C.R. (1986): Individual and combined effects of aflatoxin and deoxynivalenol (DON, Vomitoxin) in broiler chickens. *Poult. Sci.*, **65**: 1291-1298.

Huff, W.E., Harvey, R.B., Kubena, L.F., and Rottinghaus, G.E. (1988a): Toxic synergism between aflatoxin and T-2 toxin in broiler chickens. *Poult. Sci.*, **67**: 1418-1423.

Huff W.E., Kubena L.F., Harvey R.B., and Doerr J.A. (1988b): Mycotoxin interactions in poultry and swine. *J. Animal Sci.*, **66**: 2351-2355.

Huff, W.E., Kubena, L.F., Harvey, R.B., and Phillips, T.D. (1992): Efficacy of hydrated sodium calcium aluminosilicate to reduce the individual and combined toxicity of aflatoxin and ochratoxin A. *Poult. Sci.* **71**: 64-9.

Jelinek, C.F., Pohland, A.E., and Wood, G.E. (1989): Worldwide occurrence of mycotoxins in foods and feeds--an update. *J.A.O.A.C.*, **72**: 223-230.

Jonsyn, F.E. (1988): Seedborne fungi of sesame (Sesamum indicum L) in Sierra Leone and their potential aflatoxin/mycotoxin production. *Mycopathologia*, **104**: 123-127.

Kostecki, M., Golinski, P., Uchman, W., Grabarkiewicz-Szczesna, J. (1991): Decomposition of ochratoxin A by heat and gamma-irradiation. *IARC Sci. Publ.*, 109-11.

Krogh P. (1987): Ochratoxins in food. In *Mycotoxins in Food*,. Edited by P. Krogh. pp. 97-121. Academic Press, Cambridge.

Krogh P., Hald B., and Pederson E.J. (1973): Occurrence of ochratoxin A and citrinin in cereals associated with mycotoxic porcine nephropathy. *Acta Pathol. Microbiol. Scand. B*, **81**: 689-695.

Kubena, L.F., Harvey, R.B., Phillips, T.D., Corrier, D.E., Huff, W.E. (1990a): Diminution of aflatoxicosis in growing chickens by the dietary addition of a hydrated, sodium calcium aluminosilicate. *Poult. Sci.* **69**: 727-35.

Kubena, L.F., Harvey, R.B., Huff, W.E., Corrier, D.E., Phillips, T.D., Rottinghaus, G.E. (1990b): Efficacy of a hydrated sodium calcium aluminosilicate to reduce the toxicity of aflatoxin and T-2 toxin. *Poult. Sci.* **69**: 1078-86.

Kubena, L.F., Huff, W.E., Harvey, R.B., Yersin, A.G., Elissalde, M.H., Witzel, D.A., Giroir, L.E., Phillips, T.D., Petersen, H.D. (1991): Effects of a hydrated sodium calcium aluminosilicate on growing turkey poults during aflatoxicosis. *Poult. Sci.* **70**: 1823-30.

Kuiper-Goodman, T. (1990): Uncertainties in the risk assessment of three mycotoxins: aflatoxin, ochratoxin, and zearalenone. *Can. J. Physiol. Pharmacol.*, **68**: 1017-24.

Kuronen, P. (1989): High-performance liquid chromatographic screening method for mycotoxins using new retention indexes and diode array detection. *Arch. Environ. Contam. Toxicol.*, **18**: 336-348.

Lillehoj, E.B., and Ciegler, A. (1975): Mycotoxin Synergism. In *Microbiology*, Edited by D. Schlessinger. pp. 344-358. American Society for Microbiology, Washington DC.

Marquardt, R.R., Frohlich, A., and Abramson, D. (1990): Ochratoxin A: an important western Canadian storage mycotoxin. *Can. J. Physiol. Pharmacol.*, **68**: 991-9.

McLean, M. and Bach, P.H. (1993) Multiple mycotoxin exposure: Risk to domestic animals and man. *Food Chem. Toxicol.*, **Submitted**

Micco, C., Miraglia, M., Onori, R., Ioppolo, A., and Mantovani A. (1987): Long-term administration of low doses of mycotoxins in poultry. 1. Residues of ochratoxin A in broilers and laying hens. *Poult. Sci.*, **66**: 47-50.

Micco, C., Miraglia, M., Benelli, L., Onori, R., Ioppolo, A., and Mantovani, A.L. (1988): Long term administration of low doses of mycotoxins in poultry. 2. Residues of ochratoxin A and aflatoxin in broilers and laying hens after combined administration of ochratoxin A and aflatoxin B1. *Food Add. Contam.*, **5**: 309-314.

Micco, C., Grossi, M., Miraglia, M., and Brera C. (1989): A study of the contamination by ochratoxin A of green and roasted coffee beans. *Food Add. Contam.*, **6**: 333-339.

Muller, H.M., Boley, A. (1991): Einfluss von Autoklavierung und Oberflachendesinfektion von Weizen (Triticum aestivum) auf die Produktion von Ergosterin, Ochratoxin A und Citrinin durch Penicillium verrucosum. *Zentralbl. Mikrobiol.* **146**: 445-51.

Mycock, D.J., Rijkenberg, F.H.J., and Berjak, P. (1990): Infection of maize seedlings by Aspergillus flavus var. columnaris. *Seed Sci. Technol.*, **18**: 693-701.

Norred, W.P., Voss, K.A., Bacon, C.W., and Riley, R.T. (1991): Effectiveness of ammonia treatment in detoxification of fumonisin-contaminated corn. *Food Chem. Toxicol.*, **29**: 815-9.

Pathre S.V. and Mirocha C.J. (1977): Assay methods for trichothecenes and review of their natural occurrence. In *Mycotoxins in Human and Animal Health*, Edited by J.V. Rodricks, C.W. Hesseltine and M.A. Mehlman. pp. 229-253 Pathotox Publishers, Park Forest South, Illinois.

Phillips, T.D., Clement, B.A., Kubena, L.F., and Harvey, R.B. (1990): Detection and detoxification of aflatoxins: prevention of aflatoxicosis and aflatoxin residues with hydrated sodium calcium aluminosilicate. *Vet. Hum. Toxicol.*, **32**: 15-9.

Plank, G., Bauer, J., Grunkemeier, A., Fischer, S., Gedek, B., Berner, H. (1990): Untersuchungen zur protektiven Wirkung von Adsorbentien gegenuber Ochratoxin A beim Schwein. *Tierarztl. Prax.*, **18**: 483-9.

Plestina, R., Ceovic, S., Gatenbeck, S., Habazin-Novak, V., Hult, K., Hokby, E., Krogh, P., Radic, B. (1990): Human exposure to ochratoxin A in areas of Yugoslavia with endemic nephropathy. *J. Environ. Pathol. Toxicol. Oncol.*, **10**: 145-8.

Raisuddin, S., and Misra, J.K. (1991): Aflatoxin in betel nut and its control by use of food preservatives. *Food Addit. Contam.*, **8**: 707-12.

Robens, J.F., Richard, J.L. (1992): Aflatoxins in animal and human health. *Rev. Environ. Contam. Toxicol.*, **127**: 69-94.

Rogers, S.R., Pesti, G.M., Wyatt, R.D. (1991): Effect of tryptophan supplementation on aflatoxicosis in laying hens. *Poult. Sci.* **70**: 307-12.

Rousseau, D.M., and Van-Peteghem, C.H. (1989): Spontaneous occurrence of ochratoxin A residues in porcine kidneys in Belgium. *Bull. Environ. Contam. Toxicol.*, **42**: 181-186.

Scott, P.M. (1978): Mycotoxins in feeds and ingredients and their origin. *J. Food Protect.*, **41**: 385-398.

Scott, P.M., van Walbeek, W., Kennedy, B., and Anyeti, D. (1972): Mycotoxins (ochratoxin A, citrinin, and sterigmatocystin) and toxigenic fungi in grains and other agricultural products. *J. Agric. Food Chem.*, **20**: 1103-1109.

Soares, L.M. and Rodriguez-Amaya D.B. (1989): Survey of aflatoxins, ochratoxin A, zearalenone, and sterigmatocystin in some Brazilian foods by using multi-toxin thin-layer chromatographic method. *J.A.O.A.C.*, **72**: 22-26.

Sokol, P.P., Ripich, G., Holohan, P.D., and Ross, C.R. (1988): Mechanism of ochratoxin A transport in kidney. *J. Pharmacol. Exper. Therap.*, **246**: 460-465.

Stein, A.F., Phillips, T.D., Kubena, L.F., and Harvey, R.B. (1985): Renal tubular secretion and reabsorption as factors in ochratoxicosis: Effects of probenecid on nephrotoxicity. *J. Toxicol. Environ. Health*, **16**: 593-605.

Stoloff, L., Van-Egmond, HP., Park, D.L. (1991): Rationales for the establishment of limits and regulations for mycotoxins. *Food Addit. Contam.*, **8**: 213-21.

Tsubouchi, H., Yamamoto, K., Hisada, K., Sakabe, Y., and Udagawa, S. (1987): Effect of roasting on ochratoxin A level in green coffee beans inoculated with Aspergillus ochraceus. *Mycopathologia*, **97**: 111-115.

Tune, B.M. (1975): Relationship between the transport and toxicity of cephalosporins in the kidney. *J. Infect. Dis.*, **132**: 189-194.

Ueno, Y. (1987): Trichothecenes in food. In *Mycotoxins in Food.*, Edited by P. Krogh pp 123-147. Academic Press, Cambridge.

Watson, D.H. (1985): Toxic fungal metabolites in foods. *CRC Crit. Rev. Food Sci. Nutrit.*, **22**: 177-198.

WHO (1979): World Health Organisation - IPCS Environmental Health Criteria Document 11, Mycotoxins. WHO/IPCS, Geneve.

WHO (1990): World Health Organisation - IPCS Environmental Health Criteria Document 105, Selected Mycotoxins: Ochratoxin, Trichothecenes, Ergot. WHO/IPCS, Geneve.

Widstrom, N.W. and Zuber, M.S. (1983): Sources and mechanisms of genetic control in the plant. In *Southern Series Co-operative Series Bulletin 279.*, Edited by U.L. Diener, R.L. Asquith and J.W. Dickens pp. 72-76.

Xiao, H., Marquardt, R.R., Frohlich, A.A., Phillips, G.D., and Vitti, T.G. (1991a): Effect of a hay and a grain diet on the bioavailability of ochratoxin A in the rumen of sheep. *J. Anim. Sci.* **69**: 3715-23.

Xiao, H., Marquardt, R.R., Frohlich, A.A., Phillips, G.D., Vitti, T.G. (1991b): Effect of a hay and a grain diet on the rate of hydrolysis of ochratoxin A in the rumen of sheep. *J. Anim. Sci.*, **69**: 3706-14.

Yong, S., Albassam, M., and Prior, M. (1987): Protective effects of sodium bicarbonate on murine ochratoxicosis. *J. Environ. Sci. Health B*, **22**: 455-470.

Résumé

A l'heure actuelle, nous savons que les effets biologiques des mycotoxines peuvent être atténués de diverses manières permettant de protéger contre les mycotoxicoses.

Ces différentes méthodes sont compatibles avec la récolte, le stockage, la formulation et la consommation des aliments. Toutes ces stratégies peuvent être spécifiquement étudiées pour les adapter à différentes situations afin de faciliter la protection de l'animal ou de l'homme exposés aux mycotoxines.

Human ochratoxicosis and its pathologies. Eds E.E. Creppy, M. Castegnaro, G. Dirheimer. Colloque INSERM/ John Libbey Eurotext Ltd. © 1993, Vol. 231, pp. 51-58.

Ochratoxin A : its chemistry, conformation and biosynthesis

Pieter S. Steyn

CSIR, P.O. Box 395, Pretoria, 0001, Republic of South Africa

ABSTRACT

Ochratoxin A is an important naturally occurring mycotoxin, produced by *Aspergillus ochraceus* and *Penicillium verrucosum*. Its chemical properties, conformation and biosynthesis are related in this paper.

INTRODUCTION

In 1961, mycologists of the South African Council for Scientific and Industrial Research (CSIR) undertook an investigation into the microflora of domestic legume and cereal products. De B. Scott (1965) reported on the toxigenicity of these fungi and showed that 46 strains of 22 species were toxic to ducklings. *Aspergillus ochraceus Wilh* was encountered in this survey, three out of five of its strains were toxigenic. Recent mycological surveys in Southern Africa, however, showed *A. ochraceus* to be an infrequent contaminant of South African produce. It is therefore not surprising that ochratoxin was never detected on local products. (C.J. Rabie and P.G. Thiel, personal communication, 1993).

One of the most toxigenic strains of *A. ochraceus Wilh*, K-804, isolated from sorghum grain, was maintained on sterile soil and selected for the investigation of the chemistry and toxicology of its metabolites. These studies led to the discovery of the ochratoxins, the first group of important mycotoxins to be discovered after the discovery of the aflatoxins. Since then, ochratoxin A (OA) has been found to be produced by a number of both *Aspergillus* and *Penicillium* species (Holmberg, 1992). In tropical areas, OA production is associated with *Aspergillus* species, whereas ochratoxin- producing *Penicillium* species thrive and produce OA in colder climates with temperatures as low as 5°C.

The ochratoxins comprise a polyketide-derived 3,4-dihydro-3-R-methyl-isocoumarin moiety linked via its 12-carboxyl group by an amide bond to L-ß-phenylalanine. OA and its ethyl ester are the most toxic of the group of substances. Ochratoxin B(OB), the dechloro derivative of OA, is essentially non-toxic. The ochratoxins formed the substance of many reviews [Steyn (1971), Steyn (1984), Harwig et al., (1983), Kuiper-Goodman and Scott (1989), Pohland et al., (1992), as well as the thesis of Holmberg (1992)].

The occurrence of OA in several plant and animal products has been extensively reviewed (Kuiper-Goodman and Scott, 1989) - therefore, OA is a toxin of practical importance. OA contamination is typically associated with grain stored in the temperate climate of Europe and North America. It, in fact, plays a major role in poultry diseases, as well as in Danish porcine nephropathy (a serious kidney disease among pigs in Scandinavian countries).

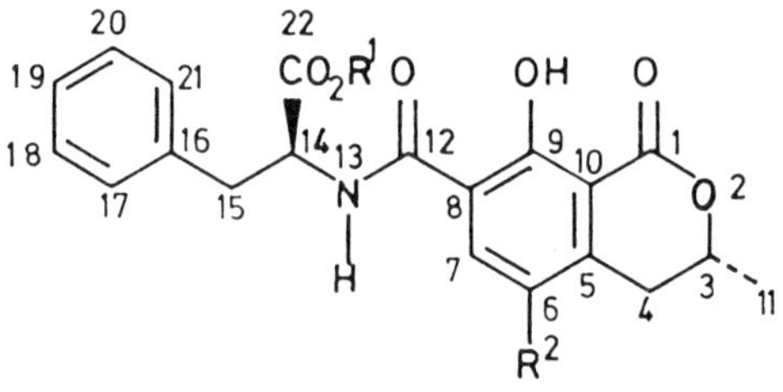

Ochratoxin A, R^1 = H; R^2 = Cl
Ochratoxin B, R^1 = H; R^2 = H
Ochratoxin C, R^1 = Et; R^2 = Cl

From experimentation in many species, OA has been clearly shown to be a toxic substance, with nephrotoxic, immunosuppressive and carcinogenic effects. The exposure of humans to OA is most disconcerting. Studies in Sweden (Breitholz et al., 1991) and Canada (Kuiper-Goodman, 1991) confirmed the presence of OA in human serum. The occurrence of OA in human serum can be attributed to the probable long half-life of OA. Some findings indicate that OA may be a factor in the etiology of Balkan Endemic Nephropahty (BEN). BEN is a serious chronic kidney disease found in high frequency in certain areas of the Balkan countries. The role of OA in BEN is currently questioned; Mantle and McHugh (1992) expressed the view that the low general natural incidence both of ochratoxigenic fungi and also of OA in foods of Bulgarian households do not support its epidemiological role in BEN.

The primary toxic effect of OA is inhibition of protein synthesis (Kuiper-Goodman and Scott, 1989). The effect is specific and involves competitive inhibition of phenylalanine $tRNA^{Phe}$ synthetase which is necessary for fundamental amino-acylation steps in protein synthesis. However, OB is non-toxic and displays no immunosuppressive activity. It is of importance to note that carboxypeptidase hydrolyses OB almost 100-fold faster than OA.

The differences in biological effects prompted us to undertake the recent study on the relationship between the structure and toxicity of OA and OB (Bredenkamp et al., 1989).

PURIFICATION AND STRUCTURE OF OA

A. ochraceus, K-804 was cultivated in bulk on sterilized maize meal as described by Scott (1965). The dried mouldy material was subjected to exhaustive extraction employing chloroform:methanol. The lipids and water-soluble products were removed from the extract by solvent partition. An "acid fraction" was obtained from the latter extract. It contained the ochratoxins and was subsequently separated on formamide-impregnated cellulose powder using mixtures of hexane and benzene as mobile phase yielding, OA and OB. Several methods are available for the analyses of the ochratoxins (Gorst-Allman and Steyn, 1984) (Sharman et al., 1992) (Nesheim et al., 1992).

OA, $C_{20}H_{18}ClNO_6$ (M^+403.8) is a white crystalline solid. Crystallization from benzene is almost quantitative; the crystals which contain 1 mole of benzene of crystallization, melt at 94-96°C (with the loss of benzene). From xylene and chloroform, OA crystallizes without solvent of crystallization (melting range: 168-173°C). $[\alpha]_D^{21}$ = -46.8° (solvent:chloroform) Ultraviolet absorption spectrum:OA displays molar absorption coefficents: $\epsilon(\lambda 214)$ = 3720 ± 55, $\epsilon(\lambda 282)$ = 89 ± 6 and $\epsilon(\lambda 332)$ = 633 ± 7 in methanol. The infrared spectrum of OA displays absorption in the carbonyl region, viz. a carboxyl group (V max 1 721 cm^{-1}), a secondary amide group (V max 1 657 cm^{-1}) and a lactone group (1 675 cm^{-1}) (Pohland et al., 1982). The mass spectrum of OA showed the following most abundant peaks (Pohland et al., 1982):

Table 1: Most abundant peaks in the mass spectrum of OA (Pohland et al., 1982)

m/z	239.0	255.0	221.1	257.1	241.0
Intensity	100.0	66.59	29.17	23.84	23.51
m/z	120.2	240.2	91.0	193.1	237.9
Intensity	19.47	17.92	14.86	11.93	10.93

The analysis of the ^{1}H nmr spectrum of OA (van der Merwe et al., 1965) was instrumental in its structure elucidation. ^{1}H nmr spectra of OA recorded at 100 MHz and at 200 MHz were depicted in the IUPAC publication of Pohland et al., (1982).

The structure of OA was derived from its physical characteristics, several degradation experiments (van der Merwe et al., 1965), its total synthesis (Steyn and Holzapfel 1967) and studies of its bio-origin employing radio-active isotopes (Steyn et al., 1970) and stable isotopes (de Jesus et al., 1980). The latter study involved the unambiguous assignment of the carbon atoms comprising OA (See Table 2). In our recent studies (Bredenkamp et al., 1989) the crystal structures of OA and OB were determined.

Table 2: ^{13}C Chemical shifts and directly bonded [^1J(C,H)] coupling constants of OA (de Jesus et al., 1980)

Carbon atom	δ_Ca	^1J(C, H)
1	169.6 S	149.7
3	75.9 D	131.9
4	32.2br	
5	141.0 S	
6	123.2 S	
7	138.9 D	168.9
8	120.3 S	
9	159.1 S	
10	110.1br S	
11	20.6 Q	128.3
12	163.0br S	
14	54.3br D	144.5
15	37.5br T	130.3
16	135.9 S	
17, 21	129.4 D	158.9
18, 20	128.6 D	159.5
19	127.2 D	160.8
22	174.9br S	

a Relative to internal Me_4Si. Capital letters refers to the pattern resulting from directly bonded protons and lower-case letters to (^{13}C, H) couplings over more than one bond. S = singlet, D = doublet, T = triplet, Q = quartet, br = broad.

A PHYSICOCHEMICAL STUDY OF THE CONFORMATIONS OF OA AND OB

OA is a potent toxin, affecting mainly the kidneys of animals, whereas its dechloro derivative, OB, is non-toxic. It is furthermore of importance to note that its hydrolysis product (ochratoxin α) is essentially non-toxic. In *in vitro* systems (bovine carboxypeptidase A) OB was hydrolysed 100-times faster than OA; this finding may explain the long half-life of OA in animals, its greater toxicity and its occurrence in human serum.

Bredenkamp et al., (1989) therefore investigated the kinetics of the HCl-catalysed hydrolysis of OA and OB. The observed relative HCl-catalysed hydrolysis rates of OA and OB are reversed compared to those observed in biological systems. The very slow rate of acid hydrolysis of the amide bond in OA and OB is very surprising (hydrolysis of OA in 6NHCl requires heating under reflux for 36 hours). The phenolic pK_a values of OA is 7 and that of OB is 8; this result may indicate that the dominantly undissociated phenolic groups of OB are less apt to interact with biological systems. The bulky chloro group in OA could also make it less accessible to the active site of the carboxypeptidase A, relative to OB.

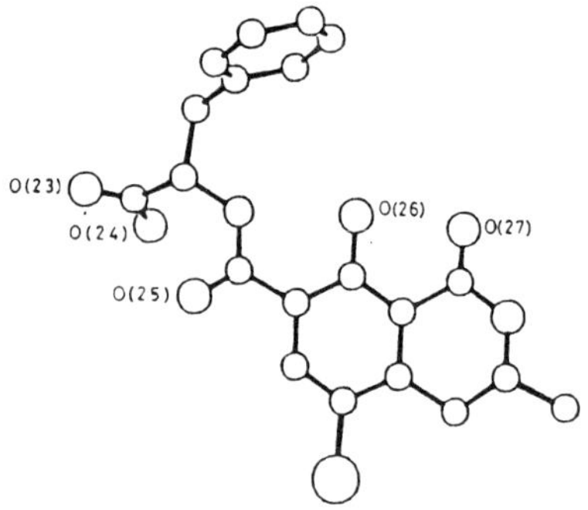

Fig.1. Perspective drawing of OA without hydrogen atoms (Bredenkamp et al., 1989).

Differences in the conformation of the OA and OB may play a role in their relative rates of hydrolysis as well as their relative toxicities. If the amide carbonyl and the phenol groups are *syn* with respect to each other, the amide carbonyl will compete with the lactone carbonyl to hydrogen-bond the phenolic proton. Additional H-bonding by the amide *per se* would have significant influence on the pK_a of the phenol group, and thus its ability to interact with vital substrates. Single-crystal X-ray crystallography was employed to determine the solid-state conformations of OA and OB. Both substances crystallize in the space group $P2_1$ (Bredenkamp et al., 1989).
From the crystallographic data it is clear that there is little difference between the conformations of the two toxins in the solid state. Fig. 1 shows the perspective drawing of OA only. The side chain dihedral angles are indicative of the similarity between the conformations. There are no significant differences in bond lengths or angles. Fig. 2 shows a view of the two molecular conformations with the phenolic carbon atoms of the dihydroisocoumarins superimposed (Bredenkamp et al., 1989).

The amide carbonyl and phenol hydroxy groups are *anti* with respect to each other, and the phenol group is only involved in hydrogen bonding with the lactone carbonyl - the interoxygen distances are 2.542 Å for OA and 2.562 Å for OB (2) ("ß-form", see Fig. 3). The amide carbonyl of both OA and OB are mainly involved in intermolecular hydrogen bonding with the hydrogen atom of the acid group. The intramolecular interoxygen distances for O(24)-O(25) are 3.153 Å and for OA and 3.090 Å for OB.

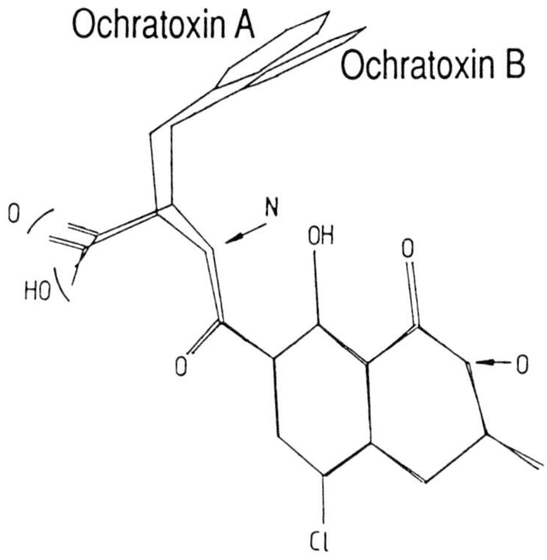

Fig. 2. Molecular graphics of OA and OB based on superimposed phenolic carbon atoms (Bredenkamp et al., 1989)

It was essential to compare the solid state conformations of OA and OB with those in solution since the biological activity relates to their conformation in solution. Hydrogen bonding of the amide and phenol protons and the orientation of the amide carbonyl in solution required special attention.

Some critically important i.r. data of OA and OB are given in the Table 3.

Table 3: Relevant i.r. bands/cm^{-1} of OA and OB in KBr and $CHCl_3$

<table>
<tr><th rowspan="2"></th><th rowspan="2">NH Stretch</th><th colspan="3">Carbonyl stretch</th><th rowspan="2">Amide II</th></tr>
<tr><th>COOH</th><th>Lactone</th><th>Amide</th></tr>
<tr><td>OA KBr pellet</td><td>3 398</td><td>1 737</td><td colspan="2">1 669</td><td>1 536</td></tr>
<tr><td>OB KBr pellet</td><td>3 372</td><td>1 728</td><td colspan="2">1 664</td><td>1 534</td></tr>
<tr><td>OA $CHCl_3$ solution</td><td>3 391</td><td>1 721</td><td>1 675</td><td>1 657</td><td>1 529</td></tr>
<tr><td>OB $CHCl_3$ solution</td><td>3 384</td><td>1 718</td><td>1 670</td><td>1 655</td><td>1 528</td></tr>
</table>

The frequencies of all four of the NH-stretch bands are within the range of the "ß-form" of a series of halogenated salicylamides (Bredenkamp et al., 1989).

Hydrogen bonding of the phenol hydrogen to the lactone carbonyl (ß-form) was verified by the value of the lactone carbonyl absorption band in ochratoxin A methyl ester (ν max 1 675 cm^{-1}) and that in O-methyl-ochratoxin A-methyl ester (ν max 1 726 cm^{-1}).

The ^{13}C nmr chemical shifts of carbonyl carbon nuclei move downfield when the carbonyl oxygen atom becomes H-bonded. The chemical shifts of the ochratoxins suggest that the phenol proton H-bonds to the lactone carbonyl in both cases, and is in agreement with the X-ray structures and i.r. findings. This is confirmed by the $^3J_{HC}$ coupling constant of the phenol proton with C-8 and C-10 in OA. The respective coupling values are 5.5 and 4.6 Hz, indicating that the phenol proton is *trans* with respect to C-8 and is situated so as to H-bond the lactone carbonyl (Bredenkamp et al., 1989). The "ß-form" allows for hydrogen bonding of the amide proton to the phenolic oxygen atom.

From these studies it is evident that the difference in hydrolysis rates (mineral acid or carboxy peptidase) of OA and OB cannot be ascribed to conformation since both toxins have the same conformation at the peptide linkage in the solid and liquid state.

THE BIOSYNTHESIS OF OA

The biosynthesis of OA has been studied using both ^{14}C- and ^{13}C-labelled precursors. DL-[1-^{14}C]-ß-phenylalanine was incorporated into OA by cultures of *A. ochraceus*. Hydrolysis of the labelled OA gave the isocoumarin acid and L-ß-phenylalanine with the amino-acid containing all the activity (Steyn et al., 1970).

Subsequent studies focussed on the origin of the isocoumarin acid. Sodium [1-^{14}C]acetate was added to resting cultures of *A. ochraceus*. The isocoumarin acid derived upon acid hydrolysis of OA contained all the activity. Kuhn-Roth oxidation and subsequent Schmidt degradation of the acetic acid provided evidence for the pentaketide origin of the dihydroisocoumarin moiety. The origin of the carboxy group at C(8) was established by Steyn et al., (1970) through the addition of DL[methyl-^{14}C] methionine to a resting culture of *A. ochraceus*. The isocoumarin acid contained all the radioactivity. Selective degradation experiments established that the C(12) was derived from methyl-methionine.

Unequivocal evidence for the acetate origin of the isocoumarin moiety in ochratoxin A was obtained by de Jesus et al., 1980. The complete assignment of the signals in the natural abundance ^{13}C NMR spectrum of OA (Table 2) allowed a study of its biosynthesis using ^{13}C-labelled precursors. The proton noise decoupled (PND) ^{13}C NMR spectrum of [1-^{13}C]acetate-derived ochratoxin A, obtained from cultures of *P. viridicatum*, showed that only C(1), C(3), C(5), C(7), and C(9) were enriched. The PND ^{13}C NMR spectrum of the [1,2-^{13}C]acetate-derived ochratoxin A showed that the signals arising from the following carbon atoms exhibited directly bonded (^{13}C, ^{13}C) couplings: C(11)-C(3), C(4)-C(5), C(6)-C(7), C(8)-C(9) and C(10)-C(1).

The involvement of a C_1 unit in the formation of the carbonyl carbon atom C(12) was verified by PND ^{13}C nmr studies of [^{13}C]formate-derived OA degradation products. The biosynthesis of OA, therefore, involves an isocoumarin moiety formed via the acetate-polymalonate pathway from one acetate and four malonate units, with the carbonyl carbon atom C(12) being derived from the C_1 pool, and the phenylalanine derived from the shikimate pathway.

REFERENCES

Bredenkamp, M.W., Dillen, J.L.M., Van Rooyen, P.H., Steyn, P.S. (1989): Crystal structures and conformational analysis of ochratoxin A and B: Probing the chemical structure causing toxicity. *J. Chem. Soc. Perkin Trans. II*, 1835-1839.

Breiholtz, A., Olsen, M., Dahlbäck A., Hult, K., (1991): Plasma ochratoxin A levels in three Swedish populations surveyed using an ion-pair technique. *Food Addit. Contam.*, 6, 183-192.

de Jesus, A.E., Steyn, P.S., Vleggaar, R. and Wessels, P.L. (1980): Carbon-13 nuclear magnetic resonance assignments and biosynthesis of the mycotoxin ochratoxin A. *J. Chem. Soc. Perkin. Trans. I.*, 52-54.

Gorst-Allman, C.P. and Steyn, P.S. (1984): Applications of chromatographic techniques in the separation, purification and characterization of mycotoxins. In *Mycotoxins - production, isolation, separation and purification*, ed. V. Betina, pp. 59-85. Amsterdam: Elsevier.

Harwig, J., Kuiper-Goodman, T., Scott, P.M. (1983): Microbial food toxicants: Ochratoxins. In *Handbook of food borne diseases of biological origin*, ed. M. Recheigl. pp. 193-238. Boca Raton F.L.: C.R.C Press

Holmberg, T. (1992): Ph.D Thesis. *Ochratoxin in cereal grain and its potential effects on animal health.* Swedish University of Agricultural Science, Uppsala, Sweden.

Kuiper-Goodman, T. (1991): Risk assessment of ochratoxin A residues in food. In *Mycotoxins, endemic nephropathy and urinary tract tumours*, eds M. Castegnaro, R. Plestina, G. Dirheimer, I.N. Chernozemsky, H. Bartsch, pp. 307-320: Lyon - IARC.

Kuiper-Goodman, T., Scott, P.M. (1989): Review: Risk assessment of the mycotoxin ochratoxin A. *Biomed. Environ. Sci.*, 2, 179-248.

Mantle, P.G., McHugh, K.M. (1992): Nephrotoxic fungi in food from nephropathy households in Bulgaria. *Mycol. Res.*: Accepted for publication.

Nesheim, S., Stack, M.E., Trucksess, M.W., Eppley, R.M. (1992): Rapid solvent-efficient method for liquid chromatographic determination of ochratoxin A in corn, barley and kidney: Collaborative study. *J. of A.O.A.C. International*, 75 (3), 481-487.

Pohland, A.E., Nesheim, S., Friedman, L. (1992): Ochratoxin A: A Review (Technical Report). *Pure and Appl. Chem.*, 64 (7), 1029-1046.

Pohland, A.E., Schuller, P.L., Steyn, P.S., van Egmond, H.P. (1982): Physicochemical data for some selected mycotoxins. *Pure and Appl. Chem.*, 54 (11), 2219-2284.

Scott, De. B. (1965): Toxigenic fungi isolated from cereal and legume products. *Mycopathol. Mycol. Appl.*, 25, 213-222.

Sharman, M., MacDonald S., Gilbert, J. (1992): Short Communication. Automated liquid chromatographic determination of ochratoxin A in cereals and animal products using immunoaffinity column clean-up. *J. Chrom.*, 603, 285-289.

Steyn, P.S., (1971): Ochratoxin and other dihydroisocoumarins. In *Microbial Toxins No. VI, Fungal Toxins*, eds. A. Ciegler, S. Kadis and S.J. Ajl, pp. 179-205. New York: Academic Press.

Steyn, P.S. (1984): Ochratoxins and related dihydroisocoumarins. In *Mycotoxins - production, isolation, separation and purification.* ed V. Betina, pp. 183 - 216. Amsterdam: Elsevier.

Steyn, P.S., and Holzapfel, C.W. (1967): The synthesis of ochratoxins A and B. Metabolites of *Aspergillus ochraceus* WIlh. *Tetrahedron*, 4449-4461.

Steyn, P.S., Holzapfel, C.S. and Ferreira, N.P. (1970): The biosynthesis of the ochratoxins, metabolites of *Aspergillus ochraceus. Phytochemistry 9*: 1977-1983.

van der Merwe, K.J., Steyn, P.S., Fourie, L. (1965): Mycotoxins. Part II. The constitution of ochratoxins A, B and C, metabolites of *Aspergillus ochraceus* Wilh., *J. Chem. Soc.*: 7083-7088.

Résumé

L'ochratoxine A est une importante mycotoxine de notre environnement, produite par Aspergillus ochraceus et Penicillium verrucosum. Cet article rassemble l'essentiel de nos connaissances sur ses propriétés chimiques, sa conformation et sa biosynthèse.

Human ochratoxicosis and its pathologies. Eds E.E. Creppy, M. Castegnaro, G. Dirheimer. Colloque INSERM/ John Libbey Eurotext Ltd. © 1993, Vol. 231, pp. 59-65.

Influence de l'ochratoxine A sur le devenir des xénobiotiques

Pierre Galtier, Gilberte Larrieu et Michel Alvinerie

Laboratoire de Pharmacologie-Toxicologie, INRA, 180, chemin de Tournefeuille, 31931 Toulouse, France

RESUME

Par ses propriétés, l'ochratoxine A est susceptible d'interférer avec la distribution, le métabolisme ou l'élimination des substances xénobiotiques. L'absorption de la toxine est réalisée par transport actif au niveau du jéjunum, cette modalité peut entraîner des interactions par compétition au niveau du transporteur. L'importance de la liaison de l'ochratoxine A aux protéines du plasma peut provoquer un déplacement de médicaments transportés par la sérum albumine. Ce type d'interaction a été mesuré avec les anticoagulants ou les anti-inflammatoires non stéroïdiens et se traduit par une toxicité accrue de la toxine. Par ailleurs, cette mycotoxine est reconnue pour réduire le métabolisme oxydatif hépatique et notamment certaines formes du cytochrome P450. L'ochratoxicose expérimentale provoque de plus un défaut d'excrétion biliaire de la tétracycline ou du chloramphénicol alors qu'une interaction avec la cholestyramine a été récemment démontrée. Enfin, la néphrotoxicité due à l'ochratoxine A peut entraîner une altération de l'élimination urinaire de certains médicaments ou encore être potentialisée comme dans le cas d'une coadministration de probénécide.

INTRODUCTION

Présente à l'état de contaminant naturel dans un grand nombre d'aliments destinés à l'Homme et aux animaux d'élevage, l'ochratoxine A se retrouve dans le sang humain de diverses populations européennes (Hald, 1991). Au delà du risque direct sur la santé des individus en raison de ses propriétés néphrotoxiques et cancérogènes, cette mycotoxine peut interférer sur le devenir habituel de substances endogènes ou exogènes. Ces dernières généralement regroupées sous le vocable de substances xénobiotiques comprennent notamment les médicaments, les additifs alimentaires et tous les contaminants présents dans l'environnement. Le cas des molécules médicamenteuses est particulièrement intéressant à considérer car la présence d'ochratoxine A dans les milieux ou les compartiments biologiques de l'organisme peut entraîner des interactions pharmacocinétiques sur le devenir particulier de ces médicaments ou encore des interactions pharmacologiques sur l'activité thérapeutique de ces mêmes molécules.

Une meilleure approche de ces interactions implique la connaissance des modalités d'absorption, de biotransformation et d'élimination des xénobiotiques envisagés. Cette connaissance est réelle dans le cas des médicaments mais peut être limitée pour les additifs ou les autres contaminants naturels. Toutefois, ces processus sont régis par la solubilité de chaque substance dans les milieux hydriques et lipidiques, leur pKa, leur poids moléculaire et leur compétitivité envers les sites d'action ; autant de facteurs qui peuvent modifier la durée et l'intensité de la réponse pharmacologique ou toxicologique du xénobiotique.
Nous envisagerons donc l'influence de l'ochratoxine A en considérant ses propriétés pharmacologiques intrinsèques. Cela nous conduira à aborder successivement les interactions possibles au niveau de l'absorption digestive, de la liaison aux protéines, des biotransformations et des excrétions biliaire et rénale.

INCIDENCE SUR L'ABSORPTION ET LA DISTRIBUTION

L'ochratoxine A possède les propriétés physico-chimiques des acides faibles confirmées par ses fonctions phénoliques et carboxyliques. Caractérisée par des pKa de 6,75 et 10,25, la toxine peut exister sous forme ionisée et non ionisée selon le pH du milieu (Galtier et coll. 1977). La diffusion passive de la forme non ionisée prépondérante au pH acide de l'estomac semble être un mécanisme majeur d'absorption de la toxine.

Tableau 1. Pourcentage de liaison de quelques médicaments acides et basiques à l'albumine plasmatique

Pourcentage	Médicaments acides	Médicaments basiques
100-98	Phénylbutazone Ibuprofène Dicloxacilline Sulfaphénazole	Diazépam
98-95	Acide nalidixique Sulfadiméthoxine Indométhacine Tolbutamide Warfarine	Bupivacaïne Pimazide
95-85	Oxacilline Acetazolamide Ethylbiscoumacétate Phénytoïne Probénécide	Amitryptiline Chlorpromazine Halopéridol Mépacrine Quinidine
85-70	Sulfaméthoxypyridazine Dapsone Acide acétylsalicylique Thiopental	Alprénolol Erythromycine Chloroquine Gentamycine

Un tel transport passif ne devrait pas influencer le franchissement par un processus similaire d'autres xénobiotiques à travers la muqueuse gastrique. Toutefois, une investigation récente (Kumagai, 1988) a montré que l'ochratoxine A était aussi absorbée à partir du jéjunum de rat anesthésié même quand sa concentration plasmatique s'avérait supérieure à celle du contenu jéjunal. Ce processus semble donc relever d'un transport actif qui serait susceptible de conduite à des compétitions avec des substances endogènes ou exogènes transitant par le même transporteur.
Une fois la mycotoxine absorbée à partir du tractus digestif, le fait majeur de sa distribution sanguine demeure sa faculté de liaison aux protéines plasmatiques. Envisagé chez diverses espèces incluant l'Homme, cette fixation est de 99-100% chez les mammifères alors qu'elle n'est que de 78% chez les poissons (Hagelberg et coll., 1989). Chez le rat, l'ochratoxine A se fixe essentiellement à la sérum albumine (Galtier 1974) et peut ainsi entrer en compétition avec les nombreux médicaments acides et basiques connus pour les mêmes propriétés (tableau 1).
Ainsi, la phénylbutazone, l'éthylbiscoumacétate et la sulfaméthoxypyridazine inhibent compétitivement la liaison de l'ochratoxine A à la sérum albumine porcine. Ceci tendrait à prouver que la toxine se fixe au même site sur la protéine et que des interactions à conséquences délétères peuvent survenir en thérapeutique humaine et vétérinaire comme cela a été démontré chez le rat (Galtier et coll. 1980).
Dans une telle hypothèse, les déplacements peuvent être réciproques en fonction de l'affinité respective entre les protéines et de la fixation préalable de l'un ou l'autre des compétiteurs. Les conséquences peuvent donc consister soit en une libération d'ochratoxine A à la suite d'une prise médicamenteuse ou une augmentation de la forme libre d'un médicament ayant une moindre affinité avec de possibles répercussions sur sa distribution et son activité pharmaco-toxicologique.

INCIDENCE SUR LES BIOTRANSFORMATIONS HEPATIQUES

Entreprises chez les rongeurs (Kane et coll. 1986 ; Fuchs et coll. 1988a) et la truite (Fuchs et coll. 1986), les études de distribution tissulaire mettent en relief la persistance particulière de l'ochratoxine A dans le tissu hépatique et la bile. Cette localisation peut être le fait d'une interaction spécifique avec des macromolécules tissulaires ou encore une participation substantielle des enzymes impliquées dans l'oxydation puis la conjugaison de la mycotoxine excrétée ensuite par voie biliaire. Concernant cette dernière hypothèse, il faut rappeler la mise en évidence d'une interaction entre l'ochratoxine A et le cytochrome P450 microsomal du foie de rat (Størmer & Pedersen, 1980). Le spectre d'interférence de type I obtenu par ces auteurs sur des fractions microsomales hépatiques de rat, démontre la liaison de la toxine sur l'apoprotéine au niveau du site actif des cytochromes. De telles liaisons ont déjà été signalées pour des médicaments comme la chlorpromazine, l'imipramine, la coumarine, les barbituriques et certaines hormones (testostérone, progestérone). Une interaction à ce niveau pourrait donc survenir et limiter le catabolisme de certains xénobiotiques.
Cette remarque se trouve confortée par l'observation de l'inhibition enzymatique survenant chez le rat soumis à un traitement quotidien d'ochratoxine A à raison de 1 mg/kg/jour (Galtier et coll. 1984). Dans ces conditions, l'expression du cytochrome P450 est réduite et certaines activités qui en dépendent sont diminuées significativement, aminopyrine déméthylase et aniline hydroxylase en particulier (Figure 1).
De tels résultats posent le problème de l'identification des cytochromes impliqués dans le métabolisme oxydatif de la toxine. Dans ce sens, on sait déjà que la 4-hydroxylation

de la toxine peut être induite chez le rat par le 3-méthylcholanthrène ou le phénobarbital (Størmer & Pedersen, 1980 ; Hietanen et coll. 1986). En outre, cette biotransformation parait indépendante du métabolisme de la débrisoquine (Castegnaro et coll. 1989).

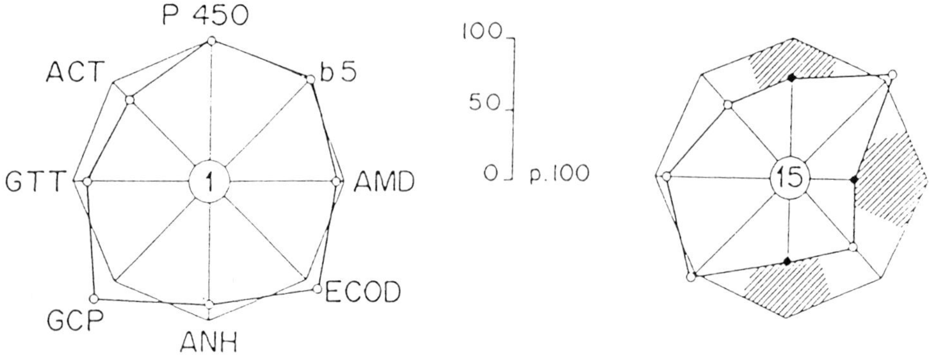

Fig. 1. Evolution des activités hépatiques de biotransformation chez des rats recevant 1 mg/kg/jour d'ochratoxine A pendant 1 jour (1) ou 15 jours (15).

P450 : cytochrome P450, b5 : cytochrome b5, AMD : aminopyrine *N*-déméthylase, ECOD : éthoxycoumarine *O-dééthylase*, ANH : aniline hydroxylase, GCP : glucuronyltransférase (*p*-nitrophénol), GTT : glutathion *S*-transférase (CDNB), ACT : acétyltransférase (isoniazide).

Les résultats sont exprimés en pourcentages des valeurs témoins.

Les points noirs et zones hachurées correspondent à des différences significatives (d'après Galtier et coll. 1984).

Il semble donc que les cytochromes P4501A (inductibles par le 3 méthylcholanthrène), 2B (inductibles par le phénobarbital), 2C (responsable de la déméthylation de l'aminopyrine) ou 2E (à l'origine de l'hydroxylation de l'aniline) pourraient être impliqués dans le catabolisme de la toxine. Toutefois, des études complémentaires sont indispensables à l'identification définitive des iso-enzymes responsables et à une meilleure appréciation des interactions métaboliques entre l'ochratoxine A et d'autres xénobiotiques.

INCIDENCE SUR L'EXCRETION BILIAIRE ET RENALE

L'ochratoxine A administrée par voie orale chez les rongeurs subit un cycle entérohépatique dû à son élimination biliaire associée à une importante réabsorption intestinale (Fuchs et coll. 1988b). Une telle mise en circulation de la toxine dans l'organisme entraîne des perturbations dans le devenir d'autres molécules. Ainsi, les concentrations hépatiques et l'excrétion biliaire de la tétracycline sont diminuées chez des rats soumis à une ochratoxicose chronique mais pour lesquels la cholérèse reste normale (Galtier & Alvinerie, 1983). Dans des conditions identiques, une réduction de

l'excrétion biliaire de la forme glucuronoconjuguée du chloramphénicol est constatée alors que les teneurs tissulaires sont inchangées. Cet exemple démontre la possible interaction de l'ochratoxine A sur la distribution et l'élimination de molécules antibiotiques largement utilisées.
A l'inverse, certains médicaments à tropisme intestinal perturbent le devenir normal de la toxine. Roth et coll. (1988) ont démontré que la cholestyramine, agent séquestrant les acides biliaires, pouvait limiter la circulation entérohépatique de la toxine. Plus récemment, Madhyastha et coll. (1992) ont prouvé la capacité de cette même substance à réduire la biodisponibilité plasmatique et l'excrétion urinaire de l'ochratoxine A en diminuant son absorption digestive et favorisant son élimination fécale. Cette expérimentation ouvre ainsi la voie à des traitements permettant de débarrasser l'organisme d'une substance reconnue pour sa forte rémanence.
L'excrétion urinaire de l'ochratoxine A a souvent été mise en parallèle avec sa néphrotoxicité. De nombreux rapports ont relaté l'existence durable de résidus dans les reins par comparaison à d'autres tissus ou organes (Krogh et coll. 1976). Le mécanisme de l'excrétion repose sur une interaction de la toxine avec les transporteurs des anions organiques et sur une filtration glomérulaire suivie de sécrétion et réabsorption tubulaires (Stein et coll. 1985 ; Sokol et coll. 1988). A ce niveau, de nombreux médicaments sont susceptibles d'entrer en compétition et de modifier l'élimination de la mycotoxine ou encore de voir leur devenir perturbé par la présence de la toxine. Le tableau 2 dresse une liste de quelques médicaments reconnus pour être activement sécrétés dans le tubule proximal.

Tableau 2. Médicaments et métabolites sécrétés par transport actif dans les tubules proximaux du rein.

Médicaments acides	Médicaments basiques
Céphaloridine	Ammonium quaternaires
Acétazolamide	Dopamine
Furosémide	Histamine
Conjugués glucuronides	Mépacrine
Conjugués à la glycine	Morphine
Conjugués sulfates	Procaïne
Indométhacine	Quinine
Pénicillines	
Phénylbutazone	
Diurétiques thazide	

Le cas du probénécide a été ainsi envisagé et sa coadministration avec l'ochratoxine A provoque une néphrotoxicité et une rémanence tissulaire accrues de la toxine (Stein et coll. 1985).

CONCLUSION

En raison de son caractère rémanent dans l'organisme animal, l'ochratoxine A peut entraîner des désordres dans le devenir de substances xénobiotiques. Ces perturbations sont souvent liées à son caractère acide faible qui lui confère des propriétés de liaison aux transporteurs intra et extracellulaires. C'est le cas des protéines plasmatiques ou des vecteurs transmembranaires de l'épithélium intestinal ou des tubules proximaux. De ce fait, les interactions les plus probables concernent d'autres composés acides tels que les anti-inflammatoires non stéroïdiens, les barbituriques, les sulfamides et les pénicillines. Nous avons également évoqué l'influence possible de la toxine sur les biotransformations hépatiques en se basant sur sa propre métabolisation par les cytochromes P450. Dans ce domaine, la compétition pourrait en particulier s'exercer vis-à-vis des coumarines et de toutes les substances médicamenteuses ou phytosanitaires rapportées pour être des inducteurs ou des inhibiteurs enzymatiques.

De telles hypothèses reposent parfois sur des études expérimentales montrant l'incidence possible de l'ochratoxine A sur le devenir d'autres xénobiotiques ou de sa réciproque. Toutefois, la réalité clinique de ces interactions reste encore à démontrer car leur survenue demeure dépendante des niveaux de concentrations respectives atteints dans les liquides et compartiments biologiques des animaux ou des hommes exposés à une alimentation contaminée.

REFERENCES

Castegnaro, M., Bartsch, H., Bereziat, J.C., Arvela, P., Michelon, J., and Broussole, L. (1989): Polymorphic ochratoxin A hydroxylation in rat strains phenotyped as poor and extensive metabolizers of debrisoquine. *Xenobiotica* 19, 225-230.

Fuchs, R., Appelgren, L.E., and Hult, K. (1986): Distribution of ^{14}C-ochratoxin A in the rainbow trout. *Acta Pharmacol. Toxicol.* 59, 220-227.

Fuchs, R., Appelgren, L.E. , and Hult, K. (1988a): Distribution of ^{14}C-ochratoxin A in the mouse monitored by whole-body autoradiography. *Pharmacol. Toxicol.* 63,: 355-360.

Fuchs, R., Radic, B., Peraica, M., Hult, K. and Plestina, R. (1988b): Enterohepatic circulation of ochratoxin A in rats. *Period. Biol.* 60, 39-42.

Galtier, P. (1974): Devenir de l'ochratoxine A dans l'organisme animal. I. Transport sanguin de la toxine chez le rat. *Ann. Rech. Vétér.* 5, 311-318.

Galtier, P., Baradat, C., and Alvinerie, M. (1977): Etude de l'élimination de l'ochratoxine A dans le lait chez la lapine. *Ann. Nutr. Aliment.* 31, 911-918.

Galtier, P., Camguillem, R., and Bodin, G. (1980): Evidence for in vitro and in vivo interaction between ochratoxin A and three acidic drugs. *Fd. Cosmet. Toxicol.* 18, 493-496.

Galtier, P., and Alvinerie, M. (1983): Incidence of experimental ochratoxicosis on hepatic disposition of [^{3}H] tetracycline and chloramphenicol in rats. *Toxicol. Lett.* 18, 263-267.

Galtier, P., Larrieu, G., and Le Bars, J. (1984): Comparative incidence of oral ochratoxicosis and aflatoxicosis on the activity of drug metabolizing enzymes in rat liver. *Toxicol. Lett.* 23, 341-347.

Hagelberg, S., Hult, K., and Fuchs, R. (1989): Toxicokinetics of ochratoxin A in several species and its plasma-binding properties. *J. Appl. Toxicol.* 9, 91-96.

Hald, B. (1991): Ochratoxin A in human blood in European countries. In *Mycotoxins, endemic nephropathy and urinary tract tumours*, eds M. Castegnaro, R. Plestina, G. Dirheimer, I.N. Chernozemsky and H. Bartsch, pp. 159-164. Lyon : IARC Sci. Publ.

Hietanen, E., Malavieille, C., Camus, A.M., Béréziat, J.C., Brun, G., Castegnaro, M., Michelon, J., Idle, J.R., and Bartsch, H. (1986): Interstrain comparison of hepatic and renal microsomal carcinogen metabolism and liver S9-mediated mutagenicity in DA and Lewis rats phenotyped as poor and extensive metabolizers of debrisoquine. *Drug Metab. Dispos.* 14, 118-126.

Kane, A., Creppy, E.E., Roth, A., Röschenthaler, R., and Dirheimer, G. (1986): Distribution of the [^{3}H]-label from low doses of radioactive ochratoxin A ingested by rats, and evidence for DNA single-strand breaks caused in liver and kidneys. *Arch. Toxicol.* 58, 219-224.

Kumagai, S. (1988): Effects of plasma ochratoxin A and luminal pH on the jejunal absorption of ochratoxin A in rats. *Fd. Chem. Toxicol.* 26, 753-758.

Roth, A., Chakor, K., Creppy, E.E., Kane, A., Röschenthaler, R., and Dirheimer, G. (1988): Evidence for enterohepatic circulation of ochratoxin A in mice. *Toxicology* 48, 293-308.

Sokol, P.P., Ripich, G., Holohan, P.D., and Ross, C.R. (1988): Mechanism of ochratoxin A transport in kidney. *J. Pharmacol. Exp. Ther.* 246, 460-465.

Stein, A.F., Philips, T.D., Kubena, L.F., and Harvey, B.B. (1985): Renal tubular secretion and reabsorption as factors in ochratoxicosis : effects of probenecid on nephrotoxicity. *J. Toxicol. Environ. Health.* 16, 593-605.

Størmer, F.C., and Pedersen, J.I. (1980): Formation of 4-hydroxyochratoxin A from ochratoxin A by rat liver microsomes. *Appl. Environ. Microbiol.* 39, 971-975.

Summary

Because of its physicochemical properties, ochratoxin A interacts with distribution, metabolism and elimination patterns of xenobiotics. This mycotoxin is actively absorbed in jejunum with possible competition with drugs at the transport level. Ochratoxin A binding to plasma proteins may cause the displacement of other drugs which are usually carried by plasma albumin. Such interactions have been described with antiinflammatory or coumarine drugs leading to an enhanced toxicity of the toxin. Ochratoxin A has been shown to decrease the liver cytochrome P450-depending oxidative enzymes and the biliary excretion of both tetracycline and chloramphenicol whereas a digestive interaction with cholestyramine has been demonstrated. The fate and toxicity of the mycotoxin in kidneys could also cause changes in the urinary elimination of certain drugs, particularly those which are actively secreted into the proximal renal tubules.

Human ochratoxicosis and its pathologies. Eds E.E. Creppy, M. Castegnaro, G. Dirheimer. Colloque INSERM/ John Libbey Eurotext Ltd. © 1993, Vol. 231, pp. 67-74.

In vitro investigations on ochratoxin A metabolism

Johanna Fink-Gremmels*, Maarten Blom and Felice Woutersen van Nijnanten

Department of Veterinary Pharmacology, Pharmacy and Toxicology, Faculty of Veterinary Medicine, Utrecht University, Yalelaan 2, P.O. Box 80.176, 3508 TD Utrecht, The Netherlands

**Corresponding author*

INTRODUCTION

Ochratoxins constitute a group of closely related derivatives of isocoumarines linked to L-phenylalanine, ochratoxin A being considered to be the most toxic secondary metabolite synthetized by various species belonging to the genera *Aspergillus* and *Penicillium*. Ochratoxin A (OA) is considered to be the causative agent in mycotoxin porcine nephropathy, which is reproducible in laboratory animal species. Furthermore, there is increasing epidemiological evidence that OA is related to the Balkan Endemic Nephropathy (BEN) and Urinary Tract Tumours (UTT) in man, predominantly observed in females showing an extreme geographic clustering and familial aggregation (*for review see IPCS 105, 1990*).

Early investigations on hepatic biotransformation of OA resulted in the detection of 4(R)- and 4(S)-4-hydroxyochratoxin A in rats as well as in 10-0-methylochratoxin A in rabbits. Furthermore, ochratoxin-alpha and phenylalanine were described to occur as cleavage product in various animal species (*Stormer et al., 1981; Hansen et al.,1982*). Based on acute toxicity studies with these metabolites, biotransformation of OA was considered to facilitate a detoxification reaction. However, in a study with *ddy* mice their was some evidence that phenobarbital treatment increased the incidence of hepatocarcinomas (*Suzuki et al., 1986*). Furthermore, in a preliminary experiment, testing the supernatant (cell culture medium) of rat hepatocytes derived from Arochlor pretreated rats and exposed *in vitro* to OA for mutagenic response in the Ames-Test, we recognized an increased incidence of point- as well as frame shift-mutations in histidin auxotroph *Salmonella typhimurium* type strains. In contrast, no mutagenic response could be found, when the assay was conducted using microsomes (S9-mix) as metabolizing system. Furthermore, hepatocytic metabolism of OA increased the number of sister-chromatid exchanges in human peripheral lymphocytes *(Hennig et al., 1991)*.

Thus, with the aim to identify potentially mutagenic and genotoxic metabolites of OA, we established a HPLC method enabling the separation and detection of OA metabolites as generated by rat and pig liver microsomal fractions and primary hepatocytes derived from rats with and without phenobarbital (PB) induction.

MATERIALS AND METHODS

Chemicals: Ochratoxin A, and Williams' Medium E, bovine serum albumin, insulin, hydrocortisone succinate and gentamicin sulfate were obtained from Sigma (St. Louis). Collagenase, glucose-6-phosphate, glucose-6-phosphate dehydrogenase and NADP were from Boehringer (Ingelheim). All other reagents were purchased from Merck (Darmstadt) and of analytical grade, except for acetonitrile which was of gradient grade (Lichrosolv).

Preparation of Microsomes: Microsomes were prepared as described by *Rutten et al.*, (1987) from male Wistar rats (225-250 grs b.w.) and castrated male pigs (35-40 kg b.w.). The final incubation mixture consisted of 0,1 ml microsomal protein (mean $\pm$ SEM = 10.175$\pm$2.042 mg/ml) dissolved in 1 ml phophate buffer (pH 7.4) containing 1.5 mM $MgCl_2$, 2.6 mM glucose-6-phosphate, 0.6 mM NADP and 0.7 IU/ml glucose-6-phosphate dehydrogenase. Ochratoxin A dissolved in DMSO was added to give a final concentration of 0.25 μM in the incubation mixture. After 30 min of incubation at 37°C the reaction was terminated by adding 50 μl 1N HCl. After centrigudation the protein-free supernatant was extracted as described below.

Preparation and culture of hepatocytes: Isolated hepatocytes were obtained from Wistar rats (260-280 grs b.w.) according to the two-step collagenase perfusion technique as described by *Seglen (1976)* with minor modifications. For the induction experiments rats were pretreated with phenobarbital administered via drinking water (0.1 %) for five consecutive days prior to liver perfusion. Briefly, the isolation of cells can be described as follows: the whole rat liver was perfused *in situ* first with Hank's balanced salt solution (HBSS). The second step was the perfusion with 50 ml of a 0.05% solution of collagenase in HBSS supplemented with 0.25 mM $CaCl_2$ in a recirculating system for 10 min. Finally, the liver capsule was carefully removed under sterile conditions and hepatocytes were dispersed in HBSS containing 2.5% BSA (bovine serum albumin). After three washing steps cells were resuspended in Williams' Medium E (WME) in a final concentration of 4 x 10^6 viable cells (as assessed by trypan blue exclusion test). Cells were cultured in a density of 10^6 cells per 6 cm Petri dish in 4 ml WME supplemented with 1 μM insulin, 10 μM hydrocortisone hemisuccinate and 50 mg/L gentamicinsulfate. Initially 3 % FCS was added to improve cell attachment to the plastic material, but final incubation in the presence of OA (2.5 μM) were performed in serum-free medium to prevent protein binding. Incubations were carried out in a humidified atmosphere of air (95% O_2 and 5% CO_2) at 37°C for 0.5, 1.0, 2.0, 4.0, 8.0, and 24 hrs, respectively. Finally, the cell supernatant was removed and acidified with 50 μl 1N HCl/ml. Cells were harvested separately and dissolved in 2.2 ml emulgen buffer for cytochrome P450 analysis. Samples were stored at -20°C (medium) and -70°C (cells), respectively, until being analysed.

Cytochrome P450 measurements: Total cytochrome P450 was measured according to the method of *Rutten et al., 1987* applying a Shimadzu spectrophotometer attached to a computerized data management system. Protein concentrations were assessed by a standard methods (Pierce) using BSA as reference.

Extraction procedure: Microsomal incubations, aliquotes of hepatocyte medium and homogenized hepatocytes (redissolved in HBSS) were lyophilised after the described acidification and finally extracted with 0.5 ml chloroform. The organic phase was separated, the solvent removed under a stream of nitrogen, and the remaining residue dissolved in 200 μl elution buffer (buffer A) of which an aliquote of 50 μl was used for HPLC analysis.

HPLC analysis: HPLC analysis of OA and its metabolites was based on a recently developed ion-pair reversed phase method applying a linear gradient. Briefly, a Spherisorb ODS 2 column (250 x 4.6 mm ID) was used and elution performed with two phosphate buffer systems (0.05 M, pH 6.6) containing tetrabutylammonium (TBA 0.025 M) as ion-pairing agent and 30 % acetonitril (eluent A) and 40 % acetonitril (eluent B), respectively. A linear gradient from 30 to 40 % acetonitril was established over a period of 10 minutes (Gynotec gradient pump), followed by an elution period of 20 minutes with eluent B only. A fluorescence detector (Iasco FP 820) set at 335 nm excitation and 450 nm emission wavelength was used as detection system and connected to an Axxiom 727 system for calculation of peak areas and data management.

RESULTS

Separation with the ion pair reversed phase method yielded a variety of metabolites of OA, with significant differences in metabolites produced by microsomal incubations and by primary rat hepatocytes. Incubations with microsomes from untreated control rats resulted in the formation of 3 different OA metabolites. Microsomes from PB pretreated rats showed an enhanced metabolic conversion of OA and in the production of measurable amounts of an additional metabolite (b). In comparable incubations with microsomes derived from livers of untreated pigs an even more complex pattern of OA metabolites was formed, showing at least 4 PB-inducible metabolites (for metabolite (e) only a non-significant induction could be observed) (table 1).

Table 1: Metabolite formation in microsomal incubations containing 0.25 mM ochratoxin A (quantities given are relative concentrations of total OA analysed; n.d. = non detectable)

	rat-non induced	rat PB induced	Pig-non induced	pig PB induced
a:	n.d.	n.d.	0.09 ± 0.01	0.16 ± 0.01
b:	n.d.	0.19 ± 0.02	0.08 ± 0.01	0.81 ± 0.04
d:	0.25 ± 0.002	0.96 ± 0.16	0.58 ± 0.02	2.78 ± 0.11
e:	0.07 ± 0.001	0.13 ± 0.08	0.06 ± 0.01	0.09 ± 0.02
h:	1.27 ± 0.110	1.36 ± 0.06	0.26 ± 0.03	1.62 ± 0.61

Comparable measurements of both, intracellular fluid and cell culture medium (extracellular compartment) indicated that cells do not retain any measurable amounts of OA biotransformation products (data not shown in detail). All metabolites could be detected extracellularly showing a time dependent increase over 24 hrs as demonstrated in figure 1 for one of the major metabolites (d). Hepatocytes from control rats produced at least 7 different metabolites, of which all were inducible by PB, with induction factors varying between 5.6 and 35 (see also table 2). After PB treatment an additional metabolite (f) appeared (see also figure 2).

Table 2: OA metabolites as produced by primary cultures of rat hepatocytes: relative metabolite concentration (as percentage of analysed OA) after 24 hrs of incubation.

metabolite	*non induced controls (n=2)*	*PB incuced (n=4)*
a:	1.28	4.06 ± 2.23
b:	0.28	3.18 ± 1,17
c:	0.065	0.61 ± 0.92
d:	3.19	105.8 ± 46.9
e:	0,51	3.13 ± 1.73
f:	n.d.	2.10 ± 1.27
g:	0.81	0.84 ± 0.43
h:	0.86	3.21 ± 2.60

After measurement of cytochrome P450 concentrations, the obtained results can be expressed as metabolite concentration per pg/ml CYP450 content. The time dependent increase of metabolite formation is presented in table 3 for the major metabolite (d).

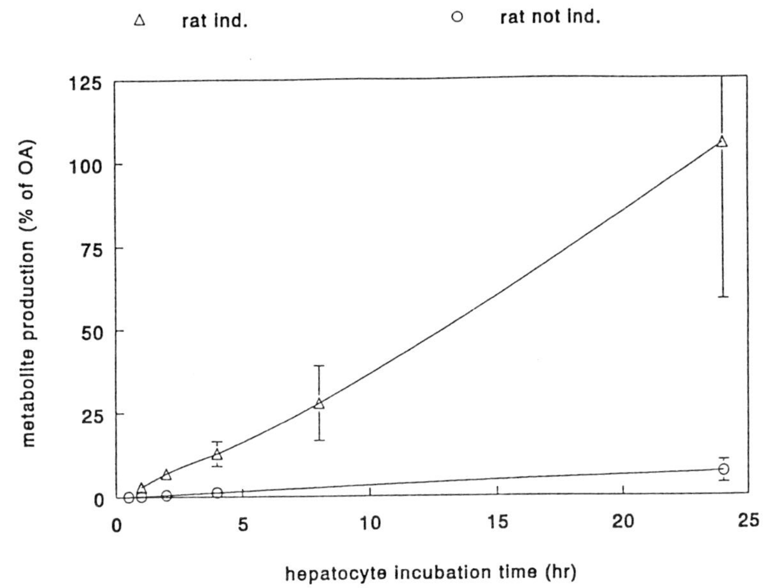

Figure 1: Time dependency of metabolite production: formation of metabolite d

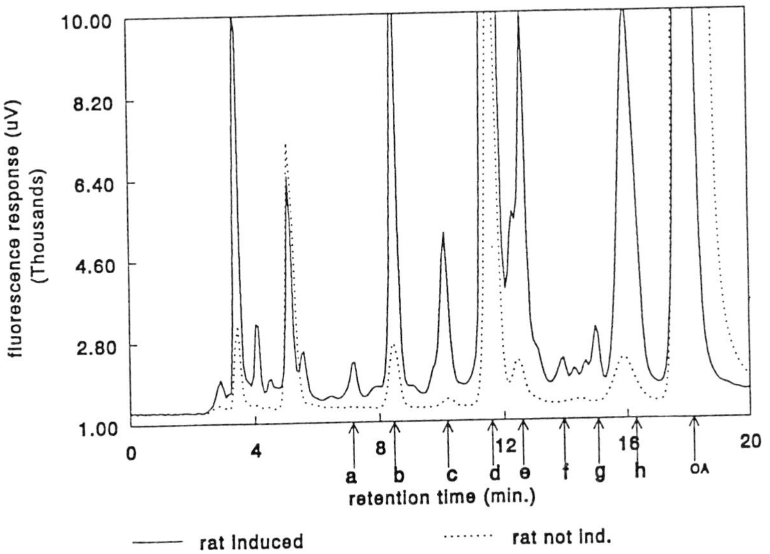

Figure 2: HPLC profile of ochratoxin A metabolites as produced by rat hepatocytes in culture (24 hrs): comparison of metabolite formation in control hepatocytes (non induced) and hepatocytes obtained from phenobarbital pretreated rats (for details of the applied method see text)

Table 3: Time dependent concentration of total cytochrome P450 in liver cell preparation obtained from PB-pretreated rats (PB3 - PB6) and relative production of metabolite d (area % of detectable OA/pmol P450/ml)

Time (hrs)	Total CYP450				rel. metabolite conc.			
	PB3	PB4	PB5	PB6	PB3	PB4	PB5	PB6
0.5	930	472	875	908	0.12	low	low	0.09
1	1083	549	711	930	0.31	low	0.44	0.26
2	717	457	694	743	1.10	0.96	1.15	1.00
4	936	579	633	398	1.72	1.35	2.18	3.42
8	904	659	750	692	2.25	2.50	4.65	5.75
24	409	115	479	263	35.28	42.20	30.02	32.81

DISCUSSION

Early studies on biotransformation of OA resulted in the detection of 4(R)- and 4(S)-4-hydroxyochratoxin A as major metabolites formed by rat microsomes and hepatocytes *in vitro* (*Hansen et al., 1982, Stormer et al., 1981*). Furthermore, *in vivo* studies indicated a cleavage of OA to ochratoxin-alpha and phenylalanine. As both, the hydroxy-metabolites and ochratoxin-alpha, are less toxic than the parent compound OA and induction of drug metabolizing enzymes decreased the acute toxicity of single injection of OA, it was concluded that biotransformation represents a detoxification reaction (*Chakor et al., 1988*). In contrast to these experiments, in which toxicological assessment was based on acute toxicity studies only, *Creppy et al., (1983)* found 4-(R)-4-hydroxyochratoxin A to be an immunosuppressor almost as highly effective as OA. Furthermore, we had demonstrated in previous experiments that metabolic transformation of OA by isolated hepatocytes resulted in mutagenic effects as demonstrated in the *Salmonella*/microsome assay and increased the incidence of sister chromatid-exchanges in human lymphocytes (*Hennig et al.,1991*). Thus, it seems relevant to study the biotransformation steps involved in OA metabolism in more detail. In this first series of experiments we developed a sensitive HPLC method for the detection of metabolites and compared the metabolic pattern of microsomes derived from rats and pigs with the pattern of metabolites generated by primary liver cells in culture. Microsomal incubations were carried out with and without pretreatment of donor rats with PB, which is known to induce the activity of various constitutive cytochrome P450 forms in the liver, including CYP 2A1, CYP 2B2 and CYP 2C6 (*Soucek and Gut, 1992*). At least 3 inducible metabolites of OA could be detected, non of which represented the cleavage product ochratoxin-alpha. As the above mentioned P450 forms not only catalyze hydroxylation reactions and all three forms show a common selectivity for demethylation reactions, it can be suggested that one of the formed metabolites contains the demethylated isocoumarin. The prominent role of cytochrome P450 in ochratoxin A induced toxicity was also demonstrated by *Omar et al.(1991),* who observed an induced lipid peroxidation in reconstituted microsomal lipid peroxidation systemes.

Furthermore, it could be demonstrated that there is a marked difference between metabolic conversion of OA in different animals species, as evidenced by the comparison of rat and pig microsomal incubations. The presence of further - as of yet unidentified - metabolites in pig liver microsomes is also described by *Oster et al. (1991)*, who found after scanning of TLC plates an unknown lipophilic metabolite. In contrast to the results of *Oster et al. (1991)* we could not detect ochratoxin-alpha in microsomal incubations.

Exposure of primary cultures of rat hepatocytes to OA resulted in at least 7 different metabolites and one additional derivative after PB induction. Based on the comparison of relative metabolite concentrations (metabolite concentration expressed as percentage of analysed OA concentration) it could be estimated that all biotransformation steps are PB inducible. Furthermore, the time dependent increase of metabolite formation indicates the generation of stable biotransformation products, which were found extracellularly only and thus may be available for circulation under *in vivo* conditions. Further studies directed to structure analysis of the metabolites are in progress. However, one of the striking effects of our previous experiments on mutagenic and cytogenetic response to OA biotransformation products was the finding, that only hepatocytic biotransformation yielded mutagenic metabolites. When comparing the presented results from this study it can be concluded that the major metabolite (d), which is present in both, microsomal incubations and in the culture medium is most likely not responsible for the mutagenic response. Further sudies are necessary to identify and isolate the obtained metabolites and to forward them to toxicological testing with special emphasis on mutagenic and genotoxic effects. In addition, the newly developed HPLC technique for the detection of OA metabolites should be used to asses circulating and excreted OA metabolites in both, laboratory animals exposed experimentally to OA, and target animal species including man exposed to OA contaminated food commodities.

REFERENCES:

IPCS (1990); Selected Mycotoxins: Ochratoxins, Trichothecens, Ergot: Environmental Health Criteria *105*. World Health Organization, Geneva.

Chakor K, Creppy EE, and Dirheimer G (1988): In vivo studies on the relationship between hepatic metabolism and toxicity of ochratoxin A. *Arch. Toxicol. Suppl. 12, 201-204.*

Creppy EE, Stormer, FC, Röschenthaler R, and Dirheimer G. (1983): Effects of two metabolites of ochratoxin A, 4-(R)-4 hydroxyochratoxin A and ochratoxin-*a*, on immune response in mice. *Infection and Immunity 39:1015-1018.*

Hansen CE, Dueland S, Drevon CA and Stormer FC (1982): Metabolism of ochratoxin A by primary culture of rat hepatocytes. *Appl. Environment. Microbiol. 43:1267-1271.*

Hennig A, Fink-Gremmels J, Leistner L (1991): Mutagenicity and effects of ochratoxin A on the frequency of sister chromatid exchanges after metabolic activation. *in:* Mycotoxins, Endemic Nephropathy and Urinary Tract Tumours. *Eds*. Castegnaro M, Plestina R, Dirheimer G, Chernozemsky IN & Bartsch H.:International Agency for Research in Cancer, Lyon. pp. 255-260.

Omar RF, Rahimtula AD and Bartsch H (1991): Role of cytochrome P-450 in ochratoxin A-stimulated lipid peroxidation. *J. Biochem. Toxicol. 6:203-209.*

Oster T, Jayyosi Z, Creppy EE, Souhili El Amri H, Batt A-M (1991): Characterization of pig liver purified cytochrome P-450 isoenzymes for ochratoxin A metabolism studies. *Toxicol. Letters, 57:203-214.*

Rutten AJJL, Falke HE, Catsburg JF, Top R, Blaauboer BJ, Van Holsteijn I, Doorn L and Van Leeuwen FXR (1987): Interlaboratory comparison of total cytochrome P-450 and protein determination in rat microsomes. *Archives Toxicol. 61:27-33.*

Seglen PO (1976): Preparation of isolated rat liver cells. *Methods in Cell Biology, XIII, 29-83.*

Stormer FC, Hansen CE, Pedersen JI, Hvistendahl G and Aasen AJ (1981): Formation of (4R)- and (4S)-4-hydroxyochratoxin-A from ochratoxin A by liver microsomes from various species. *Appl. Environment. Microbiol. 42:1051-1056.*

Soucek P and Gut I (1992): Cytochrom P450 in rats: structures, functions, properties and relevant human forms. *Xenobiotica 22:83-103.*

Suzuki S, Nowa K and Mori H (1986): Effects of drug metabolizing enzyme inducers on carcinogenesis and toxicity of ochratoxin A in mice. *Toxicol. Lett 31 (Suppl):206.*

Résumé

L' étude du métabolisme de l' ochratoxine A in vitro montre qu' après incubation d' hépatocytes de rat, on obtient au moins 7 métabolites différents. Un huitième métabolite est obtenu après induction des enzymes du microsome par le Phénobarbital.

L' augmentation du taux des métabolites en fonction du temps indique que ce sont des dérivés stables qui sont formés. Parmi ceux-ci se trouve une ou plusieurs substances mutagènes.

L' étude de la structure de ces métabollites est en cours. Mais il semble d'ores et déjà que le métabolite majeur ne soit pas mutagène.

Human ochratoxicosis and its pathologies. Eds E.E. Creppy, M. Castegnaro, G. Dirheimer. Colloque INSERM/
John Libbey Eurotext Ltd. © 1993, Vol. 231, pp. 75-82.

Détoxification de l'ochratoxine A par des moyens physiques, chimiques et enzymatiques

P. Deberghes[1], G. Deffieux[2], A. Gharbi[1], A.M. Betbeder[1], F. Boisard[3], R. Blanc[3], J.F. Delaby[3] et E.E. Creppy[1]

[1]Laboratoire de toxicologie et hygiène appliquée, UFR de Pharmacie, 3, place de la Victoire, 33076 Bordeaux Cedex; [2]Laboratoire de Botanique, UFR de Pharmacie, 3, place de la Victoire, 33076 Bordeaux Cedex; [3]Centre d'Etudes Scientifiques et Techniques de l'Armée (CESTA), 33114 Le Barp, France

INTRODUCTION

Le présent travail a été mené afin de trouver les moyens d'empêcher une moisissure toxinogène de produire de l'ochratoxine et/ou de la détruire lorsqu'elle est déjà produite aussi bien en milieu liquide que solide.

Matériels et méthodes

Dans un premier temps une fiole contenant 100 grammes de maïs et 50 ml d'eau stérile est autoclavée, puis inoculée avec un même volume de suspension de spores que dans le cas des milieux liquides. Cette première fiole est incubée pendant 2 jours à 26°C dans des conditions stériles.

Ensuite, 6 fioles, une par cas, contenant 100 grammes de maïs et 50 grammes d'eau stérile sont autoclavées puis infestées avec un même nombre de grains de maïs provenant de la fiole initiale. Les doses de radiations sont les mêmes que les milieux soient liquides ou solides.

* Congélation dans de l'azote liquide

Les inocula sont placés dans des tubes Eppendorf et plongés pendant des durées variables dans une bouteille d'azote liquide de type "Taylor-Wharton 35 VHC", avec décongélation au bain-marie à 26°C.

Deux durées d'exposition ont été testées : vingt secondes et une minute, avec le même nombre d'expositions : 2, 4, 6 et 8. Chaque traitement a été reproduit quatre fois.

* Congélation à -20°C

Les inocula sont placés dans des tubes Eppendorf et exposés 2, 4 et 6 fois pendant des durées de vingt minutes dans un congélateur à -20°C, avec décongélation au bain-marie à 26°C. Chaque traitement a été reproduit cinq fois.

b) Traitements chimiques des milieux de culture

* Ajout de carboxypeptidase A

La carboxypeptidase de pancréas de bovin a été utilisée (SIGMA); elle se présente sous la forme d'une solution aqueuse avec du toluène. Elle contient 37 mg de protéine/ml et 55 U/mg de protéine ; une unité hydrolysant 1 μmole d'hippuryl-L-phénylalanine par minute à pH 7,5 et à 25°C.

Dans ce cas on prépare une solution mère à 5 unités, que l'on filtre (filtre 0.22 μm) et à partir de laquelle on effectue des dilutions à 1 et 2 unités. Pour chaque dilution 100 μl sont injectés dans une fiole contenant du milieu de culture et préalablement inoculée, chaque traitement étant reproduit cinq fois. Les témoins reçoivent chacun 100 μl d'eau stérile.

A la fin de chaque expérience la masse mycélienne de chaque fiole est essorée et pesée avant et après séchage à l'étuve (80°C pendant 16 heures).

*Essai de destruction de la toxine déjà produite

Trois solutions d'OTA sont préparées, à 0,1, 1 et 10 μg/ml et réparties en cinq parties égales dans des tubes. Chaque dilution est exposée à 2 kGy, 2,5 kGy, 3 kGy, 4 kGy et 5 kGy, comme dans le cas des cultures.

*Protocoles d'extraction et de dosage de l'OTA

a) A partir d'un milieu liquide

Les fioles sont mises à incuber pendant au moins dix jours et l'arrêt des cultures se fait avec de l'HCl 7% (12,5 ml/fiole). On effectue une filtration sur laine de verre, avec rinçage du filtre et de la biomasse à l'eau (environ 25 ml/fiole), jusqu'à élimination de la fluorescence.

Le schéma d'extraction est alors le suivant :

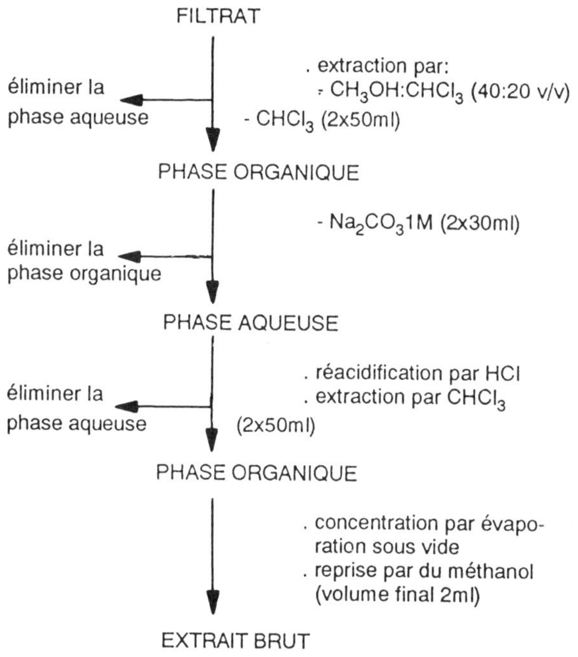

b) <u>A partir du maïs</u>

Dans un premier temps le maïs doit être broyé. Ensuite le protocole est le suivant :

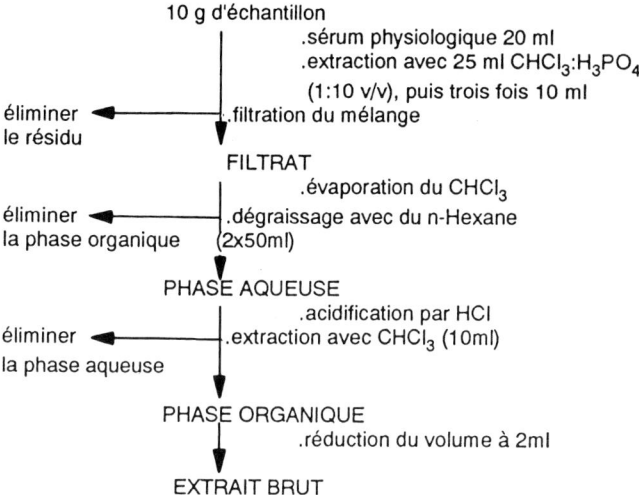

10 g d'échantillon
- .sérum physiologique 20 ml
- .extraction avec 25 ml $CHCl_3$:H_3PO_4 (1:10 v/v), puis trois fois 10 ml
- .filtration du mélange → éliminer le résidu

FILTRAT
- .évaporation du $CHCl_3$
- .dégraissage avec du n-Hexane (2x50ml) → éliminer la phase organique

PHASE AQUEUSE
- .acidification par HCl
- .extraction avec $CHCl_3$ (10ml) → éliminer la phase aqueuse

PHASE ORGANIQUE
- .réduction du volume à 2ml

EXTRAIT BRUT

Les échantillons sont ensuite purifiés par CCM préparative, l'éluant étant le suivant: acide formique:acétate d'éthyle:toluène (1:3:6 v/v/v). La silice est récupérée par grattage et épuisée avec du méthanol. La tache présentant la fluorescence de l'OTA témoin est récupérée sous UV après séchage des plaques.

Enfin l'analyse se fait en HPLC avec un détecteur de fluorimétrie, excitation (365nm) émission (421 nm). L'éluant utilisé est de l'acide phosphorique dilué à 1/1000 ; le débit est de 0,7 ml/min ; le volume d'injection est de 50 μl ; la colonne est une colonne C18 de 30 x 0,25 cm, ODS 10 μ (ICS) ; la précolonne est également une C18 ; l'injecteur est un modèle 738 universal HPLC autosampler (ICS) ; le stockage et le traitement se font par ordinateur grâce au logiciel PIC3 (ICS).

RESULTATS ET DISCUSSION

Les paramètres étudiés ont été le nombre d'amas mycéliens et la masse mycélienne totale dans chaque fiole, le pourcentage de matière sèche et la quantité d'ochratoxine produite.

Il est rapidement apparu qu'un nombre réduit d'amas mycéliens n'entraîne pas forcément la réduction de la quantité d'ochratoxine produite.

L'analyse statistique des résultats a fait appel au test de Wilcoxon (Wilcoxon Rank Sun Test)

Ce test non paramétrique est adapté aux nonombres d'essais réduits, 5 à 6 fioles et 2 à 3 séries.

<u>Effet des expositions aux UV</u>

L'exposition des inocula d'*Aspergillus ochraceus* à des UV B (10 nm) (TABLEAUX 1 et 2) induit une réduction significative aussi bien du nombre d'amas mycéliens que de la masse totale humide ou sèche par rapport aux témoins. Après 10 jours de culture la quantité d'OTA produite est nulle pour 2 heures d'exposition ou plus. Cependant après 18 jours de culture une faible quantité d'OTA est produite pour des expositions inférieures à 1 heure. Les UV B semblent donc constituer un moyen efficace et relativement peu onéreux pour réduire la production d'OTA dans les milieux liquides. Il reste à savoir ce que peuvent faire ces UV B sur des inocula solides, qui n'ont pas été testées pour l'irradiation UV à cause des difficultés techniques liées à l'exposition des grains.

Effets de l'irradiation par les rayons Gamma sur la production d'ochratoxine en milieu liquide et solide

En ce qui concerne l'effet des radiations ionisantes (TABLEAUX 3 et 4), nos résultats indiquent que ces irradiations, pour être efficaces, doivent être appliquées en quantités suffisantes. Chelak et coll. 1991 avaient abouti à la même conclusion En effet, si on n'applique que 2 ou 3 kGy la réduction de la masse mycélienne est faible, mais la production d'OTA est fortement augmentée par rapport aux témoins. Il faut aller jusqu'à 4 kGy pour obtenir une réduction de 60% de la production d'OTA, alors qu' à 5 kGy il n'y a plus du tout d'OTA produite en milieu liquide. Par contre en milieu solide l'effet est déjà très net dès 2 kGy où il y a environ 75% de réduction de la quantité d'OTA produite. Mais cet effet n'évolue pas en fonction de la quantité de radiations ionisantes. Ceci est en accord avec les conclusions de Cuero et coll. 1986, Paster et Barkaï-Golan 1986 et Aziz et coll. 1989. Ce dernier aspect du problème demanderait à être revu en tenant compte des divers paramètres (température, présence d'oxygène) et surtout l'agitation des échantillons pendant l'irradiation, qui pose un problème technique.

Lorsque les radiations ionisantes sont appliquées à des liquides contenant déjà de l'OTA, la toxine est détruite à 50% environ entre 2 et 3 kGy, et à 80% entre 4 et 5 kGy pour les concentrations élevées d'OTA (Figure 1). Cependant plus la concentration d'OTA est élevée dans le milieu et plus l'irradiation est efficace, dans nos conditions d'expérience. Compte-tenu de la stabilité de l'OTA pendant le processus de fermentation, Chu et coll. 1975

Effet de la congélation sur la production d'ochratoxine A

Lorsque les inocula sont congelés à -20°C, puis décongelés de façon répétée (TABLEAU 5), la masse mycélienne totale est réduite de façon significative dans les lots traités par rapport aux témoins, que ce soit la masse humide ou sèche, bien que la production d'OTA ne soit réduite que de 10 à 45%. Ceci pourrait être le résultat de lésions induites par les cristaux de glace qui se forment pendant la congélation lente. L'effet de cette congélation n'est pas proportionnel à la durée ni au nombre de congélations-décongélations.

Ce procédé de congélation-décongélation à -20°C, qui pourrait être appliqué en routine devra être mis au point en fonction de paramètres tels que le temps de congélation ou de décongélation, de la nature et de degré d'humidité de la denrée concernée.

Quant aux congélations dans de l'azote liquide suivies de décongélations, aucun effet n'est observé.

Effets du traitement par la carboxypeptidase sur la production d'ochratoxine A par Aspergillus ochraceus

Les traitements enzymatiques des milieux liquides par addition de carboxypeptidase sont autorisés par le ministère de l'agriculture, particulièrement en brasserie. Nos résultats indiquent une réduction très nette de la quantité d'OTA produite, environ 60% après 18 jours pour 40 U/l, et une disparition complète de l'OTA pour 100 U/l (TABLEAU 6). Compte-tenu de la stabilité de l'OTA pendant le processus de fermentation, Chu et coll. 1975, ce processus pourrait être très rapidement appliqué dans les industries agro-alimentaires. Il reste à savoir l'inocuité d'une telle quantité d'enzymes chez l'homme.

Cependant il faut noter que la masse mycélienne totale humide ou sèche, n'est pas réduite par la présence de la carboxypeptidase indiquant qu'elle agit plutôt sur l'OTA déja formée comme on pouvait s'y attendre, et pas sur son processus de formation.

Sur la base de nos résultats les traitements à retenir sont les rayonnements UV B et gamma, les lésions produites par les cristaux de glace formés lors de la congélation à -20°C, et l'utilisation d'enzyme telle que la carboxypeptidase dans le milieu liquide. Toutefois les protocoles d'application devront être étudiés plus en détail pour chaque denrée.

L'originalité de cette étude est qu'elle a porté sur des milieux liquides et solides, avec une idée d'adaptation directe aux industries agro-alimentaires : laiteries, brasseries, distilleries, stockage de céréales, préparation d'aliments pour bétail en vue de garantir une certaine qualité des produits alimentaires d'origine végétale ou animale (Kroght 1987, Golinski et Grabarkiewicz-Szczesna 1988 et Kuiper-Goodman et Scott 1989.

L'effet destructeur de la carboxypeptidase sur l'OTA dans des milieux de culture liquides peut permettre l'application directe de nos résultats en agro-alimentaire, notamment en brasserie.

Le fait que l'irradiation de milieux solides à 2 kGy permette de réduire considérablement la production d'OTA laisse espérer qu'une faible irradiation des céréales modérément contaminées peut permettre de prolonger le temps de conservation dans les conditions d'humidité et de température connues actuellement.

Une combinaison entre deux ou trois traitements de l'inoculum ou du milieu peut être envisagée avec l'espoir qu'il y ait une addition des effets, voire une synergie.

BIBLIOGRAPHIE

AZIZ N.H., REFAÏ M. & EL-FAR F., 1989, Gamma irradiation and potassium sorbate in the control of growth and ochratoxin production by *Aspergillus ochraceus*, J. Egypt. Vet. Ass., 49 (3), 951-961.

CHELACK W.S., BORSA J., MARQUARDT R.R. & FRÖHLICH A.A., Sept. 1991, Role of the Competitive Microbial Flora in the Radiation-Induced Enhancement of Ochratoxin Production by *Aspergillus alutaceus var. alutaceus* NRRL 3174, Applied and environmental microbiology, 2492-2496.

CHU F.S., CHANG C.C., ASHOOR S. & PRENTICE N., 1975, Stability of aflatoxin B_1 and ochratoxin A in brewing, Appl. Microbiol., 29, 313-316.

CUERO R.G., SMITH J.E. & LACEY J., 1986, The influence of gamma irradiation and sodium hypochlorite sterilization on maize seed mycoflora and germination, Food Microbiology, 3, 107-114.

GOLINSKI P. & GRABARKIEWICZ-SZCZESNA J., 1988, Contamination of bread and other grain products with ochratoxin A, Zanieczyszczenie ochratoksyna A pieczywa i innych przetworow zbozowych, Rocz. Panstw. Zakl. Hog., 39, 21-25.

KROGH P., 1987, Ochratoxins in food, Mycotoxins in Food, 97-122.

KUIPER-GOODMAN T. & SCOTT P.M., 1989, Risk assessment of the mycotoxin ochratoxin A, Biomed. Environ. Sci., 2, 179-248.
contamination by ochratoxin A of green and roasted coffee beans, Food. Addit. Contam., 6, 33-339.

PASTER N. & BARKAÏ-GOLAN, 1986, Heat and gamma irradiation effects on ochratoxin production by *Aspergillus ochraceus* sclerotia, Trans. Br. mycol. Soc., 87 (2), 223-228.

TABLEAU 1: Effets de diverses expositions à des UV B sur des inocula d'Aspergillus ochraceus, après 10 jours de culture en milieu liquide (50 ml).

	TEMOINS *	2 heures	3 heures	4 heures
nombre d'amas mycéliens	120	4	6	1
écart type	23.2	1.7	2.3	1.4
masse totale (g)	2.4333	1.1684	1.8753	0.323
écart type	0.613	0.648	0.477	0.4428
matière sèche (g)	0.1494	0.0482	0.0587	0.0054
écart type	0.029	0.0359	0.0296	0.0103
OTA (ng/ml)	23.7	0	0	0

* Les témoins diffèrent des cas traités au seuil 0,005 (Wilcoxon Rank Sum Test).

TABLEAU 2: Effets de diverses durées d'exposition à des UV B sur des inocula d'Aspergillus ochraceus, après 18 jours de culture en milieu liquide (50 ml).

	TEMOINS *	1/2 heure	1 heure	2 heures
nombre d'amas mycéliens	128	20	10	17
écart type	24.1	9.3	20.5	8.3
masse totale (g)	12.1942	5.877	6.1367	6.4782
écart type	3.1386	1.4678	1.3556	1.0986
matière sèche (g)	1.3818	0.8385	0.8764	0.8885
écart type	0.2471	0.1006	0.1153	0.07
OTA (ng/ml)	13.1	3.2	0	0

* Les témoins diffèrent des cas traités au seuil 0,005 (Wilcoxon Rank Sum Test).

TABLEAU 3 : Effets de diverses doses de radiations par rayons gamma sur des inocula d'Aspergillus ochraceus, après 12 jours de culture en milieu liquide (50 ml) et sur milieu solide.

	TEMOINS *	2 kGy	3 kGy	4 kGy	5 kGy
nombre d'amas mycéliens	103	7	7	1	0
écart type	8.4	6.7	7.5	1.1	0
masse totale (g)	6.6328	1.9449	1.6306	0.1437	0
écart type	0.7819	1.4565	1.2299	0.3801	0
matière sèche (g)	0.5275	0.1325	0.112	0.0252	0
écart type	0.1022	0.0956	0.0928	0.0812	0
OTA (ng/ml)	42	210.3	80.8	19.1	0
écart type	10.3	32.7	27	13.1	0
OTA sur maïs (ng/g)	6.1	4.3	2.8	3.2	4

* Les témoins diffèrent des cas traités au seuil 0,005 (Wilcoxon Rank Sum Test).

TABLEAU 4 : Effets de diverses doses de radiations par rayons gamma sur des inocula d'Aspergillus ochraceus, après 18 jours de culture en milieu liquide (50 ml) et sur milieu solide.

	TEMOINS *	2 kGy	2,5 kGy	3 kGy	4 kGy	5 kGy
nombre d'amas mycéliens	201	3	3	3	2	0
écart type	26.3	0.9	1.9	3.5	2.7	0
masse totale (g)	7.356	7.9458	5.7754	6.0491	1.267	0
écart type	1.352	1.8178	1.4272	1.7961	1.0121	0
matière sèche (g)	1.75	1.4637	1.4037	1.372	0.1197	0
écart type	0.3734	0.2199	0.4392	0.2552	0.2129	0
OTA (ng/ml)	30.3	87.6	134.9	106.7	5.6	0
écart type	14.2	21.4	17	18.6	10.4	0
OTA sur maïs (ng/g)	7.2	1.4	2.3	0.9	1.2	2.6

* Les témoins diffèrent des cas traités au seuil 0,005 (Wilcoxon Rank Sum Test).

TABLEAU 5: Effets de congélations et décongélations répétées à -20°C sur des inocula d'Aspergillus ochraceus, après 10 jours de culture en milieu liquide (50 ml).

	TEMOINS	2x20'	4x20'	6x20'
nombre d'amas mycéliens	64	35	61	56
écart type	12.9	19.1	16.2	19
masse totale (g) *	5.312	2.7104	4.5153	3.5346
écart type	1.0273	1.5228	1.3516	0.8857
matière sèche (g) *	0.5876	0.1709	0.289	0.2929
écart type	0.1564	0.104	0.0916	0.2226
OTA (ng/ml)	9	6	5.2	8

* Les témoins diffèrent des cas traités pour la masse totale au seuil 0,05, et pour la matière sèche au seuil 0,005 (Wilcoxon Rank Sum Test).

ABLEAU 6 : Effets de l'ajout de carboxypeptidase dans un milieu liquide (50 ml) contenant es inocula d'Aspergillus ochraceus, après 18 jours de culture.

	TEMOINS *	1 UNITE	2 UNITES	5 UNITES
ombre d'amas ycéliens	89	86	71	77
cart type	8.4	18.2	10.8	19.1
asse totale (g)	8.1752	8.9723	11.5608	9.0888
cart type	1.4789	0.6244	2.0009	4.1468
atière sèche (g)	0.5005	0.6056	0.6145	0.4458
cart type	0.1674	0.0474	0.0301	0.229
TA (ng/ml)	73.6	61.4	30.1	0

Les témoins ne diffèrent pas des cas traités, quel que soit le seuil choisi (Wilcoxon Rank um Test), en ce qui concerne la viabilité des mycélia et leur masse humide et sèche.

Figure 1 : Effet de rayonnements gamma sur la destruction d'ochratoxine A, pour différentes concentrations initiales d'OTA

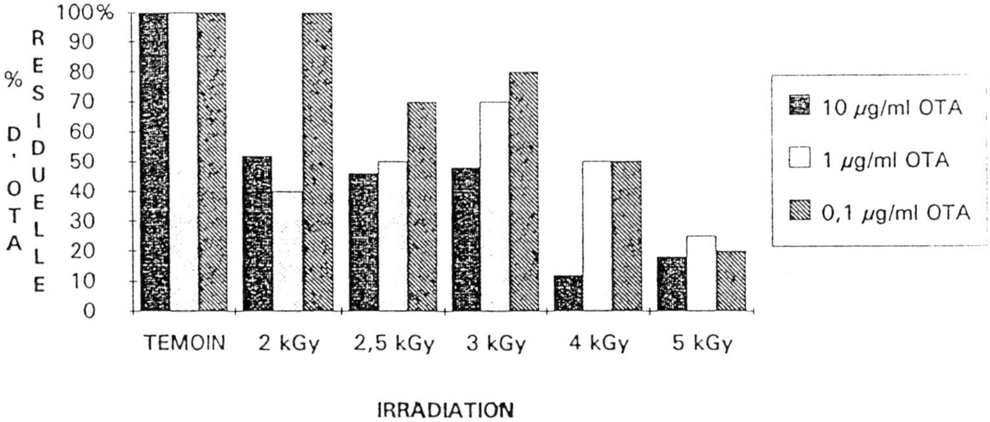

Summary:

Physical and biochemical methods have been studied to prevent a toxicogenic *Aspergillus* from producing ochratoxin A (OTA) or destroy it when already produced in a liquid or solid medium. Irradiation by UV B are efficient in a liquid medium to prevent OTA production. γ irradiation is also efficient to prevent the production of OTA or destroy it. 2 to 3 KGy and 4 to 5 KGy are needed for solid and liquid medium respectively. Repeated freezing at - 20°C and defreezing is also efficient in liquid medium to reduce the OTA production.

Carboxypeptidase (4 to 5 U/l) is very efficient to cleave the OTA already produced in a liquid medium. All these methods can be applied in food production industries and beer production.

Résumé

Des moyens physiques, chimiques et enzymatiques ont été utilisés afin d'empêcher un *Aspergillus* toxinogène de produire l'ochratoxine A (OTA) et de détruire cette toxine une fois produite dans un milieu liquide ou solide . Les UV B sont efficaces en milieu liquide pour empêcher la moisissure de proliférer et de produire l' OTA. Les rayons γ sont également efficaces pour empêcher la production d' OTA en milieu liquide ou solide et même la détruire en milieu liquide à condition d'appliquer la dose suffisante 2 à 3 KGy en milieu solide, 4 à 5 KGy en milieu liquide. La congélation répétée à - 20°C suivie de décongélation abaisse la production de la toxine en milieu liquide.

La carboxypeptidase (4 à 5 U/l) est très efficace en milieu liquide pour détruire l'OTA produite. Ces différentes méthodes ont été étudiées pour être rapidement applicables dans les industries agro-alimentaires.

II. Occurence of ochratoxin A in foodstuffs and human blood

II. Présence de l'ochratoxine A dans les aliments et le sang humain

Human ochratoxicosis and its pathologies. Eds E.E. Creppy, M. Castegnaro, G. Dirheimer. Colloque INSERM/ John Libbey Eurotext Ltd. © 1993, Vol. 231, pp. 85-100.

Worldwide ochratoxin A levels in food and feeds

G.J.A. Speijers and H.P. van Egmond

Department of Toxicology, National Institute of Public Health and Environmental Protection, P.O. Box 1, 3720 BA Bilthoven, The Netherlands

1) Introduction

Ochratoxins are metabolites produced by many Aspergillus and Penicillium species. The Aspergillus species include Aspergillus ochraceus, A. sulphureus, A. sclerotionum, A. alliacus, A. melleus, A. ostanus and A. petrakii (Maryamma and Nair, 1991; IPCS, 1990). The Penicillium species include Penicillium purpurescens Sapp, P. commune Thom, P. viridicatum Wes, P. paktans, P. cyclopium Weiling and P. variable. The production of the ochratoxins depends on environmental conditions such as water activity (moisture) and temperature (Northolt et al, 1979; IPCS, 1990; Maryamna and Nair, 1991). For A. ochraceus high temperatures (12 - 37^{0}C) are necessary, whereas frigophilic Penicillia (4 - 31^{0}C), particularly P. viridicatum , also producers of ochratoxins in areas with a colder climate (IPCS, 1990).

Ochratoxin A is the major ochratoxin which has been found in a number of countries worldwide (IPCS, 1990; Kuiper-Goodman and Grant, 1990). Ochratoxin caused nephropathy in experimental animals, and it has been associated with nephropathy in live stock animals (Kuiper-Goodman and Grant, 1990; Dalvi and Salunkhe, 1990). In addition, it has been associated with Balkan endemic nephropathy, ochratoxicoses and the occurrence of uro-genital tract tumours in animals and possibly in humans (Plestina, 1992; Ceovic et al, 1992, Radonic and Radosevic, 1992; Pohland et al, 1992).

The data on occurrence of ochratoxin A as presented in this paper have been established with a variety of analytical methods. These methods have evolved over the years. Generally, methods of analysis for ochratoxin A include the usual basic steps applied in mycotoxin methology: extraction, cleanup, separation, detection and quantification, sometimes followed by confirmation of identity (Van Egmond, 1991a). In conventional procedures, cleanup is usually achieved by liquid-liquid extraction or absorption column chromatography, followed by thin layer chromatography or high-performance liquid chromatography and UV or fluorescence detection. More recent approaches involve enzyme-linked immunosorbent assay for screening and (semi-)quantitative determination and immunoaffinity column chromatography for rapid cleanup, followed by conventional instrumental analysis. These techniques also offer possibilities for automated systems.

The worldwide occurrence of ochratoxin A is discussed in this paper by means of summarizing data per continents (Africa, America, Asia and Australia, and Europe). Distinction is made between food and feed commodities of vegetable or animal origin. Most data are shown in tables and particular matters are reported in different chapters.

The worldwide regulations for ochratoxin A as published recently by Van Egmond (1991b) are discussed briefly and finally the results of the studies on occurrence levels of ochratoxin in some countries are compared with the tolerance levels set in the respective regulations.

2) Occurrence of ochratoxin A in food and feed commodities from vegetable or animal origin.

2.1 Ochratoxin A levels in food and feed in African countries.

Very few data on the occurrence of ochratoxin in food and feed in African countries are available (table 1). Most data reported are average values if contamination was observed, and sometimes only semiquantitative determinations were available. Kane et al (1991) reported analyses in eight different food items in Senegal, performed during a period of 1984 - 1988. The only product which revealed contamination of ochratoxin A was cowpea (Vigna unguiculata), which is widely distributed in Senegal. The average level for cowpeas was 34 µg ochratoxin A/kg. In Egypt more food and feed commodities were found to be contaminated with ochratoxin A (Abdelhamid, 1990). The data available in Sierra Leone had a semiquantitative character (Jonsyn and Lahai, 1992).

Table 1. Occurrence levels of ochratoxin A in food and feed commodities from vegetable or animal origin in Africa.

Commodity	country	number of samples analyzed	% contam.	range of ochratoxin A levels in µg/kg	reference
Food					
vegetable					
Cowpeas	Senegal	31	16	34[1]	Kane et al(1991)
Maize	Egypt	3	33	12[1]	Abdelhamid(1990)
Wheat	Egypt	3	33	10[1]	,, ,, ,,
Rice germs	Egypt	3	33	577[1]	,, ,, ,,
Rice germ cake	Egypt	3	33	4[1]	,, ,, ,,
Rice bran	Egypt	3	33	9[1]	,, ,, ,,
Horse bean	Egypt	3	33	7[1]	,, ,, ,,
animal					
Smoke	Sierra	20	14	1000-	Jonsyn and Lahai,

dried fish	Leone			2000[2]	(1992)
Feed					
Broilers feed	Egypt	3	33	13[1]	Abdelhamid(1990)
Egg prod. mixed feed	Egypt	3	33	14[1]	,, ,, ,,
Milk prod. mixed feed	Egypt	3	66	19[1]	,, ,, ,,

[1] average value

[2] semiquantitative determination

2.2 Ochratoxin A levels in food and feed in American countries

A considerable amount of data on the occurrence of ochratoxin A originates from North American countries, whereas almost no data are available from South America. The data involve mainly occurrence in plant commodities. Frohlich et al. (1991) reported the occurrence of ochratoxin A in blood of slaughter pigs where exposure to ochratoxin had lead to accumulation of ochratoxin A in some cases. No data of natural occurrence in organs are available. Prior and Sisodia (1978) observed residues of ochratoxin A upto 31 µg/kg in the kidneys of the highest dose group of hens fed diets containing 0, 0.5, 1.0 or 4.0 mg ochratoxin A/kg respectively. The data on the occurrence of ochratoxin A in food and feed are summarized in table 2.

Table 2. Occurrence levels of ochratoxin A in food and feed commodities from vegetable or animal origin in America.

Commodity	Country	number of samples analyzed	% contam.	range of ochratoxin A levels in µg/kg	reference
Food vegetable					
Maize	USA	293	1	83-166	Shotwell et al. (1971)
Wheat (red winter)	USA	297	11	5-115	Shotwell et al. (1976)
Wheat (red spring)	USA	286	2.8	5-115	Shotwell et al. (1976)
Barley	USA	127	14.2	10-40	Nesheim (1971)
Coffee beans	USA	267	7.1	20-360	Levi et al (1974)
Beans	Brasil	60	0	<30	Milanez and Sabino (1989)
animal					
Pig blood	Canada	1200	0.5	100-229	Marquardt et al. (1990)
	Canada	1200	4.1	20-229	
	Canada	1200	72	5-20	,, ,, ,,
Kidneys[2] chickens	Canada			-31	Prior and Sisodia (1978)
Feed					
Wheat hay	Canada	95	7.4	30-6000	Prior (1976)

Wheat oats[1]	Canada	32	56.3	30-27000	Scott et al. (1972)
Durum wheat[3]	Canada			110-11800	Abramson et al. (1990)
Mixed feed	Canada	474	1.1	30-100	Prior(1981)
Mixed feed[1]	Canada	51	7.8	48-5900	Abramson et al. (1983)

[1] All samples suspected of containing mycotoxins

[2] Animal accumulation study

[3] Storage study

2.3. Ochratoxin A levels in food and feed in Asian and Australian countries.

There are relative few data on the occurrence of ochratoxin A in food and feed in Asia and Australia. The available data are summarized in table 3. Most of the data involve contamination of plant products, whereas data on the occurrence of ochratoxin A in commodities of animal origin are restricted to a controlled accumulation study and one study on the natural occurrence in Bhutanese cheese. Reichmann et al. (1982) recorded residue level of 3 - 10 μg ochratoxin A/kg in the kidney and 1.5 - 2.5 μg/kg in the liver of hens given 1 mg ochratoxin A/kg feed for 8 weeks. Sinha and Ranjan (1990) found nine dominant fungi in 19 samples of Bhutanese cheese, including Aspergillus ochraceus and Penicillium citrinum. In 5 samples ochratoxin A was measured (42 - 116 μg/kg), in some case citrinin was found as a co-contaminant of ochratoxin A.

Table 3. Occurrence levels of ochratoxin A in food and feed commodities from vegetable or animal origin in Asia and Australia.

Commodity	country	number of samples analyzed	% contam.	range of ochratoxin A levels in μg/kg	reference
Food vegetable					
Maize	India	22	4.5	present[1]	Pande et al. (1990)
Wheat	India	30	3.3	present[1]	Pande et al. (1990)
Spices[2]	India	108	4.5	present[1]	Saxena and Mehrotra (1989)
Rice	India	30	6.6	present[1]	Pande et al. (1990)
Dry fruits[3]	India	74	5.4	present[1]	Saxena and Mehrotra (1990)
Green coff.	Japan	68	7.3	3.2-17	Belitz and Maier (1991)
animal					
Bhutanese	India	19	27	42-116	Sinha and Ranjan

cheese					(1990)
Kidneys[4]	Australia			3-10	Reichmann et al
Liver[4]	Australia			1.5-2.5	(1982)
Feed					
Mixed feed	Australia	25	4.0	70000[5]	Connole et al (1981)

[1] No quantitaive data

[2] Included: Tumeric, coriander, fennel, cinnamon, black pepper, cardamon, India cassi, amni, cumin, chili, yellow mustard, clove and garlic.

[3] Included: Coconut, raisin,almond, walnut, cashew nut, gargon nut and pistachio nut.

[4] Accumulation study [5] Average value

2.4 Ochratoxin A levels in food and feed in European countries

By far, most data on the occurrence of ochratoxin A are obtained from Europe, especially from the Scandinavian countries, Germany and countries of the Balkan area. The data are summarized in tables 4 a-c. Determinations of ochratoxin A in human blood have also been performed (Hald, 1991). The presence of ochratoxin A in human blood has been suggested as an indicator for indirect assessment of exposure to ochratoxin A. In several countries, therefore, human blood has been collected and analysed. Analyses of human serum samples in European countries revealed that blood from healthy humans was contaminated with ochratoxin A at concentrations ranging from 0.1- 14.4 μg/l. The frequency of contamination of human blood seems to indicate continuous, widespread exposure of humans to ochratoxin A (Hald, 1991).

Tholstrup and Rasmussen (1990) analysed cereals (wheat, wheat bran, oat and barley) during the period 1986 - 1989, and observed that the humidity during the harvest season affects fungal growth and thereby the level of ochratoxin A.
This study showed unacceptable high contents of ochratoxin A in wet harvest years, leading to ochratoxin A levels in pork kidneys above the current Danish limit of 10 μg/kg, when fed to pigs.

The data on the occurrence of ochratoxin A in animal tissues, organ and eggs also include those of several accumulation studies in which pigs or chickens were fed a diet containing ochratoxin in different concentrations (table 4c).

Micco et al (1989b) determined ochratoxin A in green coffee beans, and studied the destruction of it after roasting. The beverages prepared from artificially contaminated coffee using the most common types of coffee makers showed no residues of ochratoxin A (table 4a).

Also in human milk ochratoxin A is detected, indicating that food eaten by pregnant women has been contaminated with ochratoxin A leading to carry over into the milk.

Table 4a. Occurrence levels of ochratoxin A in food commodities of vegetable origin in Europe.

Commodity	country	number of samples analyzed	% contam.	range of ochratoxin A levels in µg/kg	reference
Maize	France	463	2.6	5-200	Galtier et al.
Maize	France	461	1.3	20-200	(1977)
Maize	Yugoslavia[1]	542	8.3[2]	6-140	Pavlovic et al. (1979)
Maize	Germ.F.Rep.	40	7.5	0.1-137	Bauer & Gareis (1987)
Maize[4]	Netherlands	9	0		Van Egmond & Sizoo(1984)
Maize	UK	29	37.9	50-500	Ministry of Agriculture(1980)
Maize	Bulgaria	22	9.0	10-25	Petkova-Bocharo-
Maize	Bulgaria[1]	22	27.3	25-35	va and Castegnaro(1985)
Maize	Poland	123	1.6	25-400	Golinski et al (1991)
Maize	Bulgaria[1]	151	57.8	0.2-1418	Petkova-Bocharova
Maize	Bulgaria	113	26.7	0.2-235	et al (1991)
Maize	Italy	90	15.0	1-2	Micco et al (1989a)
Maize	Austria	27	11.1	5-100	Leibetseder(1989)
Wheat	Yugoslavia[1]	130	8.5	14-135	Pavlovic et al (1979)
Wheat	Denmark	194	37	0.8-37	Tholstrup & Ras-
Wheat(ecol)	Denmark	36	46	1.2-21	mussen(1990)
Wheat	Czechosl.	25	72	>1	Fukal(1990)
Wheat	Poland	239	11.7	5-2400	Golinsky et al (1991)
Wheat	Germ.F.Rep.	64	1.6	0.4[2]	Frank(1991)
Wheat[4]	Netherlands	38	15	0.1-4.2	Van Egmond & Sizoo(1984)
Wheat	Austria	41	9.2	5-100	Leibetseder(1989)
Wheat bran	Germ.F.Rep	84	10.7	6.8[2]	Frank(1991)
Wheat bran	Denmark	57	10.5	5-20	Pedersen & Hansen (1981)
Wheat bran	Denmark	57	68	0.5-12	Tholstrup & Ras-
Wheat bran	Denmark	15	66	0.1-26	mussen(1990)
Wheat flour	Czechosl.	27	22	>1	Fukal(1990)

Wheat flour	Poland	137	19.7	3700[2]	Golinski et al (1991)
Rye	Denmark	267	37	2.5-120	Tholstrup & Ras-
Rye (ecol)	Denmark	53	81	0.7-120	mussen(1990)
Rye flour	Poland	78	26.9	5410[2]	Golinski et al
Rye	Poland	228	27.2	5-2400	(1991)
Rye	Germ.F.Rep.	64	1.6	0.4[2]	Frank(1991)
Rye[4]	Netherlands	14	36	0.1-16.8	Van Egmond & Sizoo(1984)
Rye	Austria	41	43.9	5-100	Leibetseder(1989)
Oats	Denmark	28	43	0.8-5.6	Tholstrup & Ras-
Oats (ecol)	Denmark	9	44	0.2-4.2	mussen(1990)
Oats	Czechosl.	19	1	1-2	Fukal(1990)
Oats	Germ.F.Rep.	21	38	1.2[2]	Frank(1991)
Oats[4]	Netherlands	18	22	0.1-2.4	Van Egmond & Sizoo(1984)
Oats	Austria	48	47.3	5-1000	Leibetseder(1989)
Wheat,Rye Oats	Germ.F.Rep.	232	12.9	0.1-206	Bauer & Gareis (1987)
Barley	Poland	616	8.8	5-1200	Golinski et al (1991)
Barley	Denmark	15	33	0.2-14	Tholstrup & Ras-
Barley(ecol)	Denmark	13	54	0.2-13	mussen(1990)
Barley	Denmark	50	8	9-189	Krogh(1978)
Barley	Yugoslavia[1]	64	12.5	14-27	Pavlovic et al (1979)
Barley	Czechosl.	48	2.1	3800[2]	Vesela et al(1978)
Barley	Czechosl.	32	53	>1	Fukal(1990)
Barley	Austria	27	11.1	5-1000	Leibetseder(1989)
Rice[4]	Netherlands	11	0		Van Egmond & Sizoo(1984)
Cereals	Germ.F.Rep.	765	3.1	11.8[2]	Frank(1991)
Grain	Germ.D.Rep.	49	4.1	18-22	Fritz et al(1979)
Grain	Germ.F.Rep.	43	42	2.0-304	Hadlok et al (1989)
Grain	Poland	296	6.8	20-470	Szebiotko et al (1981)
Grain	Poland	150	5.3	50-200	Juszkiewicz & Piskorska-Pliszczynska(1976)
Corn flour	UK	13	30.8	50-200	Ministry of Agriculture(1980)
Flour	Germ.F.Rep.	93	18.2	2.2[2]	Frank(1991)
All flour	Poland	215	22.3	4370[2]	Golinski et al (1991)
Child food[3]	Czechosl.	24	0	<1	Fukal(1990)
Beans	Sweden	71	8.5	10-442	Akerstrand & Josefsson(1979)
Beans	Bulgaria[1]	24	16.7	25-27	Petkova-Bocharova
Beans	Bulgaria[1]	28	7.1	25-50	& Castegnaro (1985)

Beans	Bulgaria[1]	157	48.1	0.05-260	Petkova-Bocharova
Beans	Bulgaria	113	27.2	0.2 -285	&Castegnaro(1990)
Peas	Sweden	71	8.5	10-442	Akerstrand & Josefsson(1979)
Soy beans	UK	25	36	50-500	Ministry of
Soy flour	UK	21	19	50-500	Agriculture(1980)
Cocoa beans	UK	56	18	100-500	,, ,, ,,
Coc.roasted	UK	19	16	100	,, ,, ,,
Coffee	Italy	19	58	0.2-15	Micco et al
Cof.roasted	Italy		50	0	(1989b)
Peanuts	Poland	609	0		Golinski et al
Protein Conc.	,,	132	0		(1991)
Semolina	Germ.F.Rep.	4	50	0.5[2]	Frank(1991)
Brewery Wheat	,, , ,,	83	0		,, ,,
Malt	Germ.F.Rep.	85	1.2	12[2]	Tressl et al (1989)

Table 4a. continued.

Beer	,, , ,,	75	0		,, , , ,,
Bread	Poland	368	17.2	1360[2]	Golinski et al
Groats	Poland	35	11.4	1140[2]	(1991)
Bread	UK	50	2.0	210[2]	Osborne(1980)
Flour	UK	7	28.5	490-2900	,, ,, ,,

[1] Area with endemic nephropathy [2] Average value [3] Grain based [4] Imported grain

Table 4b. Occurrence levels of ochratoxin A in food commodities of animal origin in Europe.

Commodity	country	number of samples analyzed	% contam.	range of ochratoxin A levels in µg/kg	reference
Kidney(pig)	Denmark	686	20	2-67	Krogh et al(1973)
Kidney(pig)	Denmark	60	35	2-68	Krogh(1977)
Kidney(pig)	Denmark	25	20	>25	Buchmann and Hald (1985)
Kidney(pig)	Denmark	20	100	0.5-1955	Bauer & Gareis (1987)
kidney(pig)	Germ.F.Rep.	104	21	0.1-1.8	Bauer & Gedek (1983)
Kidney(pig)	Germ.F.Rep.	300	14	0.5-10.2	Scheuer et al (1984)
Kidney(pig)	Germ.F.Rep.	100	15	0.5-16.4	Majerus & Woller (1983)
Kidney(pig)[1]	Netherlands	46	71[7]	0.2-2.0	Van Egmond &
Kidney(pig)	Netherlands	6	100[7]	0.2-1.0	et al(1984)
Kidney(pig)[1]	Netherlands	29	7	0.2-0.4	, , ,,

tolerance level for pork kidney has been established. When the occurrence levels of ochratoxin A are compared with the tolerance levels known established in these countries the latter may be exceeded, both for foods of vegetable and animal origin. This lead to the conclusion that monitoring for the presence of ochratoxin A should be a continuous effort.

Worldwide there are many gaps in the knowledge on the occurrence of ochratoxin A. Occurrence seems to be region-dependent and climatic and storage conditions may be important factors. There is a need for more data on the occurrence of ochratoxin A in foods and feeds in many countries, particularly in those from Africa, Asia, Australia and South America, because the occurrence of ochratoxin A is suspected to be associated with health hazards in animals (including live stock) and man.

4. References

Abdelhamid, A.M. (1990): occurrence of some mycotoxins (afla toxins, ochratoxin A, citrinin, zearelenone and vomitoxin) in various Egyptian feeds. *Arch.Anim. Nutr. Berlin,* 40, 647 - 664.

Abramson, D., Mills, J.T. & Boycott, B.R. (1983): Mycotoxins and mycoflora in animal feedstuffs in western Canada. *Can. J. Comp. Med.*, 47, 23 - 47.

Abramson, D., Mills, J.T. & Sinha, R.H. (1990): Mycotoxin production in amber durum wheat stored at 15 and 19 % moisture content. *Food Additives and Contaminants*, 7, 617 - 627.

Akerstrand, K. & Josefsson, E. (1979): Fungi and mycotoxins in beans and peas. *Var Föda*, 31, 405 - 414.

Balzer, I., Bogdanic, C. & Muzic, S. (1977): Natural contamination of corn (zea Mays) with mycotoxins in Yugoslavia. *Ann. Nutr. Aliment.*, 31, 425 - 430.

Bauer, J. & Gedek, B. (1983): Proceedings of the International Symposium on Mycotoxins, march 1983, Cairo, Egypt., cited by Majerus et al (1989).

Bauer, J. & Gareis, H. (1987): Ochratoxin A in der Nahrungsmittelkette. *J. Vet. Med. B.*, 34, 613 - 627.

Bauer, J., Niemiec, J. & Scholtyssek, S. (1988): Ochratoxin A in feed for laying hens. Second communication: residues in serum, liver and eggs. *Arch.Geflgekd.*, 52, 71 - 75.

Belitz, H.D. & Maier, G.D. (1991): Zum Gehalt an Karzinogenen in Bohnenkaffee. *Dtsche. Lebensm. Rundsch.*, 87, 69 -75.

Buchmann, N.B. & Hald, B. (1985): Analysis, occurrence and control of ochratoxin A residues in Danish pig kidneys. *Food Add. Contam.*, 2, 193 -199.

Ceovic, S., Hraber, A. & Saric, M. (1992): Epidemiology of Balkan endemic nephropathy. *Fd. Chem. Toxicol.*, 30, 183 - 188.

Connole, M.D., Blaney, B.J. & Mcewan, T. (1981): Mycotoxins in animal feeds and toxic fungi in Queensland 1971 - 1980. *Aust. Vet. J.*, 57, 314 - 318.

Dalvi, R.R. & Salunkhe, D.K. (1990): Mycotoxins in foods and feeds, their potential health hazards and possible control: An overview. *J. Maharaschtra. Agric. Univ.*, 15, 36 -40.

Frank, H.K. (1991): Food contamination by ochratoxin A in Germany. In *Mycotoxins, Endemic Nephropathy and Urinary Tract tumours*, Ed. M. Castegnaro, R. Plestina, G. Dirheimer, I.N. Chernozenty & H. Bartsch, 77 - 81. Lyon: IARC.

Fritz, W., Buthig, C.L., Nonath, R. & Engst, R. (1979): Studies on the nutritionally significant formation of ochratoxin A in grain and other foodstuffs. *Lebensm. Ernhr.*, 25, 929 -932.

Frohlich, A.A., Marquardt, R.R. & Ominsky, K.H. (1991): Ochratoxin A as a contaminant in the human food chain: A Canadi- an perspective. In *Mycotoxins,Endemic Nephropathy and Urinary Tumours.* Ed. M. Castegnaro, R. Plestina, G. Dirhei mer, I.N. Chernozemsky and H. Bartsch, pp 139 - 143. Lyon:IARC.

Fukal, L. (1990): A survey of cereals, cereal products, feed stuffs and porcine kidneys for ochratoxin A by radio-immunoassay. *Food Additives and Contaminants*, 7, 253 - 258.

Fukal, L. (1991): Spontaneous occurrence of ochratoxin by immunoassay. *Dtsche. Lebensm. Rundsch.*, 87, 316 - 319.

Galtier, P., Jemmali, M. & Larrieu, G. (1977): Enquete sur la présence éventuelle d'aflatoxine et d'ochratoxine A dans des mais recolté en France en 1973 et 1974. *Ann. Nutr. Aliment.*, 31, 381 -389.

Golinski, P., Hult, K., Brabarkiewicz-Szczesna, J., Chelkowski, J., Kneblewski, P & Szebiotko, K. (1984): Mycotoxic porcine nephropathy and spontaneous occurrence of ochratoxin A residues in kidneys and blood of Polish swine. *Appl. Environ. Microbiol.*, 47, 1210 - 1212.

Golinski, P., Grabarkiewicz-Seczesna, J., Chelkowski, J., Hult, K. & Kostecki, M. (1991): Possible source of ochratoxin A in human blood in poland. In *Mycotoxins, Endemic Nephropathy and Urinary Tract Tumours*. Ed. M. Castegnaro, R, Plestina, G. Dirheimer, I.N. Chernozemsky & H. Bartsch. pp 153 - 158, Lyon: IARC.

Hadlok, R.M., Christen, U., Wiedmann, S., Moritz & Wagner, G. (1989): Mykotoxine in vom Tier stammende Nahrungsmitteln: Ochratoxin A in der Nahrungskette. Giessen: Institut für Tierärtzliche Nahrungsmittelkunde der Justus-Liebig Universität.Cited by Majerus et al (1989).

Hald, B. (1991): Ochratoxin A in human blood in European countries. In *Mycotoxins, Endemic Nephropathy and Urinary Tract Tumours*. Ed. M. Castegnaro, R. Plestina, G. Dirheimer, I.N. Chernozemsky & H. Bartsch. pp 159 - 164. Lyon: IARC.

IPCS, International Programme on Chemical Safety (1990): Selected mycotoxins: Ochratoxins, trichotecenes, ergot. *Environmental Health Criteria*, 105, 15 - 70, WHO, Geneva.

Jonsyn, F.E. & Lahai, G.P. (1992). Mycotoxic flora and mycotoxins in smoke-dried fish from Sierra Leone. *Die Nahrung*, 36, 485 - 489.

Josefsson, E. (1979): Study of ochratoxin A in pig kidneys. *Var Föda*, 31, 415 - 420.

Juszkiewicz, T. & Piskorska-Pliszczynska, J. (1976): Occurrence of aflatoxin B_1, B_2, G_1, and G_2, ochratoxin A and B,

sterigmatocystin and zearalenone in cereals. *Med. Weter.*, 32, 617 -619.
Juszkiewicz, T & Piskorska-Pliszczynska, J. (1977): Occurrence of mycotoxins in mixed feeds and concentrates. *Med. Weter.*, 33, 193 - 196.
Juszkiewicz, T., Piskorska-Pliszczynska, J & Wisnieweska, H. (1982): Ochratoxin A in laying hens: tissue deposition and passage into eggs. In *Mycotoxins and Phycotoxins.* pp 122 - 125, Vienna: Technical University Vienna.
Kane, A., Diep, N. & Diack, T.S. (1991). Natural occurrence of ochratoxin A in food and feed in Senegal. In *Mycotoxins, Endemic Nephropathy and Urinary Tract Tumours.* Ed. M. Castegnaro, R. Plestina, G. Dirheimer, I.N. Chernozemsky & H. Bartsch. pp 93 -96. Lyon: IARC.
Krogh, P., Hald, B. & Pedersen, E.J. (1973): Occurrence of ochratoxin A and citrinin in cereals associated with mycotoxic porcine nephropathy. *Acta Pathol. Microbiol. Scand.* B81, 689 - 695.
Krogh, P. (1977): Ochratoxin A residues in tissues of slaughter pigs with nephropathy. *Nord. Vet. Med.*, 29, 402 - 405.
Krogh, P. (1978): Causal associations of mycotoxin nephropathy. *Acta Pathol. Microbiol. Scand.*, A269, 1 - 28.
Krogh, P., Axelsen, N.N., Elling, F., Gyrd-Hansen, N., Hald, B., Hyldgaard-Jensen, J., Larsen, A.E,. Matsen, A., Morten sen, H.P., Möller, T., Petersen, O.K., Rarnskov, U., Rast gaard, M. & Aalund, O. (1974a). Experimental porcine nephropathy: changes of renal function and structure by ochratoxin A contaminated feed. *Acta Pathol. Microbiol. Scand.*, A246, 21 - 35.
Krogh, P., Hald, B., Englund, P., Rutqvist, L. & Swahn, O. (1974b): Contamination of Swedish cereals with ochratoxin A. *Acta. Pathol. Microbiol. Scand.*, B82, 301 - 302.
Krogh, P., Elling, F., Hald, B., Jylling, B., Petersen, V.E., Skadhauge, E., & Svendsen, C.K. (1976). Experimental avian nephropathy : changes of renal function and structure induced by ochratoxin A-contaminated feed. *Acta Pathol. Microbiol. Scand.*, A84, 215-221
Kuiper-Goodman, T. & Grant, D.L. (1990). Ochratoxin A. Toxicological evaluation of certain food additives and contaminants. *WHO Food Additives Series*, 28, 365 - 417, Geneva: IPCS, WHO.
Langseth, W. & Borgsjo, B. (1992): Ochratoksin A i plasma fra Norske slakegriser. *Norsk Veterinaertidsskrift*, 104, 205 - 207.
Leibetseder,J. (1989): Die Bedeutung der Mykotoxine für Mensch und Tier. *Ernährung*, 13, 739 - 745.
Levi, C.P., Trenk, H.L., & Mohr, H.K. (1974): Study of the occurrence of ochratoxin A in green coffee beans. *J. Assoc. Off. Anal. Chem.*, 57, 866 - 870.
Majerus, P. & Woller, R. (1983): Ochratoxin A in Schweinenieren. *5. Mykotoxin-Workshop, Detmold*, 26/27, 5.
Majerus, P., Otteneder & Hower, C. (1989): Beitrag zum Vorkommen von Ochratoxin A in Schweineblutserum. *Dtsch. Lebensm. Rdsch.*, 85, 307 - 313.

Marquardt, R.R., Frohlich, A. & Abramson, D. (1990): *Can. J. Physiol. Pharmacol.*, 68, 991 - 997.
Maryamma, K.I. & Nair, M.K. (1991): Production of ochratoxin A by Aspergillus ochraceus and Aspergillus sulphureus. *Ind. Vet. J.*, 68, 120 - 122.
Micco, C., Grossi, M., Miraglia, M., Brera, C., Libanori, A. & Faraoni, I. (1989a): Criteria evaluation of quality of national corn hybrids; Lipid profile and mycotoxins contamination. *La Rivista della Societa Italiana di Scien- za dell Alimentazione*, 18, 29 - 38.
Micco,C., Grossi, M., Miraglia, M & Brera, C, (1989b): A study of the contamination by ochratoxin A of green and roasted coffee beans. *Food Additives and Contaminants*, 6, 333 - 339.
Micco, C., Ambruzzi, N.A., Miraglia, M., Brera, C., Onori, R. & Benelli, L. (1991): Contamination of human milk with ochratoxin A. In *Mycotoxins, Endemic Nephropathy and Urinary Tract tumours*. Ed. M. Castegnaro, R. Plestina, G. Dirheimer, I.N. Chernozemsky & H. Bartsch. pp 105 - 108. Lyon: IARC.
Milanez,T.V. & Sabino, M. (1989): Ochratoxin A in beans and its stability after cooking. *Revista do Instituto Adolfo Lutz, Brasil*, 49, 131 - 135.
Ministry of Agriculture, Fisheries and Food. (1980): Survey of mycotoxins in the United Kingdom. *Food Surveillance Paper No. 4*, 35 pp. London: Ministry of Agriculture, Fisheries and Food.
Nesheim, S. (1971): Ochratoxins: occurrence, production, analysis and toxicity. Cited by IPCS 1990.
Northolt, M.D., Van Egmond, H.P. & Paulsch, W.E. (1979): Ochratoxin A production by some fungal species in relation to water activity and temperature. *J. Food Prot.*, 42, 485 - 490.
Osborne, B.G. (1980): the occurrence of ochratoxin A in mouldy bread and flour. *Fd. Cosmet. Toxicol.*, 18, 615 -617.
Pande, N., Saxena, J. and Pandey, H. (1990). Natural occurrence of mycotoxins in some cereals. *Mycoses*, 33, 126 - 128.
Pavlovic, M., Plestina, R. & Krogh, P. (1979): Ochratoxin A contamination of foodstuffs in an area with Balkan (endemic) nephropathy. *Acta Pathol. Microbiol. Scand.* , B87, 243 - 246.
Pedersen, E. & Hansen, H.N. (1981): Ochratoxin A in grain and grain products. *Stat. Levnedsm. Institute Report F81002.*. Copenhagen.
Petkova-Bocharova, T. & Castegnaro, M. (1985): Ochratoxin A contamination of cereals in an area of high incidence of Balkan endemic nephropathy in Bulgaria. *Food Addit. Contam.*, 2, 267 - 270.
Petkova-Bocharova, T., Castegnaro, M., Michelon, J. & Maru, V. (1991): Ochratoxin A and other mycotoxins in cereals from an area of Balkan endemic nephropathy and urinary tract tumours in Bulgaria. In *Mycotoxins, Endemic Nephropathy and Urinary Tract Tumours*. Ed. M. Castegnaro, R. Plestina, G. Dirheimer, I.N. Chernozemsky & H. Bartsch. pp 83 - 87. Lyon: IARC.

Petterson, H. & Kiessling, K.H. (1992): Mycotoxins in Swedish grains and mixed feeds. *J. Environm. Pathol. Toxicol. and Oncol.*, 11, 105 - 107.
Piskorska-Pliszczynska, J. & Juszkiewicz, T. (1990): Tissue deposition and passage into eggs of ochratoxin A in Japanese Quail. *J. Environm. Pathol. Toxicol. and Oncol.*, 10, 8 - 10.
Plestina, R. (1992): Some features of Balkan endemic nephropathy. *Fd. Chem. Toxicol.*, 30, 177 - 181.
Pohland, A.E., Nesheim, S. & Friedman, L (1992): Ochratoxin A: A Review. *Pure & Appl. Chem.*, 64, 1029 - 1046.
Prior, M.G. (1976): Mycotoxin in determinations on animal feedstuffs and tissues in Western Canada. *Can.J. Comp. Med.* , 40, 75 - 79.
Prior, M.G. (1981): Mycotoxins in animal feedstuffs and tissues in Western Canada 1975 to 1976. *Can .J. Comp. Med.*, 45, 116 - 119.
Prior, M.G. and Sisodia, C.S. (1978): Ochratoxicosis in white leghorns hens. *Poult. Sci.*, 57, 619 - 623.
Radonic, M. & Radosevic, Z. (1992): Clinical features of Balkan endemic nephropathy. *Fd. Chem. Toxicol.*, 30, 189 - 192.
Reichmann, K.G., Blaney, B.J., Connor, J.K. & Runge, B.M. (1982). The significance of aflatoxin and ochratoxin in the diet of Australian chickens. *Aust. Vet. J.*, 58, 211 - 212.
Rousseau, D.M. & Van Peteghem, C.H. (1989): Spontaneous occurrence of ochratoxin A residues in porcine kidneys in Belgium. *Environ. Contam. Toxicol.*, 42, 181 - 186.
Ruprich, J. Kosutsky, J., Ostry, V. & Kosutzka, E. (1991): A long-term intake of mycotoxin ochratoxin A by chicks with respect to its residues in foods and feeds. *Veter. Med. (Praha).*, 36, 685 - 693.
Rutqvist, L., Bjorklund, N.E., Hult, K. & Gatenbeck, S. (1977): Spontaneous occurrence of ochratoxin residues in kidneys of fattening pigs. *Zbl. Veterinaermed.*, A24, 402 - 408.
Sandor, G., Glavits, R., Vajda, L. Vanyi, A. & Krogh, P. (1982): Epidemiological study of ochratoxin A-associated porcine nephropathy in hungary. In *Mycotoxins and Phycotoxins, Proceedings of the 5th International IUPAC Symposium.* pp 349 - 352. Vienna: Technical University Vienna Publ.
Saxena, J. & Mehrotra, B.S. (1989): Screening of spices commonly marketed in India for natural occurrence of mycotox ins. *J. of Food Composition and Analysis*, 2, 286 - 296.
Saxena, J. & Mehrotra, B.S. (1990): The occurrence of mycotox ins in some dry fruits retail marketed in Nainital district of India. *Acta Alimentaria*, 19, 221 - 224.
Scheuer, R., Bernard, K. & Leistner, L. (1984): Rückstände von Ochratoxin A in Schweinenieren. *Mittl. Bundesanstalt für Fleischforschung, Kulmbach.*,84, 5781 - 5784.
Scheuer, R. (1989); Untersuchungen zum Vorkommen von Ochratoxin A. *Fleischwirtsch.*, 69, 1400 - 1404.
Scott, P.M., Walbeek, W., Van Kennedy, B & Anyeti, D. (1972): Mycotoxins (ochratoxin A, citrinin and sterigmatocystin) and

toxicogenic fungi in grains and other agricultural products. *J.Agric. Food Chem.*, 20, 1103 - 1109.
Shotwell, O.L., Hesseltine, C.W., Van de Graft, E.E. & Goulden, M.L. (1971): Survey of corn from different regions for aflatoxin, ochratoxin, and zearalenone. *Cereal SCi. Today*, 16, 266 - 273.
Shotwell, O.L., Goulden, M.L. & Hesseltine, C.W. (1976): Survey of US wheat for ochratoxin and aflatoxin. *J. Assoc. Off. Anal.Chem.*, 59, 122 -124.
Sinha, A.K. & Ranjan, K.S. (1990): A report of mycotoxin contamination in Bhutanese cheese. *J. Fd. Sci. Technol.*, 28, 398 - 399.
Szebiotko, K., Chelkowski, J., Dopierala, G., Godlewska, B. & Radomyska, W. (1981): Mycotoxins in cereal grain. Part I. ochratoxin, citrinin, sterigmatocystin, penicillic acid and toxicogenic fungi in cereal grain. *Nahrung*, 25, 415 - 421.
Tholstrup, B. & Rasmussen, G. (1990): Ochratoksin A i Korn 1986 -1989. *Publication Levnedsmiddelstyrelsen,* Kopenhagen.
Tressl, R., Hommel, E. & Helak, B. (1989): Bestimmung von ochratoxin A in Gerste, Malz und Bier durch Hochdruck-flüssigkeits-Chromatographie/Fluoreszenzdetektion. *Monatsschrift für Brauwissenschaft*, 8, 331 - 335.
Van Egmond, H.P. (1991a): Methods for determining ochratoxin A and other nephrotoxic mycotoxins. In *Mycotoxins, Endemic Nephropathy and Urinary Tract Tumours.* Ed. M. Castegnaro, R. Plestina, G. Dirheimer, I.N. Chernozemsky & Bartsch, H. pp 57 - 70. Lyon: IARC.
Van Egmond, H.P. (1991b): Worldwide regulations for ochratoxin A. In *Mycotoxins, Endemic Nephropathy and Urinary Tract Tumours,* Ed. M. Castegnaro, R. Plestina, G. Dirheimer, I.N. Chernozemsky & H. Bartsch. pp 331 - 336.Lyon: IARC.
Van Egmond, H.P. & Sizoo, E,A. (1984): Het voorkomen van ochratoxin A in granen en graanprodukten. *Report No. 647705001.* Bilthoven: National Institute of Public Health and Environmental Protection, The Netherlands.
Van Egmond, H.P., Van der Molen, E.J. & Paulsch, W.E. (1984): Het voorkomen van ochratoxin A in nieren van varkens met en zonder chronische nefropathie. *Report No. 647704001.* Bilthoven: National Institute of Public Health and EnvironmentalProtection, The Netherlands.
Vesela, D., Vesely, D., Jelinek, R. & Kusak, V. (1978): Detection of ochratoxin A in feed barley. *J.Vet.Med.*, 23, 431 - 436.

Résumé

Les taux de contamination des aliments de l' Homme et des animaux par l' ochratoxine A à travers le monde ont été répertoriés dans ce travail par continent: Afrique, Amérique, Asie, Australie et Europe.

Les réglementations à l'échelle mondiale publiées récemment par Van Egmond (1991) y sont discutées brièvement. Enfin une comparaison des taux de contamination par l'ochratoxine A est effectuée avec les seuils de tolérance fixés par les réglementations des pays respectifs.

Human ochratoxicosis and its pathologies. Eds E.E. Creppy, M. Castegnaro, G. Dirheimer. Colloque INSERM/ John Libbey Eurotext Ltd. © 1993, Vol. 231, pp. 101-110.

Mise en évidence de la contamination des céréales par les aflatoxines et l'ochratoxine A au Bénin

Y. Bouraïma[1], L. Ayi-Fanou[2], I. Kora[2], J. Setondji[2], A. Sanni[2] et E.E. Creppy[3]

[1]Direction de l'Agriculture, Service de Protection des Végétaux, Porto-Novo, République du Bénin; [2]Département de Biochimie et Biologie Cellulaire, Faculté des Sciences et Techniques, Université Nationale du Bénin, République du Bénin; [3]Laboratoire de Toxicologie et Hygiène Appliquée, Université de Bordeaux II, 3 ter, place de la Victoire, 33076 Bordeaux Cedex, France

Depuis un certain temps, le monde scientifique a reconnu l'existence de mycotoxines, (i.e.) métabolites toxiques produits par des moisissures lorsque les conditions de température et d'humidité sont réunies Il a admis la possibilité que ces mycotoxines contaminent les aliments de l'homme et des animaux .
L'ingestion répétée de tels aliments a été reconnue responsable d'un certain nombre de maladies humaines et animales . Ainsi vers 1930 , l'aleucémie toxique alimentaire qui ravagea les campagnes en Russie fut étroitement associée à l'infestation des céréales par des moisissures du genre Fusarium (Sarkisov 1954) dont les métabolites toxiques ont été plus tard identifiés comme étant des trichothécènes,(Mirocha et Pathre,1973)

En 1960 en Grande-Bretagne, des élevages de dindes furent décimés par des tourteaux d'arachides importés du Brésil qui contenaient du mycelium d' Aspergillus flavus , qui produit les aflatoxines B_1,B_2, G_1,G_2, métabolites toxiques et cancérogènes puissants dont la cible est surtout le foie.(Goldblatt 1969).

A la même période une mycotoxine découverte dans des moisissures des genres Aspergillus et Penicillium, en Afrique du Sud a été retrouvée dans l'alimentation des animaux et de l'homme dans les Balkans (Bulgarie, Roumanie et Yougoslavie) . Il s'agit de l'ochratoxine A (OTA). Après de longues études épidémiologiques et expérimentales , elle a été impliquée dans la néphropathie endémique des Balkans . Cette maladie est caractérisée par une tubulonéphrite interstitielle évoluant vers une insuffisance rénale chronique en quelques années chez l'homme et en quelques semaines chez l'animal
L'OTA est maintenant soupçonnée d'être aussi responsable des tumeurs du tractus urinaire et/ou de la vessie qui sont associées à la néphropathie , avec une fréquence très élevée.
Ces dernières années l'OTA est régulièrement retrouvée dans le sang humain et les carcasses d'animaux en Europe de l'Ouest et en Scandinavie , en Amérique et notamment au Canada, en Afrique par exemple en Algérie et en Tunisie . Il est vraisemblable que si elle était systématiquement recherchée, elle serait retrouvée partout.
A l'heure actuelle, tous les pays du monde se sentent concernés, et mettent au point des méthodes de détection rapides et fiables . Les chercheurs étudient tous les aspects des mycotoxicoses connues afin d'estimer les risques , et de déterminer les niveaux de contamination acceptables si le niveau zéro ne peut pas être garanti , en vue de la législation.

Dans le souci de déterminer la nature et le taux de contamination des aliments dans leur pays , des chercheurs du Bénin (Golfe du Bénin , Afrique de l'Ouest) ont recherché systématiquement sur les récoltes de 1990 et 1991. les Aflatoxines , les Ochratoxines , les Trichothécènes , la Citrinine, la Zéaralénone qui sont à l'heure actuelle les toxines les plus étudiées et aussi les plus fréquemment détectées.

MATERIEL ET METHODE

Pays et lieux de prélèvements

Le Bénin est une république de l'Afrique occidentale appartenant au Golfe de Guinée Il couvre une superficie de 15 000 km^2 · Situé entre le Nigéria et le Togo, il s'étend du 6 ° au 12 ° de latitude Nord . Il bénéficie d'un climat subtropical avec des températures oscillant entre 22°C et 34°C toute l'année avec deux saisons des pluies au sud permettant au moins une récolte de céréales par an.

Les lieux de prélèvement ont été sélectionnés selon les critères suivants:

- importance de la production et du stockage.

- proximité des marchés céréaliers nationaux et internationaux
- nécessité que l'étude couvre toute l'étendue du territoire.

Les structures de stockage incluses dans l'étude sont: des silos , des greniers traditionnels en argile, des magasins où les sacs en toile de jute contenant des grains sont empilés, et enfin, des magasins où le grain est stocké en vrac.

Au niveau des silos, les prélèvements ont été effectués suivant la méthode de l'échantillonage vertical.

Pour les stocks en sacs, les prélèvements ont été effectués au niveau des sacs situés en haut, au milieu et en bas .

Dans les magasins où le maïs est stocké en vrac, les prélèvements ont été effectués à l'aide d'une sonde à rallonge à différents niveaux des tas (en haut, au centre et à la base).

Tous les prélèvements du même silo, du même grenier ou du même magasin sont rassemblés en un échantillon unique représentatif, d'une masse de plusieurs kilogrammes.

Tous les échantillons sont rapidement transportés au laboratoire ,transvasés dans des bocaux propres et stériles et conservés dans une chambre froide et sèche à +4°C jusqu'à l'analyse.

La liste des localités prospectées avec leurs différentes caractéristiques est présentée dans le tableau 1.

Extraction - Purification - Séparation et Identification des mycotoxines

1. Pulvérisation

Une quantité de 300 grammes de chaque échantillon est pulvérisée dans un petit moulin à grain au laboratoire de la direction du Conditionnement et du Contrôle des Produits (Ministère du Développement Rural) à Cotonou. Le moulin à grain est soigneusement lavé et nettoyé après chaque échantillon.

. Les farines obtenues ont été conservées à une température d'environ 2-4°C dans une chambre froide puis transportées au laboratoire de Biochimie de la Faculté des Sciences et Techniques de l' Université Nationale du Bénin à Calavi où ont été effectués l'extraction, la purification et l'analyse qualitative.

2. Extraction

100 g de farine sont mis en suspension dans un bécher en verre contenant 100 ml d'eau physiologique (NaCl à O,9 %). A ce mélange sont ajoutés 250 ml de chloroforme et 25 ml d'acide orthophosphorique 0,1 M . Après 15 à 30 min. d'agitation mécanique, une centrifugation de 20 min. à 3000g permet d'obtenir un surnageant qui après filtration sur verre fritté sert à l'extraction des toxines éventuelles.

3. Purification

Le filtrat obtenu est débarrassé du chloroforme par évaporation dans un évaporateur rotatif à 50°C. Le résidu est lavé avec du n-Hexane dans une ampoule à décanter . La phase aqueuse est recueillie et alcalinisée avec du carbonate de sodium 0,1 N jusqu'à pH 9
Cette phase aqueuse est de nouveau extraite deux fois par le chloroforme comme l'indique le schéma ci-contre. Deux phases sont obtenues : une phase chloroformique II et une phase aqueuse II Cette dernière est acidifiée avec de l'acide chlorhydrique 6 N et est extraite deux fois au chloroforme. La phase chloroformique III acide est évaporée à sec et le résidu est repris dans 0,1 ml de chloroforme.. Cet extrait final 1 sera soumis à la recherche des ochratoxines et de la citrinine
Quant à la phase chloroformique II alcaline , elle est évaporée à sec . Le résidu est repris par l'acétonitrile et dialysé contre une solution acétone-eau (3:7,v/v).. La phase acétone-eau recueillie est extraite à nouveau avec le chloroforme et réduite à un volume de 0,1ml . C'est l'extrait final 2 qui est destiné à la recherche des aflatoxines essentiellement .

3. Séparation

Elle a été effectuée par chromatographie sur couche mince (CCM) utilisant des plaques de silice (Silicagel Merck)

Une CCM de l'extrait final I et des témoins (ochratoxine A et citrinine) est réalisée sur une première plaque avec le système solvants A : toluène : acétate d'éthyle : acide formique (30:15:5,v/v/v)

Sur une deuxième plaque de CCM , l'analyse de l'extrait final 2, avec les témoins d'aflatoxine B_1 B_2 G_1 G_2 , s'effectue dans une première dimension avec le système solvant A et dans une deuxième dimension dans le système solvant B : chloroforme acétone (9:1,v/v)

4. Révélation et Identification

L'identification des mycotoxines s'effectue sous la lumière ultra-violette à 254 et 366 nm par comparaison avec des mycotoxines de référence Sigma (fournies par le laboratoire de Toxicologie de Bordeaux)
Le schéma ci-après résume la méthodologie d'analyse des mycotoxines utilisée au cours des manipulations.
Dans une deuxième série d'étude, des essais de quantification ont été réalisés par ajout de toxines à des céréales non contaminées (ochratoxine A, Aflatoxine B_1 B_2 G_1 G_2) séparément, citrinine et zéaralénone en quantités connues .
Après extraction dans les mêmes conditions que les échantillons à doser la quantification a été réalisée par comparaison de la surface des tâches et de leur fluorescence sous des rayonnements UV ..
Une confirmation a été réalisée au laboratoire de Toxicologie et Hygiène Appliquée de Bordeaux par CLHP dans les conditions décrites pour les ochratoxines A et alpha pour les aflatoxines B_1 B_2 d'une part ,G_1 G_2 d'autre part. Elle a porté sur tous les extraits analysés par CCM pour la quantification.
La méthode utilisée dérive de celle décrite par Hietanen et coll (1986) pour les ochratoxines A et alpha .
Les conditions de CLHP ont été les suivantes:

- colonne: 20x4,2 cm, C18 .ODS 10 µm
- l'éluant , méthanol:acétonitrile:acétate de sodium 0,05mM:acide acétique (300:300:400:14 (v/v/v/v); débit 1ml/min; durée de l'analyse 10 min.
- les paramètres fluorimétriques:

pour les ochratoxines Excitation à 340 nm
Emission à 465

Les échantillons positifs sont reconfirmés par estérification dans les conditions décrites par l' AOAC(1980) (Association of Official Analytical Chemists)
: Pour les aflatoxines , les mêmes conditions de colonne, solvants, débit et durée ont été utilisées , avec les mêmes conditions de détection fluorimétrique suivantes

Excitation: 365
Emission 425
pour aflatoxines B_1B_2
Excitation : 365
Emission: 450 pour aflatoxines G_1G_2

Confirmation et quantification des Aflatoxines par CLHP:
Les extraits sont chromatographiés sur une colonne C 18 Spherisorb ODS 5 µm (25x0,40cm) . La phase mobile est constituée de méthanol et eau 50:50, v/v, avec un débit de 2 ml/min et une durée d'analyse de 10 min.. La détection est effectuée par microfluorimétrie, (excitation 365 nm, émission 418 nm, gain 10).

RESULTATS ET DISCUSSION

Une étude préalable nous a permis de déterminer la nature des moisissures présentes dans les échantillons de céréales prélevés. En effet des grains non moisis en apparence ont été prélevés au hasard dans chaque échantillon et mis à tremper quelques minutes dans du liquide physiologique.
Ce liquide a été mis en culture pendant une semaine sur de la gélose, conformément à la méthode développée par le "Commonwealth Mycological Institute " .
Après analyse au microscope des lames montées à partir des géloses, la présence de spores et de mycelium de plusieurs moisissures des genres Aspergillus, Penicillium et Fusarium a été détectée (résultats non présentés).
Dans ces conditions , l'étude a été limitée aux mycotoxines essentiellement produites par ces moisissures , à savoir aflatoxine B_1 B_2 G_1 G_2, ochratoxine A et B, citrinine et zéaralénone. Il faut rappeler en plus, que les aflatoxines, la citrinine et la zéaralénone ont déja été détectées en Afrique, au nord comme au sud du Sahara,) et que l'ochratoxine A a été découverte en Afrique du Sud Il n'est donc pas surprenant de détecter ces toxines à cet endroit du continent africain.

Trente deux échantillons ont été analysés, (3x100g de farine), en première analyse par CCM (soit 192 extraits:3x32x2=64x3). Cette première analyse au Bénin a mis en évidence des ochratoxines et des aflatoxines dans 23 échantillons /32. (soit 72 %) Ni citrinine, ni zéaralénone n'ont été détectées.
L'analyse des tâches détectées sous les rayons UV a été effectuée à Bordeaux par CLHP et détection fluorimétrique. Une première analyse a permis de détecter la présence des aflatoxines B_1 et B_2 d'une part et G_1 et G_2 d'autre part en même temps que celle des ochratoxines alpha, A, B, C, D, car ces dernières sont éluées nettement après les aflatoxines. Mais la confirmation et la quantification ont été nécessaires à cause de la découverte récente de l'existence d'analogues naturels de l'ochratoxine A dont certains pourraient avoir des temps de rétention comparables à ceux des aflatoxines , et aussi parce que dans la première analyse , il est quelquefois difficile de quantifier les aflatoxines B_1 et B_2 ou G_1et G_2 lorsqu'elles sont présentes simultanément.

Quant à la deuxième analyse CLHP, une très bonne séparation de toutes les aflatoxines avec les temps de rétention respectifs de 4,20 min - 5,06 - 6,26 et 7,66 min pour les aflatoxines G_2 G_1 , B_2 et B_1. La présence d'ochratoxine A ne constitue pas un obstacle, car elle est éluée plus rapidement.. Le logiciel PIC 3 (ICS France) (24) a permis de quantifier toutes les aflatoxines par rapport à des standards de concentrations connues chromatographiés séquenciellement au cours de l'analyse.

Après reconfirmation également par CLHP , l'essentiel des résultats de la CCM a été validé. Ainsi 21 échantillons/32 (soit 66 %) sont contaminés par des aflatoxines. (Tableau 2) à des taux variant de 0,54 ppb à 15,0 ppb d'aflatoxine B_1, de 0,50 ppb à 6 ppb d'aflatoxine B_2,de 0,09 ppb à 58 ppb d'aflatoxine G_1 et 0,05 ppb à 10 ppb d'aflatoxine G_2. Les aflatoxines G (G_1 ou G_2) sont donc les plus couramment rencontrées (15 échantillons positifs sur 21) dont 6 fois en association ($G_1 + G_2$) . L'aflatoxine B_2 est retrouvée seule 4 fois Quant à l'aflatoxine B_1, elle est retrouvée dans 8 échantillons, 2 fois seule et les autres fois en association avec soit l'aflatoxine B_2 soit l'aflatoxine G_2. Les concentrations les plus élevées sont 58 ppb d'aflatoxine G_1 et 15 ppb d'aflatoxine B_1.

Une comparaison des taux d'aflatoxines déterminés au Bénin avec les concentrations maximales tolérées dans certains pays du monde où une législation est en place permet de tirer quelques conclusions .

31% des échantillons sont au-dessus du seuil de 5 ng/g soit 5 ppb Ils seraient impropres à l'alimentation des bébés en France et rejetés de toute alimentation humaine en Russie selon la règlementation en vigueur.

8 échantillons/32 (supérieurs à 10 ppb) seraient exclus de l'alimentation humaine dans la CEE. Cela représente des tonnages importants, car il s'agit des stocks qui approvisionnent les plus gros centres urbains du pays: Cotonou, Parakou etc ...

Les Etats-Unis admettent entre 15 et 20 ppb dans l'alimentation humaine selon les cas. Dans ces conditions seulement 2 échantillons sur 32 auraient été rejetés. Il pourrait sembler à première vue que la situation de la contamination des céréales par les aflatoxines ne soit pas très alarmante puisqu'un seul échantillon dépasse les 50 ppb qui constituent la concentration maximale tolérée dans l'alimentation du bétail en France. Mais il est important de se rappeler d'une part que ces céréales sont destinées essentiellement à l'alimentation humaine et d'autre part que la fréquence des contaminations par l'aflatoxine est supérieure à 50 % dans le contexte du Bénin où les céréales constituent la base de l'alimentation journalière d'une très grande partie de la population. Le risque n'est donc pas négligeable puisqu'une fois sur deux le consommateur risque d'être exposé aux aflatoxines.

En ce qui concerne les ochratoxines, seuls cinq échantillons (soit 15%) se sont révélés contaminés par l'ochratoxine A entre 15 et 45 ppb (15 - 45 µg/kg) . Deux autres échantillons qui avaient été considérés positifs par CCM se sont avérés contenir moins de 0,2 ppb en première analyse CLHP et négatifs à la confirmation après estérification et CLHP.

Ces 5 échantillons contaminés par l'ochratoxine A auraient été rejetés de l'alimentation humaine dans de nombreux pays d' Europe et d' Amérique Même si des taux de contamination plus élevés ont déjà été trouvés dans d'autres pays du monde , ceux trouvés au Bénin constituent un risque pour la santé, surtout en association avec des aflatoxines comme c'est quelquefois le cas.

Il apparaît que le nombre d'échantillons positifs pour les aflatoxines est largement supérieur à celui des échantillons positifs pour l'OTA, malgré la présence de moisissures des genres Aspergillus et Penicillium dans près de 60 % des échantillons. Une telle constatation avait été faite par Kane et coll. (1991) qui avaient trouvé après une étude de ce type au Sénégal, que la présence d'Aspergillus flavus parmi d'autres espèces d'Aspergillus conduit vers la production d'aflatoxines essentiellement , et que les ochratoxines dans ces conditions sont très peu produites ou totalement absentes.. Ce phénomène peut s'expliquer par le fait que dans la biogénèse des aflatoxines, l'Aspergillus flavus utilise la phénylalanine du support sur lequel il se développe . Or l'ochratoxine A est un analogue de la phénylalanine, qui ne peut pas être produite sans phénylalanine.

Il y aurait donc une compétition entre Aspergillus flavus et Aspergillus ochraceus et autres. A. flavus plus abondant, détournerait toute la phénylalanine à son seul profit. Il serait donc souhaitable de rechercher d'autres ochratoxines ayant des acides aminés différents à la place de la phénylalanine. Trois de ces ochratoxines ont été récemment isolées et identifiées . Certaines de ces ochratoxines , homologues de l' OTA sont aussi toxiques que cette dernière , comme cela a été montré avec des analogues synthétiques .

Il est vraisemblable qu'une association aflatoxines-ochratoxines (avec un acide aminé autre que la phénylalanine) existe au Bénin. En effet dans cette étude une partie des produits fluorescents détectés en première analyse n'a pu être identifiée faute de substances de références, et surtout parce que le comportement en CLHP des analogues de l'ochratoxine A est mal connu. Cela peut être préoccupant, car il y a une synergie entre OTA et aflatoxine pour de nombreux effets toxiques (tératogénèse, génotoxicité et cancérogénèse).

Cette étude devrait se répéter annuellement , et se prolonger par la détermination des mycotoxicoses humaines par la recherche d' OTA et d' aflatoxine dans le sang humain pour évaluer le risque réel encouru. Car des concentrations de 15 ppb et plus (6 % des échantillons au Bénin entre 1991-1992) peuvent entraîner un fort taux de carcinome hépatique chez l'animal d'expérience après 50 semaines .

Cette étude a toutefois prouvé que lorsque les céréales produites localement sont si conservées dans de bonnes conditions de température et d'humidité, elles sont très peu contaminées par les mycotoxines. A l'inverse les céréales d'importation achetées ou données , transportées pendant de nombreuses semaines par bateaux doivent faire l'objet de soins particuliers en cas de distribution non immédiate. Faute de quoi la contamination est fréquente.

Il est à remarquer que les échantillons du Nord du pays résultent de vieux stocks constitués à partir des récoltes de la dernière campagne; tandis que les échantillons du Sud proviennent de la grande saison pluvieuse , de la campagne en cours au moment de l'étude.Or des mycotoxines sont détectées dans tous les échantillons prélevés au Nord,cela confirme que la durée du stockage joue un rôle important dans la prolifération des moisissures et dans la production des mycotoxines.

Un seul échantillon prélevé directement au champ après la maturation physiologique du maïs (quelques jours avant la récolte) a montré la présence d'aflatoxines et d'ochratoxine.. Ceci vient confirmer l'idée que la production de mycotoxine peut démarrer depuis le champ.

L'étude a permis par ailleurs de constater que la plupart des échantillons prélevés dans les magasins du Programme Alimentaire Mondial (P.A.M.) ont révélé la présence de mycotoxines:ochratoxine A (Bohicon:maïs et sorgho) et les aflatoxines B_1 et B_2 (INSAE Cotonou:maïs et sorgho).

Par contre dans les structures de stockage où les conditions de conservation et les règles d'hygiène (propreté du grenier, débarrassé des résidus de la dernière campagne et nettoyé avant remplissage) sont bien respectées, les échantillons n'ont pas révélé la présence de mycotoxines. C'est le cas de la chambre froide de la Ferme Semencière du Carder Ouémé et les silos métalliques de l'Office National des Céréales.

REMERCIEMENTS

Ce travail a été rendu possible grâce à l'appui logistique de la Deutsche Gesellschaft für Technische Zummenarbeit (GTZ) GmbH et du laboratoire de Toxicologie et Hygiène Appliquée de l'Université de Bordeaux II .

REFERENCES

ARORA R.G. (1985): Experimental mycotoxicosis : Some observations on the teratopathological effects of a single dose of aflatoxin B_1 and ochratoxin in CBA /Ca mice. *In* Trichothecenes and Other Mycotoxins (J.Lacey, Ed.)537-544. Wiley, Chichester

BAUER J. & GAREIS M. (1987): Ochratoxin A in der Nahrungsmittelkette *Z.Veterinärmed.B.* ,*34*, 613-627.

BENDELE A.M. , CARLTON W.W., KROGH P. & LILLEHOJ E.B. (1985) : Ochratoxin A carcinogenesis in the (C57BL/6JxC3H) F1 mouse.*J. Natl. Cancer Inst.*,*75*, 733-742

BREITHOLTZ A., OLSEN M., DAHLBACK A. & HULT K. -(1991) Plasma ochratoxin A levels in three Swedish populations surveyed using an ion-pair HPLC technique. *Fd Addit and Cont* .(.Accepted for publication.)

BUCHI G. & WEINREB S.M. (1969): *J.Am.Chem.Soc*,*91*,5408-5409.

CASTEGNARO M., CHERNOZEMSKY I.N., HIETANEN E. & BARTSCH H (1990): Are mycotoxins risk factors for endemic nephropathy and associated urothelial cancers? *Arch. Geschwulstforsch.60*, 4,295-303.

CHERNOZEMSKY I.N., STOYANOV I.S., PETKOVA-BOCHAROVA T. ,NOCOLOV I.G., DRAGANOV I., STOICHEV I.N., TANCHEV Y., NAIDENOV D. & KALCHEVA N.D. (1977): Geographic correlation between the occurence of endemic nephropathy and urinary tract tumours in Vratza district , Bulgaria;. *Int..J . Cancer* ,*19*,1-11.

CREPPY E.E., KERN D., STEYN P.S., VLEGGAAR R., RöSCHENTHALER R. & DIRHEIMER G.(1983): Compartive study of the effect of ochratoxin A analogues on yeast aminoacyl-tRNA synthetases and on growth and protein synthesis in hepatoma cells.*Toxicol Letters 19*,217-224.

CREPPY E.E., BETBEDER A.M., GHARBI A., COUNORD J.,CASTEGNARO M., BARTSCH H., MONCHARMONT P., FOUILLET B., CHAMBON P.& DIRHEIMER G.(1991):Human ochratoxicosis in France. *in* Mycotoxins, Endemic Nephropathy and Urinary Tract Tumours. Lyon IARC Castegnaro M., Plestina R., Dirheimer G., Chernozemsky I.N. & Bartsch H. eds.,pp145-151.

FROHLICH A.A., MARQUARDT R.R. & OMINSKI K.H.(1991): Ochratoxin A as a contaminant in the human food chain : a Canadian perspective. *in* Mycotoxins, Endemic Nephropathy and Urinary Tract Tumours Castegnaro M., Plestina R., Dirheimer G., Chernozemsky I.N. & Bartsch H. eds Lyon IARCs.pp 139-143.

GOLDBLATT L.A. (1969) : *in* Aflatoxin, Academic Press. New-York,13-46,223-224.

GOLUMBIC C. & KULIK M.M.(1969):.Fungal spoilage in stored crops and its control. *In* Aflatoxin. Academic Press New York Ed. L A.Goldblatt, .307-332.

HADIDANE R., ROGER-REGNAULT C., ELLOUZE F., BACHA H., CREPPY E.E. & DIRHEIMER G (1988):. - Monitoring and identification of fungal toxins in food products and cereals in Tunisia . *J.Sored.Prod. Res.*,*24*,199-206.

HADIDANE R., CREPPY E.E., HAMMAMI MI, ELLOUZE F., BACHA H. & DIRHEIMER G.- (1992) :Isolation and structure determination of natural analogues of the mycotoxin ochratoxin A produced by Aspergillus ochraceus. *Toxicology* ,(sous presse).
HENNIG A., FINK-GREMMELS J. & LEISTNER L. (1991) : Mutagenicity and effects of ochratoxin A on the frequency of sister chromatid exchange after metabolic activation . in : Mycotoxines, Endemic Nephropathy and Urinary Tract Tumours Castegnaro M., Plestina R., Dirheimer G., Chernozemsky I.N. & Bartsch H. eds Lyon IARC pp255-260.
HULT K., PLESTINA R., HABAZIN-NOVAC V., RADIC B. & CEOVIC S. (1982): Ochratoxin A in human blood and Balkan endemic nephropathy. *Arch.Toxicol.,51*,313-321.
HULT K., PLESTINA R., CEIVIC S., HABAZIN-NOVAK V. & RADIC B. (1982) : Ochratoxin A in human blood: Analytical results and confirmational tests from a study in connection with Balkan endemic nephropathy. In Proceedings V International IUPAC Symposium Mycotoxins and Phycotoxins, september 1-3 Vienna, Austria ,338-341 *Austrian Chem. Soc. Vienna.*
KANE A., CREPPY E.E., RöSCHENTHALER R. & DIRHEIMER G. (1986) Changes in urinary and renal tubular enzymes caused by subchronic administration of ochratoxin A in rats. *Toxicology,42*,233-243.
KANE A., DIOP N. & DIACK T.S. (1991): Natural occurrence of ochratoxin A in food and feed in Senegal.. *in* Mycotoxins, Endemic Nephropathy and Urinary Tract Tumours.Castegnaro M., Plestina R., Dirheimer G., Chernozemsky I.N. & Bartsch H. eds Lyon IARC.,pp 93-96.
KANISAWA M. and SUZUKI S. (1978) : Induction of renal and hepatic tumors in mice by ochratoxin A, a mycotoxin *Gann ,69*,599-600.
KROGH P., HALD B., PLESTINA R. & CEOVIC S. (1977): Balkan (endemic) nephropathy and foodborn ochratoxin A : Preliminary results of foodstuffs. *Acta Pathol. Microbiol. Scandj. section B.*1977 *85*,238-240.
KUIPER-GOODMAN T. & SCOTT P.M. (1989): Risk assessment of the mycotoxin ochratoxin A. *Biomed.Environ. Sci.2,*179-248.
LE BARS J.(1976)- Mycotoxines:Ecologie des Moisissures Toxinogènes. *Cah.Nutr.Diét.,2,* 23-27.
MIROCHA C.J. , PATHRE S.(1973): *Appl. Microbiol.26.*719-724.,
NEWBERNE P.M. (1974): The new world of mycotoxins, animal and human health *Clin. Toxicol. .7.* 161-177
NTP(1989) - NTP Technical Report on the Toxicology and Carcinogenesis Studies of Ochratoxin A (CAS N° 303-47-9) jn/N Rats (Gavage Studies) (G. Boorman, Ed.), NIH Publication N° 89-2813, U.S. Department of Health and Human Services , National Institute of Health, Research Triangle Park, NC.
PETKOVA-BOCHAROVA T., CHERNOZEMSKY I.N. & CASTEGNARO M.(1988): Ochratoxin A in human blood in relation to Balkan endemic nephropathy and urinary system tumours in Bulgaria. *Food Addit. Contam ., 5*,299-301.
PFOHL-LESZKOWICZ A., CHAKOR K., CREPPY E.E. & DIRHEIMER G (1991): DNA-Adduct(s) formation after treatment of mice with ochratoxin A.*in* Mycotoxins, Endemic Nephropathy and Urinary Tract Tumours.Castegnaro M., Plestina R., Dirheimer G., Chernozemsky I.N. & Bartsch H. eds Lyon IARC.pp .245-253
PFOHL-LESZKOWICZ A., GROSSE Y., CASTEGNARO M., PETKOVA-BOCHAROVA T., NICOLOV I.G., CHERNOZEMSKY I.N., BARTSCH H., BETBEDER A.M., CREPPY E.E. & DIRHEIMER G.(1992) Ochratoxin A related DNA adducts in urinary tract tumours of bulgarian subjects.,Castegnaro M., Plestina R., Dirheimer G., Chernozemsky I.N. & Bartsch H. eds Lyon IARC (sous presse.)
SARKISOV, A.C. (1954) : Mycotoxicoses (Fungal Infections) Moscow State Publishing House for Agricultural Literature .
VAN DER MERWE K.J., STEYN P.S. , FOURRIE L., SCOTT D.B. & THERON J.J. (1965): Ochratoxin A, a toxic metabolite produced by Aspergillus ochraceus Wilh. *Nature205,*1112-1113.
VAN EGMOND H.P.(1991): Worldwide regulations for ochratoxin A. in Mycotoxins, Endemic Nephropathy and Urinary Tract Tumours Castegnaro M., Plestina R., Dirheimer G., Chernozemsky I.N. & Bartsch H. eds Lyon IARC . pp,331-336.
WOGAN G.N. - Aflatoxin carcinogenesis. *In* Methods in Cancer Research Academic Press (H.Busch, ed.) 1973, Vol. VII,309-344. .

Tableau 1 - Liste des échantillons prélevés et indications					
des lieux ,	des structures de stockage et de la température du grain				
	au moment du prélèvement				
N° Réf	LOCALITES		Nature	Structures	T°Grain °C
			Denrée	Stockage	
1	ALLADA	(Agbanou)	maïs	Chambre	26
				vrac au sol	
2	AZOVE		maïs	Chambre	28
				vrac au sol	
3	BASSILA		maïs	Crib sac	30
4	BEMBEREKE	(SRCV INA)	maïs	Grenier en	31
				argile	
5	BOHICON	MagasinPAM	maïs	Magasin	29
				PAM(sacs)	
6	LOKOSSA		maïs	MagasinPAM	27
				piles de sacs	
7	BOHICON		maïs	Magasin	29
				vrac au sol	
8	BOHICON		sorgho	MagasinPAM	32
				(sacs)	
9	COTONOU	(Dantokpa)	maïs	Magasin	32
				sac	
10	COTONOU	INSAE	maïs	Magasin	33
			jaune	PAM(sac)	
11	COTONOU	INSAE	sorgho	Magasin	31
				PAM (sacs)	
12	COTONOU	INSAE	maïs	Magasin	36
				PAM (sac)	
13	COTONOU	(O.N.C.)	maïs	silo métal-	28
				lique (sac)	
14	DJIDJA		maïs	Chambre	25
				sac	
15	DJOUGOU	(Donga)CPR	maïs	Chambre	31
				vrac	
16	DJOUGOU	(Sosso)	maïs	Champ en	33
				maturation	
17	DOGBO	(Dévé	maïs	Magasin	28
	Ferme	semencière		vrac au sol	
18	GLAZOUE		sorgho	Hangar(sac)	29
				air libre	

19	GLAZOUE		maïs	Hangar	29
				sac	
20	KETOU		maïs	Magasin	29
	Ferme	semencière	épi	vrac au sol	
21	KETOU		maïs	Chambre	29
				vrac au sol	
22	KETOU	(Iloulofin):	maïs	Chambre	27
	Ferme	semencière		froide(sac)	
23	LOKOSSA		maïs	MagasinPAM	27
				piles sacs	
24	MALANVILLE	(marché)	sorgho	Hangar	32
				sac	
25	MALANVILLE	(marché)	maïs	Magasin	35
				sac	
26	NATITINGOU		maïs	Chambre	29
				sac	
27	PARAKOU	(Alafiarou)	maïs	Crib	31
	Ferme	semencière			
28	PARAKOU		sorgho	Magasin	32
				PAM	
29	TOFFO	(Agbotagon)	maïs	Magasin	27
	Ferme	semencière		vrac au sol	
30	TORI-BOSSITO		maïs	Chambre	28
				vrac en épi	
31	TOUKOUNTOUNA		sorgho	Grenier	32
	Tampégre			en argile	
				complète	
32	ZAKPOTA		maïs	Crib	27
	Ferme	semencière			
N°réf : Numéro d'enregistrement					
INSAE					
ONC:Office	national	des Céréales			
PAM: Programme Alimentaire Mondial					
SRVC INA					

Tableau 2 : Liste des échantillons positifs
Concentrations en ppb des mycotoxines identifiees et quantifiées par CLHP.

N° Enregistrement	Nature de la denrée	Localité	MYCOTOXINES AFLATOXINES B1	B2	G1	G2	Ochratoxine
1	maïs	ALLADA	0,54	-	-	-	-
2	maïs	AZOVE	-	-	10,00	-	-
3	maïs	BASSILA	-	-	-	-	15
4	maïs	BEMBEREKE	-	-	-	-	33
5	maïs	BOHICON	15,00	-	0,09	-	-
10	maïs	COTONOU	-	0,5	-	0,75	-
11	sorgho	COTONOU	10,30	-	4,4	-	-
12	maïs	COTONOU	-	-	0,25	-	45
15	maïs	DJOUGOU	-	-	1,20	0,06	-
16	maïs(champ)*	DJOUGOU	-	-	-	0,22	44
19	maïs	GLAZOUE	5,7	-	2,30	-	-
20	sorgho	KETOU	-	-	11	0,29	-
21	maïs	KETOU	1,16	-	3,50	-	-
23	maïs	MALANVILLE	-	0,30	0,08	-	42
24	maïs	MALANVILLE	2,66	2,40	-	0,40	-
25	maïs	MALANVILLE	-	6,00	-	-	-
26	sorgho	NATITINGOU	-	-	2,15	0,05	-
27	maïs	PARAKOU	9,10	3,30	-	-	-
28	sorgho	PARAKOU	-	-	57,90	10	-
29	maïs	TOFFO	-	-	10,90	0,54	-
30	sorgho	TORI-BOSSITO	-	-	10,20	-	-
31	sorgho	TOUKOUNTOUNA	-	2,10	-	0,36	-
32	sorgho	ZAKPOTA	10,30	-	-	-	-

*Prélevé avant la récolte

Summary:

Food and feeds contaminations by mycotoxins have been reported in several countries of Africa in both north and south of Sahara. The present paper concerns a mycotoxin survey of the main cereals all over the country of Benin. It shows contamination by ochratoxin A , 15 % in the range of 15 to 45 ppb and also by aflatoxins , 66 % in the range of 0,5 to 58 ppb. In some cases (10%), ochratoxin A and aflatoxins are simultaneously present.
These two mycotoxins are known to induce synergistic effects in animal. So the determination of their prevalence in human blood and related pathologies is going on.

Résumé

La contamination des aliments par des mycotoxines a déjà été observée dans plusieurs pays d' Afrique au nord comme au sud du Sahara.
La présente étude qui a porté sur les céréales les plus consommées au Bénin (du Sud au Nord) montre une contamination par l' ochratoxine A, 15 % avec des taux compris entre 15 et 45 ppb et des aflatoxines 66 % avec des taux allant de 0,5 à 58 ppb. Dans 10 % des cas, il s'agit d'une contamination multiple par l'ochratoxine A et une ou plusieurs aflatoxines.
Compte-tenu des effets synergiques possibles entre ces deux mycotoxines, la détermination des fréquences de ces mycotoxicoses chez l' Homme est entreprise de même que la mise en évidence de pathologies qui leur sont imputables.

Human ochratoxicosis and its pathologies. Eds E.E. Creppy, M. Castegnaro, G. Dirheimer. Colloque INSERM/ John Libbey Eurotext Ltd. © 1993, Vol. 231, pp. 111-121.

Ochratoxines et ochratoxicoses humaines en Tunisie

H. Bacha[(1)], K. Maaroufi[(1)], A. Achour[(2)], M. Hamammi[(3)], F. Ellouz[(1)] et E.E. Creppy[(4)]

[(1)]*Laboratoire de Biochimie et de Toxicologie Moléculaire, Faculté de Médecine Dentaire, rue Avicennes, 5019 Monastir, Tunisie;* [(2)]*Service de Néphrologie, CHU, rue Avicennes, 5019 Monastir, Tunisie;* [(3)]*Laboratoire de Biochimie, Faculté de Médecine, rue Avicennes, 5019 Monastir, Tunisie;* [(4)]*Laboratoire de Toxicologie et d'Hygiène Appliquée, UFR des Sciences Pharmaceutiques de l'Université de Bordeaux II, 33076 Bordeaux, France*

Les mycotoxines sont sécrétées par des champignons microscopiques contaminant diverses denrées alimentaires. Lors de leur prolifération, ces champignons se développent sous forme de moisissures blanches, bleues, rouille, vertes ou noires.

Les mycotoxines n'ont pas d'unité structurale ni pathologique ni la même origine fongique. Plus de 200 espèces de champignons toxinogènes ont été recensées. Les espèces de champignon sécrétant les mycotoxines sont ubiquitaires, le problème des mycotoxines se présenta comme une question d'importance mondiale, tant sur le plan économique que sur le plan de la santé publique.

Bien que la situation soit différente entre pays développés et pays en voie de développement, plusieurs d'entre eux, comme la Tunisie, ont pris conscience de l'ampleur de la question et accordent une attention grandissante aux problèmes posés par les mycotoxines.

Nos travaux en Tunisie ont montré que plusieurs mycotoxines, aflatoxines B1, B2, G1, G2, ochratoxine A, citrinine zéaralénone, stérigmatocystine et certains trichothécènes, contaminent de manière importante les produits céréaliers locaux et importés, les produits couramment consommés des circuits commerciaux, les produits traditionnels consommés quotidiennement, ainsi que les aliments fabriqués industriellement et destinés aux animaux d'élevage (poulets, bovins, poissons), (Hadinane et coll., 1985 ; Bacha et coll., 1988).

Notre étude épidémiologique avec une enquête alimentaire a établi clairement la relation entre la consommation d'aliments contaminés par des mycotoxines et l'apparition de pathologies spécifiques (Bacha et coll., 1986 et 1988).

Lors de cette étude épidémiologique, une toxine, l'ochratoxine A, omniprésente, a particulièrement retenu notre attention.

L'Ochratoxine A (OTA), mycotoxine produite par des moisissures du genre *Aspergillus* et *Penicillium* (Van Der Merwe et coll., 1965), peut se retrouver dans le sang de l'homme et des animaux, apportée par l'alimentation (Krogh, 1987).

Plusieurs travaux ont montré que l'OTA provoque in vivo une néphropathie tubulo-interstitielle, ainsi que des tumeurs au niveau du tractus urinaire

(Chernozemsky et coll., 1987 ; Petkova-Bocharova et coll., 1988 ; Castegnaro et coll., 1990).

L'OTA se lie aux protéines sériques et peut être métabolisée en métabolites de toxicité analogue (Creppy et coll., 1983a).

L'OTA est un analogue structural de la phénylalanine. Elle est capable d'entrer en compétition avec cet acide aminé au cours de l'aminoacylation de $tRNA^{phe}$ (Bunge et coll., 1978 ; Creppy et coll., 1979a ; Creppy et coll., 1983b), cette compétition se traduit par l'inhibition de sunthèse protéique in vitro et in vivo (Creppy et coll., 1979b et 1984).

La structure de l'OTA montre qu'elle est constituée de deux parties : un noyau dihydroisocoumarinique chloré (OTα) lié par liaison amide à la phénylalanine (figure 1).

Fig. 1 : Structure de l'Ochratoxine A ou OTA (Phe-Ochratoxine)

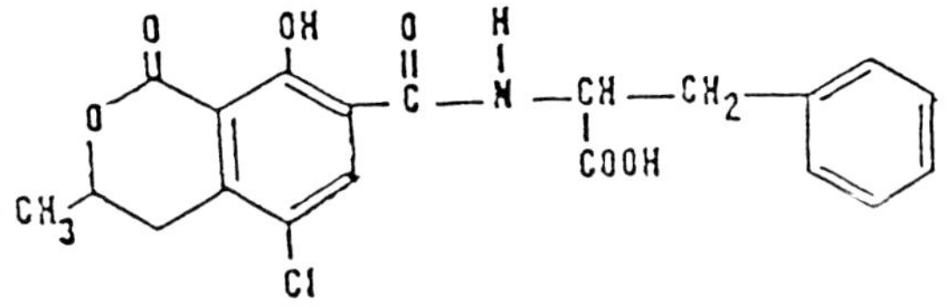

Actuellement, il semble certain que l'OTA soit l'agent causal principal dans la Néphropathie Endémique des Balkans (Krogh et coll., 1977 ; Hult et coll., 1982 ; Castegnaro et coll., 1987).

Nos résultats préliminaires et ceux obtenus en Algérie (Creppy, 1992) concernant l'ochratoxicose humaine, montrent qu'il s'agit d'un problème dont l'étendue dépasse les régions Balkaniques.

Dans le présent article, nous exposons des résultats qui appuient la relation entre la présence d'OTA dans le sang et les problèmes de néphropathie en Tunise. Nous montrons également que l'effet néphrotoxique n'est pas dû uniquement à l'OTA (phénylalanine-OTA) mais très probablement à la somme des effets d'un ensemble d'ochratoxines (analogues naturels de l'OTA dans lesquels la phénylalanine est remplacée par d'autres acides aminés).

Matériels et méthodes

1. Produits :

- Ochratoxine A (Phe-OTA) : Sigma Chemicals, St Louis Mo, USA. L'OTA est dissoute dans du méthanol, sa concentration est déterminée spectrophotométriquement à $\lambda = 330$ nm avec $\Sigma = 5500$.

- L'Ochratoxine alpha (OTα) est purifiée à partir de Phe-OTA après hydrolyse acide (Creppy et coll., 1983a).
- Acides aminés témoins : Sigma Chemicals, St Louis Mo, USA. Ils sont dissous dans le mélange éthanol-eau-triethylamine (2:2:1, v/v/v/).
- Les plaques de chromatographie en couches minces de gel de silice et de cellulose : Merck, Darmstadt, Allemagne).
- Les solvants et autres produits chimiques : Merck, Prolabo ou Sigma et sont tous de qualité pour analyse.

2. Echantillons de sérum :

- Les échantillons de sérum de sang humain ont été collectés en Tunisie dans les hôpitaux de Sfax (pour le sud), Sousse-Monastir (pour le centre), Tunis (pour la région nord) et Jendouba (pour le nord ouest). La collecte a été faite dans les services de néphrologie et dans les centres d'hémodialyse.
- Pour les témoins sains, la collecte a été réalisée dans les banques de sang des différents hôpitaux.

3. Extraction et analyses par chromatographies sur couches minces et par CLHP :

Dans un premier temps, les sérums collectés sont purifiés sur des cartouches C_{18} Sep-Pack Waters ; après lavage à l'eau, l'OTA est élué par du méthanol contenant 1% d'acide chlorydrique. Les différents extraits sont alors divisés en plusieurs fractions F1, F2, F3 :

Les fractions (F1) sont chromatographiées sur gel de silice avec le système de solvant : Toluène : acétate d'éthyl : acide formique (6:4:1, v/v/v). Les taches fluorescentes sont visualisées sous U.V. Pour chaque tache visualisée on détermine le Rf ou mobilité relative (rapport entre la distance de migration de la tache fluorescente et la distance du front du solvant). Les différents Rf sont comparés à celui de l'OTA témoin.

Les fractions (F2) du même éluat subissent une hydrolyse en milieu acide ; ces fractions séchées, sont additionées de HCl 6N contenant 5 % de phénol saturé et 2 % de β mercaptoétahnol, dans des tubes scellés sous azote. L'hydrolyse se déroule alors 24h à 100°C. Les hydrolysats séchés et repris avec un minimum de méthanol (100 µl) subiront 2 types de chromatographies :

i- Une chromatographie sur cellulose dans le système de solvant pyridine : n butanol : acide acétique : eau, (50:75:15:60, v/v/v/v). Les révélations se font par vaporisation de ninhydrine sur les plaques et les Rf des taches ninhydrines positives (couleur bleu foncé, jaune pour la proline), sont comparés aux Rf des différents acides aminés témoins ayant migré dans les mêmes conditions.

ii- Une chromatographie sur gel de silice dans le système de solvant toluène : acétate déthyle, acide formique (6:4:1, v/v/v). Les révélations se font sous U.V. et les Rf des taches fluorescentes visualisées sont comparés au Rf de l'ochratoxine alpha (OTα).

* Fractionnement par CLHP

Les fractions (F3) obtenues après purification sur Sep-Pack C_{18} sont traités selon la technique décrite (Figure 2) ; les extraits finaux sont fractionnés par CLHP sur une colonne C_{18} hypersil 5µM (8 x 250 mm). L'élution s'effectue par le mélange : méthanol : acétonitrile : acétate de sodium 5mM : acide acétique (300:300:300:13, v/v/v/v) avec un débit de 1 ml/min ; l'excitation a lieu à λ = 340 nm et l'émission à λ = 465 nm.

Fig. 2 MÉTHODE D'EXTRACTION, DE SEPARATION, DE PURIFICATION DE L'OTA

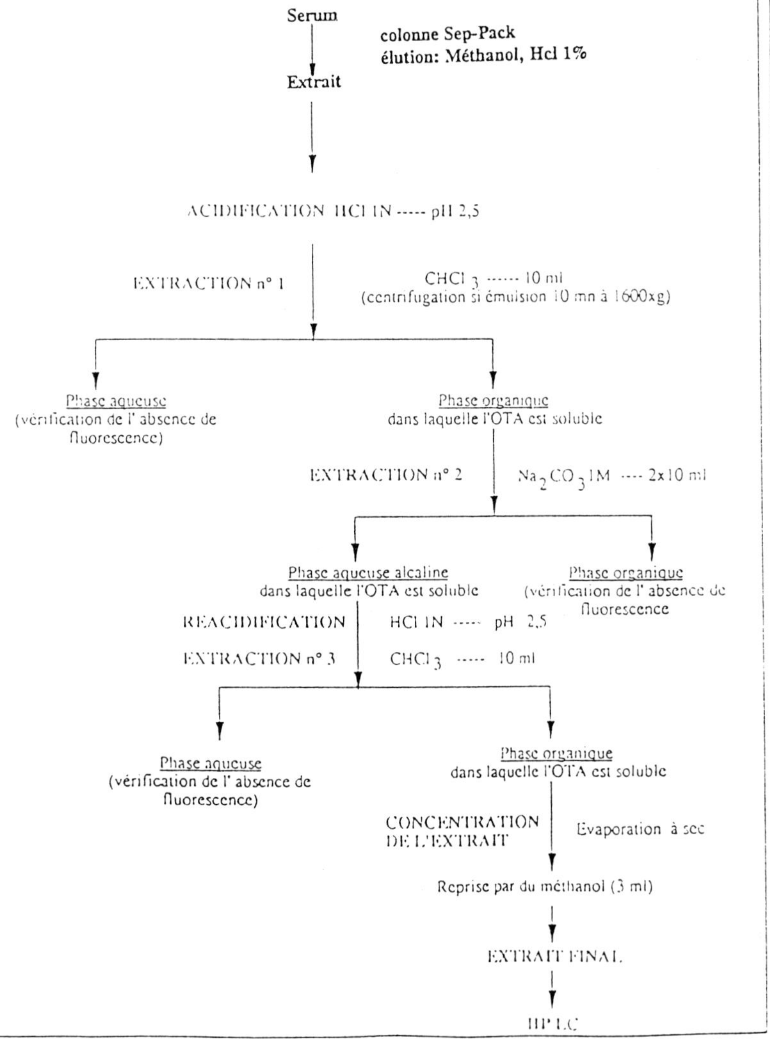

Résultats et discussion

Au cours de nos précédentes études (Bacha et coll., 1986) nous avons montré l'omniprésence de l'OTA parmi les mycotoxines contaminant les céréales locales ou importées, les denrées alimentaires couramment consommées ainsi que les composants industriels d'aliments destinés à des animaux d'élevage (Bacha et coll., 1986 et 1988). L'étude des réponses aux questionnaires remplis par les malades et l'analyse de l'OTA nous ont révélé une relation entre l'existence de l'OTA dans les aliments et l'existence de néphropathies (Néphropaties Interstitielles Chroniques N.I.C.) à éthiologie inexpliquée.

Nos premiers résultants montrent que sur 442 échantillons de la population générale et à la limite de la détection de 2 ng/ml, environ 70 % sont OTA positifs avec des taux d'OTA qui vont de 2 à 100 ng/ml. Si on abaisse la limite de détection à 0,1 ng/ml comme c'est le cas pour des études effectuées en Allemagne (Bauer et Gareis, 1987), 82 % des échantillons de la population générale sont alors OTA positifs (Tableau 1).

Tous les échantillons de patients présentant des problèmes rénaux (N.I.C. ou autres) sont OTA positifs avec des taux d'OTA allant de 25 à 100 ng/ml, avec une moyenne de 56 ± 19 ng/ml.

Pour la population apparemment saine, l'OTA est présente à un taux moyen de 3,5 ± 6,8 ng/ml. L'importance de l'écart type vient du fait que 2 témoins sains présentent des taux d'OTA anormalement élevés, 110 et 42 ng/ml, cela est dû sans doute à une consommation d'un aliment fortement contaminé juste avant les prélèvements de sang. Mis à part ces deux cas, tous les autres échantillons de ce groupe présentent des taux d'OTA inférieurs à 10 ng/ml (avec 12 % : inférieurs à 1,5 ng/ml).

La comparaison par le test (Wilcoxon Rank Sum Test) des deux groupes montre que le groupe des néphropathes et le groupe témoin (population générale) sont significativement différents à p=0,005 (Tableau 1).

TABLEAU 1:

OTA détectée dans des échantillons de sang (population du Sahel Tunisien).

Echantillons	Nombre	limite de détection	OTA positifs	Taux ng/ml
Population générale	442	2ng/ml	70 %	2 à 100
Population générale	442	1ng/ml	82 %	1 à 100
Population néphropathes	310	2ng/ml	100%	25 à 100

Population Apparemment saine ——> Taux moyen = 3.5 ± 6.8 ng/ml
avec 2 cas aberrants (42 et 110 ng/ml)

Population N.I.C. et Hémodialysées–> Taux moyen = 56 ± 19 ng/ml
(N.I.C. = Néphropathie Interstitielle Chronique)

Population apparemment saine
et
population N.I.C. et Hémodialysées

Comparées par le test "Wilcoxon Rank Sum Test"
sont significativement différentes à p = 0,005

Les comparaisons par le même test des différents groupes présentant des problèmes rénaux entre eux, montrent des taux de 61 ng/ml ± 36 pour le groupe présentant une néphropathie tubulointersticielle, 45 ng/ml ± 7 pour le groupe présentant une hypertension artérielle d'origine rénale.

Cela montre clairement que le taux d'OTA le plus important concerne les cas de néphropatie tubulo-intersticielle (Tableau 2). Le groupe des tubulo-néphrites interstitielles est significativement différent des autres, p = 0.05.

Tableau 2 : Concentration d'OTA dans le sang, par groupe de néphropathies

Néphropathies tubulo-interstitielles	→	61 ± 36 ng/ml
Néphropathies glomérulaires	→	45 ± 7 ng/ml
Hypertension d'origine rénale	→	49 ± 6 ng/ml

Il semble donc y avoir en Tunisie une relation entre les cas de néphropathies et la présence de l'ochratoxine A dans le sang des patients.

Il est clair que l'OTA présente dans le sang de la population tunisienne ne peut avoir comme origine qu'une contamination alimentaire. Cela est parfaitement confirmé puisque l'OTA est omniprésente dans les céréales et les produits de consommation courante comme nous l'avons montré à l'occasion d'enquêtes sur la contamination alimentaire. La situation des ochratoxicoses en Tunisie semble donc endémique.

Une étude similaire a été menée en parallèle en Algérie (Creppy, 1992), elle montre une corrélation analogue à celle que nous observons en Tunisie.

Ces résultats montrent donc que des ochratoxicoses et des néphropathies avec implication directe de l'OTA existe en Afrique du Nord. La situation semble endémique et parfaitement comparable à celle observée dans la région des Balkans (Hult et coll., 1982).

Si nous appliquons la méthode de Hagelberg et coll. (1989) pour calculer la quantité d'OTA prise chez les patients à insuffisance rénale (Ko) :

TABLEAU 3: Valeurs comparatives de "Daily Intake" (Ko)

$$Ko = Cpl \times \frac{Cp}{A}$$

Avec: Cpl = Clairance du plasma
Cp = concentration de l'OTA dans le plasma
A = Biodisponibilité
Ko s'exprime en ng/Kg corporel/jour

<u>Pays Scandinaves:</u>

Ko = 9 ng / Kg corporel / jour (Hagerberg et coll. 1989)

<u>Algérie:</u>

Ko = de 3,8 à 80,4 ng / Kg corporel /jour

<u>Tunisie:</u>

Ko = de 4,7 à 130 ng / kg corporel / jour

Ko Max. toléré et fixé en pays Scandinaves: 5 ng / Kg corporel / jour

En Tunisie, le "daily intake" se situe à des valeurs de 4,7 à 130 ng/kg corporel/jour.

En Algérie, le "daily intake" est de 3,8 à 80,4 ng/kg corporel/jour. Sachant que le plus fort "daily intake" mesuré par Hagelberg en pays scandinave correspond à 9 ng/kg corporel/jour, on voit que les valeurs les plus élevées trouvées en Tunisie sont de 14 fois plus fortes. Il faut encore savoir que les valeurs maximales tolérées ont été fixées en régions scandinaves à 5 ng/kg corporel/jour. La Tunisie est largement au-dessus (Tableau 3).

Au vu de ces résultats, nous pouvons dire qu'il existe un lieu certain entre des cas de néphropathie et les ochratoxicoses en Tunisie. Il reste à démontrer plus largement que ces néphropathies induites par l'OTA sont endémiques.

La question que nous nous posons actuellement est de savoir pourquoi le champignon synthétise l'OTA (phe - OTA) de façon préférentielle et non pas d'autres ochratoxines où lapartie phénylalaline serait remplacée par d'autres acides aminés.

Nos récents travaux (Hadidane et coll., 1993) ont montré que le champignon cultivé dans des conditions proches de ses conditions naturelles de prolifération, synthétise d'autres ochratoxines naturelles. Notre méthode de culture, d'extraction et de purification nous ont permis de caractériser 3 autres ochratoxines dont les structures ont été déterminées par Spectrométrie de Masse. Nous pensons que le champignon est capable de synthétiser 20 ochratoxines (autant que les 20 acides aminés) et que seules les conditions de culture d'extraction et de purification n'ont pas permis de les caractériser toutes.

Plus récemment, nous venons de découvrir, dans les échantillons de sérums de néphropathes, chromatographiés sur gel de silice, plusieurs bandes de forte fluorescence, dont l'une a été identifée comme étant l'OTA (phe - OTA). Les autres bandes traitées selon la même méthodologie décrite (Figure 3), subissent séparément une élution.

Fig. 3 : Hydrolyse acide de fractions fluorescentes extraites de sérums de néphropathes.

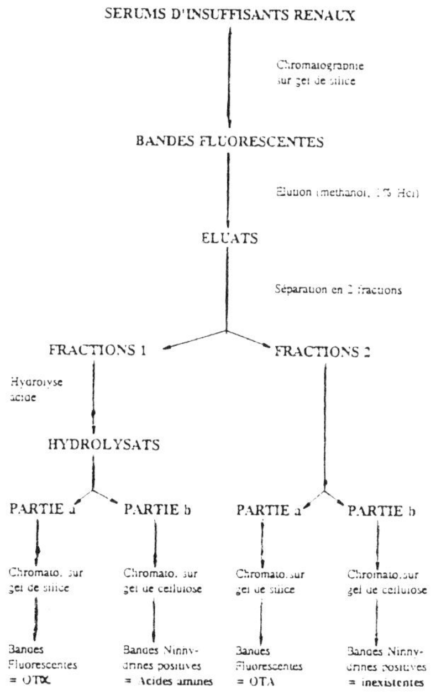

Chaque éluat est divisé en 2 fractions. Une des fractions subit une hydrolyse acide, l'autre ne subit aucune hydrolyse. Après chromatographie sur cellulose et révélation à la ninhydrine des fractions hydrolysées en milieu acide et des fractions non hydrolysées; nous avons observé que seules les fractions hydrolysées montrent l'apparition d'acides aminés (visualisés par des taches ninhydrines positives), alors que les mêmes fractions non hydrolysées ne montrent rien.

Les fractions hydrolysées, chromatographiées sur gel de silice montrent toutes une bande fluorescente identifiée sous UV comme étant l'OTα. Cela veut donc dire que l'hydrolyse acide a rompu la liaison pseudopeptidique entre l'OTα et l'acide aminé.

Nous avons trouvé après analyse d'environ 50 échantillons de forte fluorescence de sérum d'insuffisants rénaux, 12 ochratoxines naturelles où l'OTA (phe - OTA), bien qu'elle soit importante, n'est ni exclusive ni la plus dominante (figure 4). Leurs structures seront déterminées par Spectrométrie de Masse.

A la lumière de ces résultats, nous pouvons dire que les ochratoxicoses en Tunisie seraient induites par un ensemble d'ochratoxines naturelles. Le détournement des structures d'acides aminés en ochratoxines serait pour le champignon un moyen de mieux coloniser le substrat. La contamination humaine impliquerait l'ensemble des ochratoxines dont l'action pourrait être synergique.

Fig. 4 : Révélations après hydrolyse acide de (a) d'acides aminés; (b) d'ochratoxines. à partir de fractions fluorescentes extraites de sérums de néphropates.

(a)

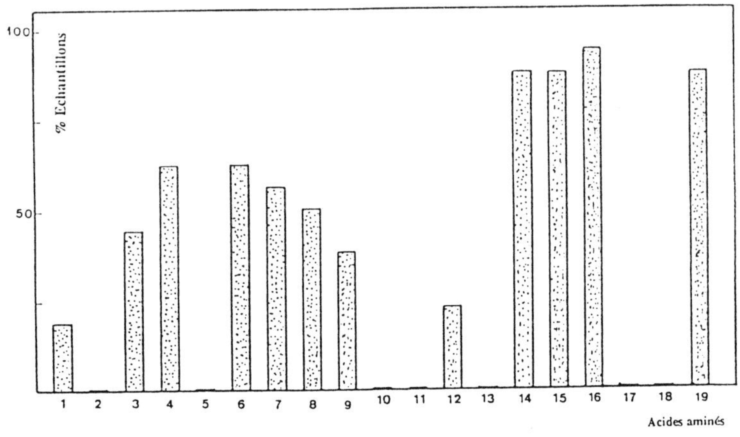

(b) OCHRATOXINES REVELEES
(Présence simultanée d'OTα)

1- Gly ------------> Gly-Ochratoxine
2- Ala
3- Val ------------> Val-Ochratoxine
4- Leu ------------> Leu-Ochratoxine
5- Ile
6- Ser ------------> Ser-Ochratoxine
7- Thr ------------> Thr-Ochratoxine
8- Cys ------------> Cys-Ochratoxine
9- Met ------------> Met-Ochratoxine
10- Asp
11- Asn
12- Glu ------------> Glu-Ochratoxine
13- Gln
14- Phe ------------> Phe-Ochratoxine
15- Tyr ------------> Tyr-Ochratoxine
16- Lys ------------> Lys-Ochratoxine
17- Arg
18- His
19- Pro ------------> Pro-Ochratoxine
20- Trp (détruit)

REFERENCES :

BACHA H., HADIDANE R., REGNAULT C.R., ELLOUZ F., CREPPY E.E. and DIRHEIMER G. (1986).
Mycotoxines et Mycotoxicoses en Tunisie. Cahiers Médicaux de Tunisie - Nutrition et Santé, 49, 34-35.

BACHA H., HADIDANE R., CREPPY E.E., REGNAULT C.R., ELLOUZ F. and DIRHEIMER G. (1988).
Monitoring and identification of fungal toxins in food products, animal feel and cereals in Tunisia. J. Stored Prod. Res., 24, N°4, 199-206.

BAUER J. and GAREIS M. (1987).
Ochratoxin A in der Nahrungsmittelkette. Z. Veterinärmed. B. 34, 613-627.

BUNGE I., DIRHEIMER G. and RÖSCHENTALER R. (1978).
In vivo and in vitro inhibition of protein synthesis in Bacillus stearothermophilus by ochratoxin A. Biochem. Biophys. res. Commun., 83, 398-405.

CASTEGNARO M., BARTSCH H. and CHERNOZEMSKY I. (1987).
Endemic nephropathy and urinary tract tumors in the Balkans. Cancer Res., 47, 3608-3609.

CASTEGNARO M., CHERNOZEMSKY I.N., HIETNANEN E. and BARTSCH H. (1990).
Are mycotoxins risk factors for endemic nephropathy and associated urothetial cancers ? Arch. Geschwulstforsh. 60, 4, 295-303.

CHERNOZEMSKY I.N., STOYANOV I.S., PETKOVA-BOCHAROVA T., NOCOLOV I.G., DRAGANOV I., STOICHEV I.N., TENCHEY Y., NAIDONEV D. and KALCHEVA N.D. (1977). Geographic correlation between the occurence of Endemic Nephropathyl and Urinary Tract Tumours in Vratza District, Bulgaria. Int. J. Cancer, 19, 1-11.

CREPPY E.E. (1992).
Human Ochratoxicosis in northern Africa, correlation with casses of nephropathy. Report of a network from Tunisia and Algeria. 14th Mykotoxin Workshop Giessen, pp. 68-69.

CREPPY E.E., LUGNIER A.A.J., FASOLO F., HELLER K., RÖSCHENTALER R. and DIRHEIMER G. (1979a).
In vitro inhibition of yeast phenylalanyl-tRNA synthetase by ochratoxin A. Chem. Biol. Interact., 24, 257-262.

CREPPY E.E., LUGNIER A.A.J., BECK G., RÖSCHENTHALER R. and DIRHEIMER G. (1979b).
Action of ochratoxin A on cultured hepatome cells-reversion of inhibition by phenylalaline. FEBS Lett., 104, 297-290.

CREPPY E.E., STORMER F.C., RÖSCHENTALER R. and DIRHEIMER G. (1983a).
Effects of two metabolites of ochratoxin A, (4R)-4-hydroxy ochratoxin A and ochratoxin-alpha, on immune response in mice. Infect. Immun. 39,1015-1018.

CREPPY E.E., KERN D., STEYN P.S., VLEGAAR R., ROSCHENTHALER R. and DIRHEIMER G. (1983b).
Balkan endemic nephropathy. Arch. Toxicol., 51, 313-321. Comparative study of the effect of ochratoxin A analogues on yeast aminoacyl-tRNA synthetases and on the growth and protein synthetis of hepatoma cells. Toxicol. Lett., 19, 217-224.

CREPPY E.E., RÖSCHENTHALER R. and DIRHEIMER G. (1984).
Inhibition of protein synthesis in mice by ochratoxin A and its prevention by phenylalaline. Food Chem. Toxicol. 22, 883-886.

FROHLICH A.A., MARQUARDT R.R. and OMINSKI K.H. (1991).
Ochratoxin A as a contaminant in the human food chain: a Canadian perspective. In mycotoxins, Endemic Nephropathy and Urinary tract tumours. IARC n° 115, Scientific Publication, pp. 139-143.

HADIDANE R., REGNAULT C.R., BOUATTOUR H., ELLOUZ F., BACHA H., CREPPY E.E.and DIRHEIMER G. (1985).
Correlation between alimentary micotoxin contamination and specific diseases. Human Toxical., 4, 491-501.

HADIDANE R., BACHA J., CREPPY E.E., HAMMANI M., ELLOUZ F. and DIRHEIMER G. (1993).
Isolation and structure determination of natural analogues of the mycotoxin ochratoxin A produced by Aspergillus Ochraceus.
Accepté pour publication dans Human and Experimental Toxicology.

HAGELBERG S., FUCHS R., HULT K. (1989).
Toxicokinetics of ochratoxin A in several species and its plasma binding properties. J. Appl. Toxicol. 9, 91-96.

HULT K., PLESTINA R., CEOVIC S., HABAZIN-NOVAK V and RADIC B. (1982).
Ochratoxin A in human blood: Analytical results and confirmational tests from a study in connection with Balkan endemic nephropathy. In Proceedings, V International IUPAC Symposium Mycotoxins and Phycotoxins, september 1-3, Vienna, Austria, pp. 338-341.

HULT K., PLESTINA R., HABAZIN-NOVAK V., RADIC B. and CEOVIC S. (1982).
Ochratoxin A in human blood and Balkan endemic nephropathy. Arch. Toxicol., 51, 313-321.

KLASSEN C.D. (1986).
Distribution, excretion and absorption of toxicants. Casarett and Foull's Toxicology.
The Basic Science of Poisons, edited by C.D. Klaassen, M.O. Amdur, and J. Doull (New-York: Macmillian Publishing Company), third edition, 33-66.

KROGH P. (1987).
Ochratoxins in food. In: Mycotoxins in food, Ed. P. Krogh (Academic Press, New-York), 97-121

KROGH P., HALD B., PLESTINA R. and CEOVIC S. (1977).
Balkan (endemic) nephropathy and foodborn ochratoxin A: Preliminary results of foodtuffs. Acta Pathol. Microbiol. Scandi., section B, 85, 238-240.

PETKOVA-BOCHAROVA T., CHERNOZEMSKY I.N. and CASTEGNARO M. (1988).
Ochratoxin A in human blood in relation to Balkan Endemic Nephropathy and Urinary System Tumours in Bulgaria. Food Addit. Contam., 5, 299-301.

Van DER MERWE K.J., STEYN P.S., FOURRIE L., SCOTT D.B. and THERON J.J. (1965).
Ochratoxin A, a toxic metabolite produced by Aspergillus Ochraceus Wilh. Nature, 205, 1112-1113.

Summary :

The prevalence of human ochratoxicosis is being determined in Tunisia. It seems higher than in Europe (55 to 80 % at the detection limit of 0.1 ng / ml).

95 % of people suffering from nephropathy are OTA positive with blood concentrations higher than 90 ng / ml in several cases.

In addition some natural analogues of OTA with different aminoacids are found in their blood extracts. This confirms the natural occurence of these analogues in cultures of *Aspergillus*.

Résumé

La fréquence de l' ochratoxicose humaine a été déterminée en Tunisie. Elle semble plus élevée qu' en Europe (55 - 80 % au seuil de 0,1 ng / ml). Les personnes souffrant de néphropathie sont à plus de 95 % OTA positives, avec des concentrations qui dépassent 90 ng / ml. De plus d'autres ochratoxines avec des acides aminés différents de la phénylalanine sont également mis en évidence dans le sang humain, ce qui confirme l' existence d' analogues naturels de l' ochratoxine A dans les extraits d' *Aspergillus*.

Human ochratoxicosis and its pathologies. Eds E.E. Creppy, M. Castegnaro, G. Dirheimer. Colloque INSERM/ John Libbey Eurotext Ltd. © 1993, Vol. 231, pp. 123-127.

Ochratoxicose humaine en Algérie

A. Khalef(1), C. Zidane(2), A. Charef(2), A. Gharbi(3), M. Tadjerouna(2), A.M. Betbeder(2) et E.E. Creppy(3)

(1)*UGTA, Maison du Peuple, Place du 1er mai, Alger, Algérie;* (2)*CHU Parnet, Hussein Dey, Alger, Algérie;* (3)*Laboratoire de toxicologie et hygiène appliquée, 3, place de la Victoire, 33076 Bordeaux Cedex, France*

RESUME

La fréquence de l'ochratoxicose humaine a été déterminée en Algérie. Les échantillons de sang ont été prélevés d'une part sur des ouvriers travaillant dans des usines d'état sur toute l'étendue du pays, et qui prennent deux repas/jour sur place, et d'autre part sur des néphropathes à la clinique néphrologique d'Alger.

95 % des 84 échantillons de néphropathes contenaient plus de 0,1 ng d'OTA par ml. A cette même limite de détection, 66 % des échantillons provenant des autres personnes (346) étaient également positifs.

De part la fréquence de l'ochratoxicose humaine et les taux trouvés, l'Algérie semble avoir un taux de contamination supérieur à ceux trouvés en France et en Europe.

La principale source de contamination est sans doute les céréales, vu les habitudes alimentaires et la contamination par diverses mycotoxines et l'ochratoxine A en particulier mise en évidence en Afrique du Nord.

La recherche se poursuit pour déterminer la dose journalière d'OTA ingérée en Algérie.

L'ochratoxine A (OTA) est une mycotoxine produite par des moisissures des genres Aspergillus et Penicillium (Van der Merve et coll. 1965). Elle est retrouvée dans une grande variété d'aliments pour l'homme et le bétail, notamment les céréales, les oléagineux, les fruits et les légumes secs (Krogh 1987). L'ochratoxine A est néphrotoxique pour toutes les espèces animales testées jusqu'ici (Krogh et coll. 1974, Krogh 1980,, Hult et coll. 1982).

Elle est impliquée dans les néphropathies endémiques des Balkans comme agent causal principal (Krogh et coll. 1977). En plus de cette néphropathie, l'OTA est immunosupressive (Tor et coll. 1989, Dwivedi et Burns 1984, Creppy et coll. 1983), tératogène (Arora et Frölen 1981, Mayura et coll. 1981), génotoxique (Creppy et coll. 1985, Pfohl-Leszkowicz et coll. 1991, Malaveille et coll. 1991) et cancérogène (Kanisawa et Suzuki 1978, Bendele et coll. 1985, NTP 1989).

En Algérie, la détermination des taux d'OTA dans l'alimentation ou dans le sang de l'homme et des animaux n'avait jamais été entreprise. Compte-tenu des habitudes alimentaires (forte consommation de céréales) et du dépistage fréquent des nombreux cas de néphropathie sans étiologie définie, il est apparu opportun de procéder à la détermination de la fréquence de l'ochratoxicose humaine en Algérie.

MATERIELS ET METHODES :

Les prélèvements ont eu lieu à 3 endroits différents : Est, Ouest et Centre de l'Algérie. Les personnes prélevées sont des travailleurs en bonne santé apparente employés par l'entreprise nationale des industries métalliques (fabrications des emballages métalliques). Ces personnes effectuent un travail posté dans la tranche horaire de 5 à 13 h et prennent 2 repas sur 3 à l'usine. Les prélèvements sont effectués le matin dans des tubes secs. Après coagulation du sang (10 ml) et centrifugation (2000 g pendant 15 mn), le sérum (5 ml) est recueilli dans des tubes à hémolyse, il est immédiatement congelé jusqu'à l'extraction de l'OTA puis l' analyse.

La technique d'extraction dérive de celles de Bauer et Gareis 1987 avec les modifications préconisées par Hietanen et coll. En bref, l'OTA est extraite par le chloroforme à partir du sérum acidifié par un mélange de chlorure de magnésium et de l'acide chlorhydrique ajusté au pH 2,5. L'OTA est alors extraite de la phase chloroformique avec une solution de bicarbonate. Après acidification de cette nouvelle solution à un pH à 2,5, l'OTA est extraite à nouveau par du chloroforme. Cette phase chloroformique est évaporée à sec sans pression réduite à 40°C. L'extrait sec est repris par le méthanol et séparé en 3 parties aliquotes : une sert à l'analyse directe pour HPLC et fluorimétrie et les deux autres sont conservées à -20°C pour confirmation ultérieure si nécessaire.

Les conditions d'HPLC sont les suivantes :

- colonne C18, 20 x 0,64 cm, 10 μm ODS2. L'éluant est méthanol : acétonitrile : acétate de sodium 5mM : acide acétique (300:300 : 400 : 14 v/v/v/v).

Les paramètres fluorimétriques sont excitation 340, émission 465.

Tous les échantillons positifs sont confirmés à la fois par estérification selon la méthode décrite par l'AOAC (Assiation of Official Analytical Chemist 1980) et par l'action de la carboxypeptidase (Hult et coll 1982). Le rendement de l'extraction de l'OTA est d'environ 90 % de telle sorte que les résultats sont exprimés directement en ng/ml (nanogrammes par millilitre). Par cette méthode et avec le fluorimétre (JASCO-821 FP) il est possible d'arriver à une limite de détection inférieure à 0,1 ng/ml.

RESULTATS ET DISCUSSION

430 échantillons dont 84 provenant de personnes souffrant de diverses néphropathies ont été analysés. Sur les 346 échantillons provenant de la population générale, 57 % sont positifs avec un taux supérieur à 2 ng/ml. Au seuil de 0,1 ng/ml, le pourcentage d'échantillons positifs passe à 66,9 % (tableau I). Parmi les néphropathes au seuil de 0,1 ng/ml, il y a 95 % d'échantillons positifs confirmés et au seuil de 2 ng/ml il y a 90 % (tableau I). Il faut noter que les échantillons de sang des autres néphropathes considérés comme négatifs au seuil de 0,1 ng/ml contiennent des traces d'OTA.

Une comparaison des 2 populations (néphropathes et population générale) fait apparaître une différence significative pour $\leq$ 0,005. En effet les moyennes et écarts-types des taux d'OTA sont de 2,8 ng/ml ± 2,6 avec une valeur maximale de 9 ng/l et de 7,4 ± 7,1 ng/ml pour les néphropathes avec une valeur maximale de 46 ng/ml. Le choix des 2 seuils de 0,1 et de 2 ng/l nous permet de comparer les résultats obtenus en Algérie avec ceux obtenus dans les pays scandinaves et les pays de l'Europe de l'Ouest d'une part ainsi qu'avec ceux des Balkans d'autre part.

Au seuil de détection de 0,1 ng/ml, l'Algérie semble avoir un taux de contamination plus élevé que le taux français (66,9 % contre 22 %) (Creppy et coll. 1991). En comparaison avec l'ochratoxicose humaine en Allemagne on peut constater que l'Algérie a une fréquence légèrement supérieure mais pour ce qui concerne la population générale les taux semblent plus élevés en Allemagne qu'en Algérie (14,4 ng/ml contre 9 ng/ml pour l'Algérie) (Bauer et Garies 1987), (Kuiper-Goodman et Scott 1989). D'autres pays d'Europe tels que la Suède ou le Danemark ont des taux d'OTA dans le sang comparables à ceux trouvés en Algérie même si la fréquence y est légèrement plus faible (Breitholz et coll. 1991).

La comparaison avec les résultats d'ochratoxicose humaine dans les Balkans fait apparaître des taux nettement plus élevés (10 à 40 ng/ml avec une fréquence plus faible dans les zones des néphropathies endémiques (Petokova-Bocharova et coll. 1988). Il faut noter toutefois que les taux trouvés dans les zones non endémiques de l'ex-Yougoslavie (7-8 ng/ml) sont exactement du même ordre que ceux trouvés en Algérie (Petkova-Bocharova et coll. 1988). On pourrait s'attendre alors à ce que les taux d'OTA dans le sang humain en Algérie soient sans conséquence, or il semble que le nombre de cas de néphropathies répertoriés en Algérie pour 100000 habitants soit nettement plus élevé que la moyenne des pays européens où il ne semble pas y avoir de problèmes particuliers.

L'origine de la contamination de l'homme est sans doute essentiellement céréalière compte-tenu des importations massives de céréales provenant de pays où l'OTA a été identifiée dans les céréales (Amérique du Nord).

De plus les conditions de transport maritime, de stockage dans les circuits de distribution et du stockage à domicile sont telles qu'il n'est pas surprenant que des moisissures toxinogènes se développent dans ces céréales. Cependant la recherche systématique de l'OTA et des toxines telles que la citrinine devra être entreprise afin de déterminer l'ampleur de la contamination. Une autre source de contamination est constituée par les volailles, les fruits et légumes secs qui ont déjà fait l'objet d'étude en Tunisie. Les résultats des chercheurs tunisiens montrent que la contamination par l'OTA est fréquente et quelque peu alarmante. Compte-tenu de la similitude des situations climatiques et des habitudes alimentaires on peut penser que la situation est identique en Algérie.

En attendant de déterminer avec précision les quantités réelles dans l'alimentation de l'algérien moyen, il est possible à partir des résultats de l'ochratoxicose humaine en Algérie de calculer la dose journalière ingérée suivant la formule de Klaassen 1986

$$Ko = Clp \times C_{P/A}$$

dans laquelle Clp est la clearance plasmatique en ml/kg de poids corporel, Cp la concentration plasmatique en ng/ml et A la biodisponibilité. Dans ces conditions la dose journalière ingérée serait comprise entre 3,8 et 80 ng/kg/jour de poids corporel. En comparaison avec ce qui est retenu en Scandinavie comme dose journalière acceptable (5 ng/kg Pc/jour), il est clair que l'exposition à l'OTA est plus importante en Algérie que dans les pays d'Europe où cette détermination a été faite. Il reste à confirmer ce calcul théorique par la détermination pratique qui est en court sur le terrain .

La question peut se poser alors de savoir s'il existe en Algérie une néphropathie humaine liée à l'OTA.

BIBLIOGRAPHIE

Arora R.G. and Frölen H. (1981). Interference of mycotoxins with prenatal development of the mouse. Ochratoxin A induced teratogenic effects in relation to the dose and stage of gestation. Acta Vet. Scand., 22, 535-552

Bauer J. and Gareis M. (1987). Ochratoxin A in der Nahrungsmittelkette. Z. Veterinämed B., 34, 613-627

Bendele A.M., Carlton WW., Krogh P. and Lilihoj E.B. (1985). Ochratoxin A carcinogenesis in the (C57BL/6JxC3H) F1 mouse. J. Nat. Cancer Inst., 75, 733-742

Breitholz A., Olsen M., Dahlback A. and Hult K. (1991). Plasma ochratoxin A levels in three swedish populations surveyed using an ion-pair HPLC technique.Fd. Addit. and Cont. 8, 2, 183-192

Creppy E.E., Stormer F.C., Röschenthaler R. and Dirheimer G. (1983). Effects of two metabolites of ochratoxin A, (4R)-4-hydroxy-ochratoxin A and ochratoxin α on immune response in mice. Infect. Immun., 39, 1015-1018.

Creppy E.E., Kane A., Dirheimer G., Lafarge-Frayssinet C., Mousset S. and Frayssinet C (1985). Genotoxicity of ochratoxin A in mice : DNA single-strang break evaluation in spleen, liver and kidney. Toxicol. Lett. 28, 29-35

Creppy E.E., Betbeder A.M., Gharbi A., Counord J., Castegnaro M., Bartsch H., Moncharmont P., Fouillet B., Chambon P. and Dirheimer G. (1991). Human ochratoxicosis in France. In Mycotoxins, endemic nephropathy and urinary tract tumours, IARC Lyon, 145-151

Dwivedi P. and Burns R.B. (1984). Effect of ochratoxin A on immunoglobulins in broiler chicks. Res. Vet. Sci. 36, 117-121

Hult K., Plestina R., Habazin-Novac V., Radic B. and Ceovic S. (1982). Ochratoxin A in human blood and Balkan endemic nephropathy. Arch. Toxicol., 51, 313-321

Kanisawa M. and Suzuki S. (1978). Induction of renal and hepatic tumors in mice by ochratoxin A, a mycotoxin. Gann 69, 599-600

Krogh P., Axelson N.H., Elling F., Gyrd-Hansen N., Hald B., Hyldgaard-Jensen J., Larsen A.E., Madsen A., Mortensen H.P., Moller T., Petersen O.K., Ravnskov U., Mortensen H.P., Moller T., Petersen O.K., Ravnskov U., Rostgaard M. and Aalund O. (1974). Experimental porcine nephropathy : changes of renal function and structure induced by ochratoxin A-contamined feed. Acta Pathol. Scandi., section A, suppl. 246, 1-21

Krogh P., Hald B., Plestina R. and Ceovic S. (1977). Balkan endemic nephropathy and foodborn ochratoxin A : preliminary results of foodsssstuffs. Acta Pathol. Microbiol. Scandi., section B, 85, 238-240

Krogh P. (1980). Ochratoxins : occurence biological effects and causal renal disease in : Eaker D. and Wadstrom T., ed. Natural toxins, Oxford Pergamon Press, 673-680

Krogh P. (1987). Ochratoxins in food. Mycotoxins in food. Ed. P. Krogh (Academic Press, New York), 97-121

Kuiper-Goodman T. and Scott P.M. (1989). Risk assessment of the mycotoxin ochratoxin A. Blomed Environ. Sci. 2, 179-248

Malaveille C., Brun G. and Bartsch H. (1991). Genotoxicity of ochratoxin A and structurally related compounds in Escheria Coli strains : studies on the mode of action in : Mycotoxins, endemic nephropathy and urinary tract tumours, IARC, scientific publication n°115, Castegnaro M., Plestina R., Dirheimer G., Chernozemski I.M., Bartsch H., ed. Lyon IARC, 261-266

Mayura K., Edwards J.F., Maull E.A., Phillips T.D. (1989). The effects of ochratoxin A on post implantation rat embryos in culture. Arch. Environ. Contam.Toxicol., 18, 411-415

NTP technical report on the toxicology and carcinogenesis studies of ochratoxin A (1989). CAS n° 303-47-9 in F344/N rats C. gavage studies. (NIH publication n° 89-2813). Boorman G. ed. Research Triangle Park. North Carolina, National Toxicology Program. US dept of Health and human services

Petkova-Bocharova T., Chernozemsky I.N. and Castegnaro M. (1988). Ochratoxin A in human blood in relation to Balkan endemic nephropathy and urinary system tumours in Bulgaria. Food Addit. Contam., 5, 299-301

Pfohl-Leszkowicz A., Chakor K., Creppy E.E. and Dirheimer G. (1991). DNA-adducts formation in mice treated with ochratoxin A, in : Mycotoxins, endemic nephropathy and urinary tract tumors, IARC, scientific publication n°115, Castegnaro M., Plestina R., Dirheimer G., Chernozemski I.M., Bartsch H., ed. Lyon IARC, 245-253

Tor L., Steien K., Stormer F.C. (1989). Mechanism of ochratoxin A induced immunosuppression . Mycopathologia 107, 107, 153-159

Van der Merve K.J., Steyn P.S., Fourrie L., Scott D.B. and Theron J.J. (1965). Ochratoxin A, a toxic metabolite produced by Aspergillus ochraceus. Nature, 205,1112-1113

Tableau 1 : Fréquence de l'ochratoxine en Algérie dans la population générale et les personnes atteintes de néphropathie chronique

	Origine des échantillons			
nombre d'échantillons analysés	Population générale 346		Nephropathes 84	
Niveau de contamination (ng/ml)	nombre d'échantillons positifs	%	nombre d'échantillons positifs	%
0.1-0.2	34	9.9	4	4.7
> 2	197	57.0	76	90.4
Total	231	66.9	80	95.0
Taux maximum	9		46	

Summary

The prevalence of human ochratoxicosis has been determinated in Algeria. The blood samples were from workers in governmental factories all over the country and from the nephrology department in Alger.

95 % of the 84 samples from nephropathy patient were positive at the detection limit of 0.1 ng/ml. At the same detection limit 66 % of the samples from the workers (346) were also positive.

According to the prevalence of human ochratoxicosis and the level found, Algeria seems to have a level of human exposure and blood of OTA higher than those observed in Europe.

The main contamination source is likely the cereals according to the eating habits and food contamination by mycotoxins observed in northern Africa. The determination of the daily intake is under the way.

Human ochratoxicosis and its pathologies. Eds E.E. Creppy, M. Castegnaro, G. Dirheimer. Colloque INSERM/ John Libbey Eurotext Ltd. © 1993, Vol. 231, pp. 129-139.

Ochratoxin A in italy : status of knowledge and perspectives

M. Miraglia, C. Brera, S. Corneli and R. De Dominicis

Istituto Superiore di Sanitá, Laboratorio Alimenti, Viale Regina Elena, 299, 00161 Rome, Italy

The problem of contamination by ochratoxin A (OTA) in Italy has been faced only in the last few years and will gain, in the future, increasing attention in view of the well known widespread occurence of this toxin in temperate climates.

The first strategies adopted to assess the relevance of OTA in the human food chain in Italy were first, the development of programs to check the status of food and feed contamination and then, for some matrices, the verification of the possible destruction of the toxin with technological procedures.

The other strategy was to evaluate the human exposure to ochratoxin A by its analysis in biological fluids.

Occurence of OTA in agricultural commodities

Many researchers have reported on the heterogeneous distribution of mycotoxins in grain. Therefore the criteria that we have adopted in selecting samples for our preliminary survey programs was to take samples from large bulk of materials by representative sampling instead of analyzing a large number of randomly collected samples.

The lack of a specific sampling plan for OTA analysis prompted us to adopt in this study those developed for aflatoxin surveys.

In this investigation, the commodities tested for OTA contamination were selected mainly on the basis of the consumption levels and on the potential for contamination by mycotoxins.

The results obtained are reported in Table 1, they show a widespread contamination even if at a low level.

The methods adopted for the determinations of OTA was generally a modification of that of Tsubouchi et al. (1987). The limit of detection ranged from 0.1 to 0.2 µg/kg depending on the commodities.

As for durum and soft wheat, samples of conventionally and organically grown wheat, were evaluated for OTA content together with their milling fractions.

Higher levels of OTA was detected in the organically grown wheat as compared to the conventionnaly grower ones. Furthermore a slight penetration of OTA in the inner parts of the grain was noted with a concentration in the peripheral part of the kernel of 4 µg/kg and of 0.2 µg/kg in the flour.

It is noteworthy that in some cases (bran and extruded food) the upper values for OTA contamination exceeded the levels established by some European countries (van Egmond, 1991).

No conclusions have been drawn yet in consideration of the limited number of analyzed samples and more extensive survey programs are planned in order to better scrutinize OTA contamination in agricultural commodities.

Fate of OTA during technological procedures

Detailed information on the fate of OTA during several technological procedures such as milling, heating, breadmaking, and beer manufacturing has been already reviewed (Scott, 1984).

Since a widespread contamination by OTA in coffee beans was shown in our preliminary investigations, researches were carried out in order to investigate the fate of OTA in samples of contaminated coffee beans during the usually employed technological procedures (Micco et al., 1989b).

Both naturally and artificially contaminated samples were roasted at different operation times (5-6 min.) to verify the percentage of destruction of the toxin.

The reduction ranged from 48 to 87 % and from 90 to 100 % in artificially and naturally contaminated samples respectively.

The apparently greater stability in the artificially contaminated beans could be a result of the preliminary wetting process performed with the aim of making the grains more able to absorb the mycotoxin, and, as has been already claimed for cereals, OTA appears to be more rapidly destroyed in dry products than in the presence of water.

Furthermore in order to verify the reduction of OTA by combined decaffeinization and roasting processes, a sample of naturally contamined coffee (OTA level 6 µg/kg) was decaffeinated by using pilot equipment and procedures which included steaming and extraction with methylene chloride and then, roasted.

The reduction of OTA by decaffeination was 60 % and the following roasting procedures causes a further loss of up to 100 %.

Also the possibility of migration of OTA into the beverage was verified. The trial performed on artificially contaminated and roasted coffee at an OTA level after roasting of 6 µg/kg revealed no OTA in the beverages.

On the basis of one study reporting the occurence of OTA in cocoa beans up to 500 µg/kg (MAFF, 1980) and of a slight diffusion of contamination even if it was found at a very low level (our unpublished data), we also carried on researches aimed at investigating the fate of OTA in cocoa beans during the different technological procedures leading to main final products (milk and dark chocolate).

Samples of inoculated cocoa beans were mixed with "blank" beans in order to give contaminated batches (mean contamination level 10 μg/kg). A single batch was used for each of following processes:

1) Technological procedures leading to milk-chocolate
2) Technological procedures leading to alkali treated-chocolate
3) Warehouse stock under usual storage conditions for 45 days then step 1).
4) Warehouse stock under usual storage conditions for 45 days then step 2).

The technological procedures were performed in a pilot plant and are summarized below:

- Winnowing: is a simple process that entails separating the nib from the inedible shell. In our experiment the conditions used in winnowing were: T=600°C, duration 40 sec.
- Alkali treatment (only for dark chocolate): soaking into $NaHCO_3$ solution (5%, ph=10) at 70°C.
- Roasting: treatment at T=130°C, duration 3 mn.
- Liquor grinding: the grinding of the nib of the cocoa beans leading to the chocolate liquor production. Grinding liberates the fat locked within the cell-wall, while producing temperatures as high as 110°C.

The analysis of the chocolate produced after every single step revealed that the whole technological process usually employed in the processing both of milk and alkali treated chocolate caused a destruction of the toxin up to 99%.

Alkali treatment resulted, as expected, in beeing the most effective step in destroying OTA, while during winnowing the percentage of destruction was up to 22%.

In this step a marked migration of the toxin to the inner part of the beans was also observed.

Although the whole technological process causes an almost exhaustive destruction of the toxin, precautions should be taken in order to prevent raw cocoa beans from high contamination level, in consideration also of the observed increase of the concentration of OTA in batches stored for 45 days at warehouse conditions.

The analytical method employed for OTA determinations in this study, was suitable for all the products under analysis (raw cocoa beans, shell, nibs, alkali treated beans, chocolate liquor, raosted beans). It basically consisted in a modification of that of Tsubouchi et al. (1987).

OTA in human fluids

The organization of the programs to evaluate the incidence of contamination by mycotoxins is problematic because of the lenght of time required and great economical efforts. Furthermore, the potential of OTA contamination is influenced by various factors such as favourable climatic conditions, inadequate storage techniques. Consequently, fluctuation of the incidence of contamination and unreliable assessment of the real exposure to human, can occur.

A program for the evaluation of the occurence of OTA in biological fluids could overcome the above uncertainty. The first action undertaken in our country to evaluate the presence of OTA in bilogical fluids, has been a study performed on human milk. This secretion fluid could also be, in view of the high chemical affinity of OTA for the fatty nature of milk, reflective of the possible intake of the toxin, in addition to being the most important food for a very critic group of people.

Two studies aimed at the evaluation of OTA levels in human milk were performed in our laboratory (Micco et al., 1991, 1992).

All together, 111 samples of human milk were collected from mother donors, 30 of whom being hospitalized. The overall contamination levels of OTA in human milk are shown in Table 2.

In the mean time, data on the alimentary habits of the mothers were collected in order to obtain information on the type and amount of food usually consumed. No correlation could be established between the diet and the contamination of milk.

The group of hospitalized mothers showed a higher percentage of contaminated samples even if at a lower level. A possible explanation of the above differences could be suggested in the preparation of the meals within the two groups. In the hospitalized mothers a dilution factor acting during the preparation of the common diet provides a more diffuse and a lower level of OTA contamination with respect to the domestically prepared food.

Furthermore, in order to be able to follow any possible trend of OTA levels over a short period of time, six mothers not hospitalized were tested during a week. Each day, a sample of milk coming from a single such was drawn at the same time. Very interesting results for OTA contamination were observed (Table 3).

Particularly in two cases a very hight peak of concentration (21.9 and 8.5 ng/ml) preceded and followed by very low levels were noted, giving a possible, even if only indicative, route of metabolization of this toxin by women.

The OTA level was determined by using an HPLC method (Breitholtz et al. 1991) adapted to human milk. The limit of detection was 0.1 ng/ml. The confirmation of identity of OTA peak in positive samples was performed in two different ways: first, a conversion of OTA to its methyl ester by pre-column derivatization with BF_3-methanol solution and second, a formation of ammonia derivative of OTA by a post column derivatization with 10% solution of ammonia. The detection of ammonia treated samples was made at wavelengths (370-460 nm) different from those used in the not-treated samples (333-470 nm) because of the much higher sensitivity of ammonia OTA derivative.

These studies represent the first investigation on OTA contamination of human milk in Italy. In the near future further researches on human fluids (milk, serum, urine) are planned based on multidisciplinary collaboration in order to obtain a better overview of the Italian situation and to contribute in establishing the health risk to human deriving from OTA contamination.

REFERENCES

Breitholtz A., Olsen M., Dahlbach A. and Hult K., 1991, Plasma ochratoxin A levels in three Swedish population surveyes using an ione-pair HPLC techniques. Food Additives and Contaminants 8, 183-187.

Cantafora A., Grossi M., Miraglia M. and Benelli L., 1983, Determination of ochratoxin A in coffee beans using reversed-phase high performance liquid chromatography. La Rivista della Società Italiana di Scienza dell'Alimentazione, anno 12, n°2, 103-108.

van Egmond H.P., 1991, Worlwide regulations for ochratoxin A. In Mycotoxins, Endemic Nephropathy and Urinary Tract Tumors, Ed. M. Castegnaro, R. Plestina, G. Dirheimer, I.N. Chernozemsky, and H. Bartsch. IARC scientific publications, n°115, pp.331-336.

Micco C., Ambruzzi M.A., Miraglia M., Brera C., Benelli L. and Corneli S., 1992, Evaluation of ochratoxin A level in human milk in Italy. In Proceedings of VIII International IUPAC Symposium on Mycotoxins and Phycotoxins, Mexico City, Mexico, 6-13 nov.

Micco C., Ambruzzi M.A., Miraglia M., Brera C., Onori R. and Benelli L., 1991, Contamination of human milk with Ochratoxin A. In Mycotoxins, Endemic Nephropathy and Urinary Tract Tumours. IARC Scientific Publication, n°115, pp.105-108.

Micco C., Gross M., Miraglia M., Brera C., Libanori A. and Faraoni I., 1989 a, La qualità degli ibridi nazionali di mais: corredo lipidico e contaminazione da micotossine. La Rivista della Società Italiana di Scienza dell'Alimentazione, anno 18, n°1, 29-38.

Micco C., Grossi M., Miraglia M. and Brera C., 1989 b, A study of the contamination by ochratoxin A of green and roasted coffee beans. Food Additives and Contaminants, 6, 333-339.

Micco C., Grossi M., Onori R., Chirico M. and Brera C., 1986, Aflatossina B_1 Ochratossina A e Zearalenone in mais nazionale: monitoraggio della produzione relativa agli anni 1982, 1983 e 1984. La Rivista della Società Italiana di Scienza dell'Alimentazione, anno 15, n°3, 113-116.

Ministry of Agriculture, Fisheries and Food, 1980 Survey of mycotoxins in the United Kingdom, Food Surveillance paper, n° 4.

Scott P.M., 1984, Effets of food processing on mycotoxins, Journal of Food Protection, 47, 489-499.

Tsubouchi H., Yamamoto K., Hisada K., and Udagawa S., 1987, Effect of roasting on ochratoxin A level in green coffee beans inoculated whith Aspergillus ochraceus. Mycopathologia, 97, 11-115.

Table 1 - Occurrence of Ochratoxin A in food in Italy

Food commodity	No. of samples analysed	Range of contamination (µg/kg)	No. of positive samples	Reference
Corn	111	0.1 - 1.0	39 -35%)	Micco et al. (1986)
Corn	90	1.0 - 2.0	14 (15%)	Micco et al. (1989a)
Rice	15	0.3 - 1.0	8 (53%)	Unpublished data
Durum Wheat and derived products	10	0.3 - 5.6	9 (90%)	Unpublished data
Soft Wheat and derived products	10	0.3 - 2.6	10 (100%)	Unpublished data
Bran	35	1.0 - 11.0	5 (14%)	Unpublished data
Extruded Foods	15	0.2 - 15.0	8 (53%)	Unpublished data
Green Coffee	40	0.5 - 23.0	9 (23%)	Cantafora et al. (1983)
Green Coffee	29	0.2 (15.0	17 (59%)	Micco et al. (1989b)

Table 2 - OCHRATOXIN A LEVELS IN HUMAN MILK SAMPLES AVERAGE LEVELS (No. OF REPLICATES=3)

STATUS	NOT HOSPITALIZED	HOSPITALIZED
No. of samples analyzed	81	30
No. of positive samples	13 (16%)	9 (30%)
OTA range (ng/ml)	0.7-12.0	0.1 - 1.0

RELATIVE STANDARD DEVIATION = 1.9%

Table 3 -OCHRATOXIN A LEVELS (ng/ml) IN SAMPLES COLLECTED FOR SIX DAYS (No. OF REPLICATES=3).

PATIENT N°	Days M	T	W	Th	F	S
1	2.7	1.7	21.9	0.8	0.7	0.9
2	1.6	1.3	1.8	0.7	0.3	1.0
3	0.2	0.1	ND	0.7	0.3	0.2
4	1.1	0.7	0.1	0.3	0.2	1.4
5	0.2	ND	ND	ND	ND	ND
6	ND	ND	8.5	0.1	ND	ND

ANALYZED SAMPLES N = 6 - RSD = 2.1% DL=0.1 ng/ml

Summary

The strategies adopted to assess the relevance of OTA in the human food chain in Italy were the following: a) development of programs to check the status of food contamination b) verification of the possible degradation of OTA when using the technological procedures c) analysis of biological fluids.

A preliminary survey on agricultural commodities revealed a widespread contamination, generally at low level. The technological procedures used in the processing of cocoa and coffee beans are able to destroy almost completely the toxin. The study carried out on human milk samples (each obtained from a single suck), showed a meaningful incidence of OTA contamination especially in non hospitalized donor mothers. Milk samples collected over six days consecutively revealed in two cases peaks of OTA contamination (21.8 and 8.6 ng/ml) preceded and followed by very low levels.

The results obtained encourage us to intensify multidisciplinary researches both on food and on biological fluids in order to obtain better overview on the Italian situation and, to contribute in establishing the health risk to humans occasioned by OTA contamination.

Résumé

La stratégie adoptée pour apprécier la contamination de la chaîne alimentaire de l' Homme en Italie est la suivante: mise au point de programmes de recherche pour:

- déterminer le taux de contamination de la nourriture ,
- vérifier les possibilités de destruction de la mycotoxine par des procédés industriels,
- analyser les fluides biologiques.

Une étude préliminaire sur les produits agricoles a révélé une contamination générale mais à des taux faibles. Les procédés industriels de préparation de certains aliments détruisent presque entièrement la toxine.
Les analyses de lait maternel (d'une seule traite) ont révélé la présence de l'ochratoxine A surtout chez les femmes non hospitalisées. L'analyse du lait durant 6 jours consécutifs a montré des pics d'OTA à 8,6 ng/ml et 21,8 ng/ml , précédés et suivis de valeurs nettement plus basses.
Une recherche multidisciplinaire sera nécessaire pour déterminer les risques de l'exposition de l' Homme à l'ochratoxine A en Italie.

Human ochratoxicosis and its pathologies. Eds E.E. Creppy, M. Castegnaro, G. Dirheimer. Colloque INSERM/ John Libbey Eurotext Ltd. © 1993, Vol. 231, pp. 141-145.

Human ochratoxicosis in Germany updating 1993

R.M. Hadlok

Institute of Veterinary Meat- and Foodhygiene, Justus-Liebig-University of Giessen, Frankfurter Strasse 92, Giessen, Germany

Summary

Scientific research in the Federal Republic of Germany during the 1980's showed the presence of ochratoxin A in human kidneys, milk and blood. In the kidneys, levels between, 0.1 and 0.3 μg/kg were found; in mothers milk they ranged from 0.017 to 0.03 μg/l. 56.5 to 68.3 % of blood serum samples were found to be ochratoxine A positive; (0.1 to 14.4 μg/l). So far results obtained from human material are to be considered insufficient. According to available information, there is a deficit in the field of ochratoxin A research in humans by professionals of human medicine. Research is especially necessary since the presence of ochratoxin A may be directly related to habits of consumption and clinical symptoms in humans.

In the Federal Republic of Germany, test results over the past years showed the presence of ochratoxin A in kidneys, mothers milk, and blood.
Three out of 46 human kidneys showed ochratoxin A levels between 0.1 and 0.3 μg/kg, according to Bauer et al. (1986).
Gareis et al. (1987/88) indentified ochratoxin A in 4 out of 36 samples of mothers milk; concentrations between 0.017 and 0.030 μg/l were discovered.
In human blood serum from Bavaria, examined between 1977 and 1985, 56.9 % of the samples (173 out of 306) showed concentrations from 0.1 to 14.4 μg/l. The average was 0.6 μg/l. 158 of the positive samples were found to have values between 0.1 and 0.9 μg/l (Bauer et al. 1986).

A study of 211 whole human blood samples, conducted in 1986 in Niedersachsen determined 96 (or 45.5 %) samples to be positive; ochratoxin A levels ranged from 0.1 to 0.4 μg/l (Scheuer and Leistner, 1986/87).

In Bavaria 1990 and 1991, Märtlbauer and Straka reexamined the serum of 25 people (n = 70). This study took habits of consumption into consideration. Ochratoxin A concentrations were shown to range from 0.05 to 1.4 µg/l; the median value was about 0.38 µg/l.

In Hessia 1988 (Hadlok et al. 1989) 208 samples of human blood were tested for ochratoxin A; 142 (68.3 %) tested positive for this mycotoxin. The values of the positive serum samples ranged from 0.1 to 8.4 µg/l, the median value was 1.1 µg/l. Men showed an average level of 1.0, women of 1.3 µg/l (see table 1).

Table 1: Detection of Ochratoxin A in serum of human blood

Origin Hessia	Number of exa-minated samples	Number of positive samples	Concentration (µg/l) x_{pos}	range
total	208	142 (68.3 %)	1.1	0.1 - 8.4
male	124	85 (68.5 %)	1.0	0.1 - 8.4
female	84	57 (67.9 %)	1.3	0.1 - 7.2

Limit of detection: 0.1 µg/l

The distribution pattern of ochratoxin A in the blood serum of men and women is shown in table 2.

Table 2: Prevalence of Ochratoxin A in serum of blood of men and women

Ochratoxin A µg/l	Men Number	% of total samples n = 124	% of positive samples n = 85	Women Number	% of total samples n = 84	% of positive samples n = 57
nn*	39	31.5		27	32.1	
0.1 - 0.9	37	29.8	43.5	21	25.0	36.8
1.0 - 1.9	28	22.6	32.9	22	26.2	38.6
2.0 - 2.9	7	5.7	8.2	8	9.5	14.0
3.0 - 3.9	6	4.8	7.1	2	2.4	3.5
4.0 - 4.9	4	3.2	4.7	1	1.2	1.8
≥ 5	3	2.4	3.5	3	3.6	5.3

*nn: not detectible (Limit of detection 0.1 µg/l)

In both groups, concentrations between 0.1 and 1.9 µg/l prevailed; 52.4 % of samples taken from men and 51.2 % of samples from women fell in this range.

Table 3 lists the results according to sex and age.

Table 3: Occurence of Ochratoxin A in serum of blood of men and women depending on age

Ochratoxin A µg/l	Men Number	% of total samples	% of positive samples	Women Number	% of total samples	% of positive samples
< 25 years	n = 33			n = 41		
nn*	8	24.2		16	39.0	
0.1 - 0.9	12	36.4	48.0	11	26.8	44.0
1.0 - 1.9	5	15.2	20.0	9	22.0	36.0
2.0 - 2.9	3	9.1	12.0	2	4.9	8.0
3.0 - 3.9	3	9.1	12.0	1	2.4	4.0
4.0 - 4.9	2	6.1	8.0	1	2.4	4.0
≥ 5.0	-	-	-	1	2.4	4.0
25 - 39 years	n = 65			n = 30		
nn*	22	33.8		9	30.0	
0.1 - 0.9	19	29.2	44.2	6	20.0	28.6
1.0 - 1.9	13	20.0	30.2	10	33.3	47.6
2.0 - 2.9	3	4.6	7.0	4	13.3	19.0
3.0 - 3.9	3	4.6	7.0	-	-	-
4.0 - 4.9	2	3.1	4.7	-	-	-
≥ 5.0	3	4.6	7.0	1	3.3	4.8
≥ 40 years	n = 26			n = 13		
nn*	9	34.6		2	15.4	
0.1 - 0.9	6	23.1	35.3	4	30.8	36.4
1.0 - 1.9	10	38.5	58.8	3	23.1	27.3
2.0 - 2.9	1	3.8	5.9	2	15.4	18.2
3.0 - 3.9	-	-	-	1	7.7	9.1
4.0 - 4.9	-	-	-	-	-	-
≥ 5.0	-	-	-	1	7.7	9.1

*nn: not detectable (Limit of detection 0.1 µg/l)

This table shows that as men get older the percentage of positive samples (in relation to all samples) declines, while for women the percentage increases with age. Also, as men get older ochratoxin A concentrations increasingly fall into the 0.1 to 1.9 µg/l range (increase from 68.0 to 94.1 %), while the opposite holds true for women (decrease from 80.0 to 63.7 %).

These results emphasize the necessity to pay closer attention to ochratoxin A occurence in humans, including the possible clinical and pathologic-anatomical consequences. Patients with renal symptoms should be studied in particular. It could also be of interest to organize the experiments according to habits of consumption. The German Research Association (1990) estimated that most of the ochratoxin A contamination comes from plant foods, and only about 2 % from animal foods. Meat products from pork, especially products containing pigs blood, constitute the main source of contamination. Studies from 1987 to 1988 (Hadlok et al. 1989) showed that 58.8 % of 908 slaughtered pigs had ochratoxin A in their blood serum. The concentrations ranged from 0.1 to 50.0 g/l, the median being 0.7 g/l.

Literature

Bauer, J., Gareis, M., Gedek, B. (1986): Incidence of Ochratoxin A in Blood Serum and Kidneys of man and animals. In: Proceedings of the 2nd World Congress Foodborne Infections and Intoxications, 1986, Berlin.

Deutsche Forschungsgemeinschaft (1990): Ochratoxin A, Vorkommen und toxikologische Bewertung; VCH Verlagsgesellschaft mbH, Weinheim.

Gareis, M., Märtlbauer, E., Bauer, J., Gedek, B. (1987/1988): Bestimmung von Ochratoxin A in Muttermilch. Z. Lebensm. Unters. Forsch. 186, 114-117 und 9. Mykotoxin-Workshop 1987 in Braunschweig, Biologische Bundesanstalt für Land- und Forstwirtschaft; Institut für Pflanzenschutz in Ackerbau und Grünland

Hadlok, R.M. (1989): Ochratoxinvorkommen in Blut von Schlachtschweinen und von Menschen. 11. Mykotoxin-Workshop, 1989, Berlin, Bundesgesundheitsamt Berlin, Max-von-Pettenkofer-Institut

Hadlok, R.M., Christen, U., Wiedemann, S., Moritz, Angela, Wagner, Gabriele (1989): Institut für Tierärztliche Nahrungsmittelkunde der Justus-Liebig-Universität; Gießen. Forschungsbericht, P. 1-90

Märtlbauer, E. and Straka, Margit (1992): Ochratoxin A in menschlichem Blutserum. 14.Mykotoxin-Workshop, Gießen/Schloß Rauischholzhausen; Institut für Tierärztliche Nahrungsmittelkunde der Justus-Liebig-Universität Gießen. Proceedings, P. 64-66

Scheuer, R. and Leistner, L. (1986/1987): Nachweis von Ochratoxin A in Humanblutproben aus der Bundesrepublik Deutschland. Jahresbericht 1986, C-21, Bundesanstalt für Fleischforschung Kulmbach und 9. Mykotoxinworkshop 1987 in Braunschweig. Biologische Bundesanstalt für Land- und Forstwirtschaft; Institut für Pflanzenschutz in Ackerbau und Grünland

Résumé

La recherche scientifique au cours des années 80 a montré la présence d'ochratoxine A dans les reins, le lait et le sang humain. Dans les reins, des taux compris entre 0,1 et 0,3 μg/kg ont été trouvés, alors que dans le lait maternel ces taux étaient de 0,017 à 0,03 μg/l. La fréquence de l'ochratoxicose humaine était de 56,5 à 68,3 % de la population avec des taux compris entre 0,1 et 14,4 μg/l.

Les résultats obtenus sur le matériel humain doivent être considérés comme insuffisants, car des informations de sources purement médicales font défaut.
D'autres recherches seront nécessaires étant donné que la présence d'ochratoxine dans le sang peut être directement liée aux habitudes alimentaires et à une symptomatologie chez l'homme.

Human ochratoxicosis and its pathologies. Eds E.E. Creppy, M. Castegnaro, G. Dirheimer. Colloque INSERM/ John Libbey Eurotext Ltd. © 1993, Vol. 231, pp. 147-158.

Etude de l'ochratoxicose humaine dans trois régions de France : Alsace, Aquitaine et région Rhône-Alpes

E.E. Creppy[(1)], M. Castegnaro[(2)], Y. Grosse[(3)], J. Mériaux[(4)], C. Manier[(5)], P. Moncharmont[(6)], C. Waller[(7)] et coll.

La liste complète des auteurs et leurs adresses figurent en fin d'article

Résumé: La fréquence de l'ochratoxicose humaine a été déterminée dans trois régions de France: l'Alsace, l'Aquitaine et la région Rhône-Alpes de janvier 1991 à mai 1992.
Il apparaît que cette fréquence est plus faible en France que dans les autres pays européens (moins de 20 % au total, au seuil de détection de 0,1 ng/ml). La raison de cette fréquence faible peut être la sécheresse des deux dernières années et aussi les différences possibles entre les espèces de moisissures impliquées dans les contaminations des aliments.
Cette étude a aussi révélé que les populations rurales sont plus exposées que les populations urbaines (19 à 33 % au lieu de 4 à 18 %). Cette situation est similaire à celle observée dans les régions rurales de Suède, indiquant que les habitudes alimentaires et les aliments disponibles constituent un facteur déterminant dans l'exposition aux mycotoxines en général et à l'ochratoxine en particulier.

INTRODUCTION

L'ochratoxine A (OTA) est une mycotoxine produite par Penicillium verrucosum (Frisvad et Filtenborg, 1989) et de nombreuses espèces d'Aspergillus dont la plus connue et la plus toxinogène est Aspergillus ochraceus (IARC 1993). Cette mycotoxine est néphrotoxique chez toutes les espèces animales sur lesquelles elle a été testée (Krogh 1987), immuno-suppressive (Haubeck et coll., 1981; Creppy et coll., 1983; Dwivedi et Burns, 1984;Tor, et coll. 1989); tératogène (Arora et Froëlen 1981; Hoshino et coll., 1988; Mayura et coll., 1989); génotoxique (Creppy et coll.,1985; Malaveille et coll., 1991; Pfohl-Leszkowicz et coll.,1991; Hennig et coll.,1991; Manolov et coll., 1991); cancérogène chez les rongeurs (Kanisawa et Suzuki, 1978; Bendele et coll., 1985; Boorman, 1989). Elle est suspectée d'être à l'origine de la néphropathie endémique des Balkans (Krogh, 1977) et des tumeurs du tractus urinaire qui y sont associées (Castegnaro et coll.,1987). elle est actuellement classée dans le groupe 2B (Produits considérés comme cancérogènes possibles pour l'homme) par le groupe de travail des monographies du CIRC (IARC, 1993).

L'ochratoxine A est un contaminant naturel des céréales, haricots blancs, flageolets etc... et se retrouve dans les aliments de l'homme et des animaux d'élevage (volailles, porcs etc...) (pour une revue, voir IARC, 1993).

L'ochratoxine A se lie fortement aux protéines sanguines (Hagelberg et coll.,1989) et entre dans la circulation entéro-hépatique chez l'animal (Roth et coll., 1988) ce qui allonge sa durée de vie dans l'organisme. Elle a été trouvée dans le sang non seulement des populations balkaniques dont l'exposition est très documentée (Hult et coll., 1982; Petkova-Bocharova et coll., 1988; Fuchs et coll., 1991; Petkova-Bocharova et Castegnaro; 1991) mais aussi dans le sang de nombreuses personnes d'autres pays d'Europe (Scandinavie, Allemagne, Pologne, France) ou d'Amérique du Nord (Canada) et d' Afrique (Tunisie et Algérie) (Hald, 1991; Breitholtz et coll., 1991; Bauer et Gareis, 1987; Golinski et Grabarkiewicz-Szczësna, 1989; Creppy et coll., 1991; Frohlich et coll., 1991; Creppy, 1992).

Les résultats préliminaires de l'étude de la fréquence de l'ochratoxicose humaine en France avaient montré, sur un petit nombre d'échantillons, environ 18 % d'échantillons positifs à un seuil de 0,1 ng/ml. Une étude portant sur un plus grand nombre d'échantillons a donc été entreprise dans les mêmes régions de France : Alsace, Aquitaine et Rhône-Alpes, qui nous paraissaient à titres divers, les plus exposées à une contamination par l'ochratoxine A.

La région Alsace a été sélectionnée du fait des habitudes alimentaires qui sont proches de celles de la population allemande dans laquelle une forte proportion d'ochratoxicose a été déterminée (Bauer et Gareis,1987). Dans cette région, plusieurs types de céréales sont produites qui sont utilisées pour la nourriture de porcs et de volailles et pour la fabrication de bières, qui sont des candidats potentiels pour la contamination par l'ochratoxine A. Dans la région Aquitaine (Dordogne, Gironde, Landes, Lot-et-Garonne, Pyrénées-Atlantiques) et le département voisin du Gers sont produites de grandes quantités de maïs utilisées pour nourrir les volailles de consommation directe ou destinées à la production des foies gras , magrets, etc... . Enfin en région Rhône-Alpes, l'Ain et le Rhône ont été sélectionnés comme représentatifs d'une population purement urbaine (Lyon et ses environs) ou à dominante rurale (l' Ain) qui a une forte concentration d'élevages de volailles nourries aux grains.

Materiel et Méthodes

Les échantillons de sang ou de sérum ont été obtenus dans les centres suivants: pour la région Alsace, le Centre de Transfusion Sanguine de Strasbourg; pour la région Aquitaine, le Centre des Bilans de Santé de Bordeaux et la Caisse Mutuelle Agricole de Bordeaux ainsi que le Centre de Transfusion Sanguine de Bordeaux; pour la région Rhône-Alpes, le Centre de Transfusion Sanguine de Lyon-Beynost. Tous ces échantillons ont été codés et conservés à -25°C jusqu'à l'analyse par Chromatographie Liquide Haute Performance (CLHP) avec détection spectrofluorimétrique.

Pour l'extraction, après avoir prélevé 5 ml de sérum et ajouté 40 ml de mélange $MgCl_2$/HCl (0,1 M/0,05 M)(v/v), le pH a été ajusté à 2,5 (±0,1) par ajout de HCl 1M. Le tout a été extrait par 10 ml de chloroforme en agitant vigoureusement environ 3 minutes puis en centrifugeant pour séparer les phases (environ 10 minutes à 1600 g), 8 ml de chloroforme ont été collectés. Cette phase chloroformique a été extraite par deux fois avec 8 ml d'une solution de bicarbonate de sodium 0,1 M. Les deux extraits ont été combinés et ont été ajustés à pH 2,5 (±0,1) par ajout de HCl 1M. L'OTA a été extraite de la solution précédente par 10 ml de chloroforme. Cette dernière phase chloroformique a été évaporée à sec à l'aide d'un évaporateur rotatif sous pression réduite à environ 40°C.

L'extrait sec a été repris dans 3 ml de méthanol et divisé en 3 parties aliquotes de 1 ml (la première partie servira à l'analyse de l' OTA par HPLC, la seconde à la confirmation par formation du dérivé méthylé, la troisième à la confirmation après traitement par la carboxypeptidase)

L'une des parties aliquotes a été analysée par CLHP en utilisant les conditions chromatographiques suivantes:

- colonne: 30 cm x 1/4' ODS2, 10µm
- solvant:méthanol:acétonitrile: acétate de sodium 0,005 M :acide acétique (300:300:400:14).

Ce solvant sera ajusté en fonction de la colonne pour donner un temps de rétention pour l' OTA de 8 à 10 min.

- débit de solvant: 1,5 ml/min.
- volume d'injection 50 µl
- détection par fluorimétrie: excitation (340 nm) émission (465 nm)

La quantification a été effectuée par rapport à un standard contenant de 1 à 10 ng d'ochratoxine A/ml.
Tous les résultats positifs sont à confirmer par les deux méthodes suivantes:
a) par formation de l'OTA méthyl ester et analyse par CLHP. Pour cela, évaporer le méthanol de la seconde partie aliquote, ajouter 1 ml de la solution BF_3 (trifluorure de bore), (14 %) dans le méthanol, fermer le flacon, placer dans un bain-marie à 60°C et laisser agir 15 min. Analyser en utilisant les conditions chromatographiques précédentes avec le solvant dont les proportions respectives seront de 300:300:300:13.
Comme pour l'OTA, le solvant sera ajusté en fonction de la colonne pour donner un temps de rétention de l'OTA méthylester de l'ordre de 8 à 10 min.
La quantification sera effectuée par rapport à un standard dérivé dans des conditions analogues.

b) par dégradation de l' OTA par la carboxypeptidase. Pour cela, évaporer le méthanol de la 3ème partie aliquote, et reprendre par 0,9 ml de tampon Tris-HCl 0,04 M (pH 7,5) contenant 1 mole de NaCl par litre, ajouter 100 µl d'une solution de carboxypeptidase (100U/ml) et fermer le flacon.
Incuber pendant 2 heures à 37°C puis ramener à température ambiante.
Analyser en utilisant les conditions décrites pour l'analyse de l' OTA.
Le pic de l'OTA doit avoir disparu du chromatogramme.

Résultats:

Les analyses et confirmations ont été réalisées en aveugle avant décryptage des origines rurale ou citadine des échantillons.
Les résultats sont présentés dans les tableaux de 1 à 4. Dans ces tableaux, les échantillons ont été classés suivant leurs origines en ruraux purs, citadins purs et citadins dans de petites localités rurales dans le cas de la région Rhône-Alpes où la distinction a pu être faite dans les types de citadins (tableau 4).
Dans le tableau 2 les valeurs maximales observées sont cependant des cas isolés: par exemple dans la population rurale la seconde valeur la plus élevée est de 4 ng/ml et parmi les citadins elle est de 28 ng/ml. Pour le tableau 3 (Aquitaine) la seconde valeur pour les échantillons ruraux est de 3,7 ng/ml et pour les citadins de 20 ng/ml.

Discussion:

Les résultats ci-dessus confirment ceux de l'étude préliminaire et indiquent une présence de l'ochratoxicose humaine en France.

En étudiant les résultats, région par région, nous remarquons que:

- pour la région Alsace

(i) 94 % des échantillons positifs ont des concentrations inférieures ou égales à 2 ng/ml

(ii) il y a 2,4 fois plus de ruraux que de citadins qui sont positifs

(iii) toutes les valeurs dépassant 2ng/ml sont obtenues dans les échantillons ruraux , avec un maximum de 11,8 ng/ml.

- pour la région Aquitaine: Les résultats de l'analyse des échantillons d' origine rurale non-ambiguë (Caisse mutuelle Agricole, tableau 2) montrent une prépondérance d'ochratoxicose, surtout dans les valeurs les plus élevées (> 2 ng/ml) par comparaison aux taux trouvés dans les échantillons d'origine citadine certaine (Centre de Bilan de Santé de Bordeaux, tableau 2) .

Cette différence n'est pas observée dans les résultats des analyses des échantillons provenant du Centre de Transfusion sanguine de Bordeaux qui regroupent cependant des personnes habitant aussi bien à la campagne qu'à la ville (tableau 3). Il faut cependant noter que les ruraux définis dans le tableau 2 sont des personnes travaillant à la campagne et vivant, en partie du moins, de leur propre production (céréales, volailles etc...)

Dans le tableau 3, cette catégorie est pour l'instant moins bien définie et recouvre probablement des personnes travaillant en ville, mais demeurant dans de petites agglomérations.

Une telle situation est également observée en région Rhône-Alpes (tableau 4) où on retrouve une prépondérance d'ochratoxicose en milieu purement rural par rapport au milieu purement urbain avec des taux et incidence intermédiaires pour les petites zones rurales en milieu urbain. En effet, la région Rhône-Alpes offre tous les cas de figure avec en plus des ruraux et des citadins, les citadins vivant dans les petites agglomérations rurales.

On constate en effet que:

(i) 98 % des échantillons ruraux sont compris entre 0,1 et 2 ng/ml, dont 10 % entre 1 et 2 ng/ml,

(ii) aucun citadin n'a plus de 1 ng/ml;

(iii) de tous les ruraux citadins, (tous inférieurs à 2 ng/ml), 75 % ont entre 1 et 2 ng/ml.

Les habitudes alimentaires rurales ou citadines semblent influencer très nettement le taux d'ochratoxicose humaine en France. Ceci est à rapprocher des résultats obtenus en Suède (Breitholz et coll. 1991) où le comté purement rural de Visby dans lequel les gens vivent de leur production a une prépondérance d'ochratoxicose humaine par rapport aux deux autres comtés étudiés.

La comparaison des régions entre elles fait ressortir une fréquence plus élevée des taux les plus forts en Aquitaine par rapport aux 2 autres régions en ce qui concerne les populations aussi bien rurales que citadines.
Les fréquences globales d'ochratoxicose humaine en milieu rural sont similaires dans deux régions (situées autour de 20 %) Aquitaine et Rhône-Alpes. Elles sont plus fortes cependant en région Alsace (33,8 %). Pour les populations citadines Alsace et Rhône-Alpes sont proches (environ 13 à 14 %) alors que l' Aquitaine a un taux plus élevé (17,8 %)
Les fréquences globales, toutes populations confondues, indiquent que l'Alsace a 19,6 % de positifs contre 18,7 % en Aquitaine et 14,6 % en région Rhône-Alpes. Néanmoins, l'Alsace et la région Rhône-Alpes ont la majeure partie de leurs échantillons positifs compris entre 0,1 et 2 ng/ml alors que l' Aquitaine se situe dans les valeurs plus élevées. Une étude des habitudes alimentaires dans ces régions ainsi que des taux de contamination des nourritures de base seraient nécessaires pour discuter plus en détail ces différences.
Il est cependant important de remarquer que, à niveau de détection comparable (0,1 ng/ml), la France est de tous les pays d' Europe où la fréquence de l'ochratoxicose humaine a été déterminée, celui où elle est la plus basse. Ceci peut être lié à plusieurs faits.
Les échantillons ont tous été prélevés dans les années 1991, début 1992 où une sécheresse globale prévalait en France. D'autre part, il est à noter une amélioration des conditions de séchage et de stockage des grains depuis quelques années, effort qu'il est indispensable de poursuivre. Enfin, les sources de l'ochratoxine A en France sont peut-être différentes de celles des pays de l 'Europe du Nord où les contaminations sont dues au Penicillium en général.
Une recherche des sources de contamination en France est donc nécessaire en relation avec les études des taux sanguins.

Bibliographie:

- Arora, R.G. & Fröelén, H. (1981) : Interference of mycotoxins with prenatal development of the mouse. II. Ochratoxin A induced teratogenic effects in relation to the dose and stage of gestation. - Acta Vet. Scand., 22, 535-552.
- Bauer, J. & Gareis, M. (1987) Ochratoxin in the food chain. Z. Veterinarmed.B, 34, 613-627.(in German)
- Bendele , A.M., Carlton, W.W., Krogh, P. & Lillehoj, E.B. 1985: Ochratoxin A carcinogenesis in the (C57B1/6JxC3H)F mouse. J.Natl Cancer Inst., 75,733-742.
- Boorman, G., ed. (1989) NTP Technical Report on the Toxicology and Carcinogenesis Studies of Ochratoxin A (CAS N° 303-47--9) in F344/N Rats (Gavage Studies) (NIH Publication N° 89-2813), Research Triangle Park, North Carolina, National Toxicology Program

, US Department of Health and Human Services .
- Breitholz, A., Olsen, M., Dahlback, A., et Hult, K. (1991) : Plasma ochratoxin A levels in three Swedish populations surveyed using an ion-pair HPLC technique. Food Addit. Contam., 8, 183-192.
- Castegnaro, M., Bartsch, H. & Chernozemsky, I.N., (1987) : Endemic nephropathy and urinary-tract tumors in the Balkans. Cancer Res., 47, 3608-3609.
- Creppy, E.E., Betbeder , A.M., Gharbi, A., Counord, J., Castegnaro, M., Bartsch, H., Montcharmont, P., Fouillet, B., Chambon, P., Dirheimer, G. (1991): Human ochratoxicosis in France. In : Mycotoxins, Endemic nephropathy and urinary tract tumours IARC Scientific publications N°115 Castegnaro M., Plestina R., Dirheimer G., Chernozemsky I.N. & Bartsch H. eds. Lyon:IARC pp145-151.
- Creppy , E.E., Kane , A. , Dirheimer , G. , Lafarge-Frayssinet , C., Mousset , S et Frayssinet C. (1985): Genotoxicity of ochratoxin A in mice: DNA single strand break evaluation in spleen , liver and kidney. Toxicol. Lett. 28, 29-35.
- Creppy, E.E., Stormer , F.C., Röschenthaler, R. & Dirheimer, G. (1983): Effects of two metabolites of Ochratoxin A, (4R)-4-hydroxyochratoxin A and ochratoxin alpha on immune response in mice . Infect. Immunol., 39, 1015-1018.
- Creppy, E.E. (1992) : Human ochratoxicosis in Northern Africa. correlation with cases of nephropathy. 14. Mykotoxin Workshop J. Liebig Universität Giessen.
- Dwivedi P. & Burns R.B. (1984) : Effect of ochratoxin A on immunoglobulins in broiler chicks. Res. Vet. Sci., 36, 117-121.
- Frisvad, J.C. & Filtenborg O. (1989) : Terverticilatte penicillia: chemotaxonomy and mycotoxin production. Mycologia, 81, 837-861.
- Fröhlich, R.A., Marquardt, R.R., Omainski , K.H.(1991): Ochratoxin A as a contaminant in the human food chain : a canadian perspective In : Mycotoxins, Endemic nephropathy and urinary tract tumours IARC Scientific publications N°115 Castegnaro M., Plestina R., Dirheimer G., Chernozemsky I.N. & Bartsch H.. eds. Lyon:IARC pp139-143.
- Fuchs , R., Radic B., Ceovic, S., Sostaric, B. & Hult, K. (1991) Human exposure to ochratoxin A.In : Mycotoxins, Endemic nephropathy and urinary tract tumours IARC Scientific publications N°115 Castegnaro M., Plestina R., Dirheimer G., Chernozemsky I.N. & Bartsch H.. eds. Lyon:IARC pp131-134.
- Golinski, P. & Grabarkiewicz-Szczesna, J. (1985): The first polish cases of the detection of ochratoxin A residues in human blood. Rocz. Panstw. Zakl. hyg., 36, 378-381.
- Hagelberg, S., Fuchs, R. & Hult, K., (1989): Toxicokinetics of ochratoxin A in several species and its plasma-binding properties. J. Appl.Toxicol. 9, 91-96.
- Hald, B. (1991) : Ochratoxin A in human blood in European countries In : Mycotoxins, Endemic nephropathy and urinary tract

tumours IARC Scientific publications N°115 Castegnaro M., Plestina R., Dirheimer G., Chernozemsky I.N. & Bartsch H.. eds. Lyon:IARC pp159-164.

- Haubeck, H.D., Lorkowski, G., Kölsh, E. & Röschenthaler, R. (1981) : Immunosuppression by ochratoxin A and its prevention by phenylalanine. Appl. environ. Microbiol., 41: 1040-1042.

- Hennig , A., Fink-Gremmels, J. et Lestner, L. (1991): Mutagenicity and effects of ochratoxin A on frequency of sister chromatid exchange after metabolic activation In : Mycotoxins, Endemic nephropathy and urinary tract tumours IARC Scientific publications N°115 Castegnaro M., Plestina R., Dirheimer G., Chernozemsky I.N. & Bartsch H. eds. Lyon:IARC pp255-260.

- Hietanen, E., Malaveille, C., Camus A.M., Béréziat, J.C., Brun, G., Castegnaro, M., Michelon, J., Idle, J.R. & Bartsch, H. (1986) Interstrain comparison of hepatic and renal microsomal carcinogen metabolism and liver S9-mediated mutagenicity in DA and Lewis rats phenotyped as poor and extensive metabolizers of debrisoquine. Drug Metab. Disposition, 14, 118-126.

- Hoshino, K. , Fukui, Y., Mayasaka , I., & Kameyama, Y. (1988) : Developmental disturbance of cerebral cortex of mouse offspring from dams treated with ochratoxin A during pregnancy. Cong. Anom. 28, 287-294.

- Hult, K., Plestina, R., Habazin-Novak, V., Radic, B & Ceovic , S. (1982) Ochratoxin A in human blood and Balkan endemic nephropathy. Arch. Toxicol., 51, 313-321.

- IARC (1993): IARC Monograph on the evaluation of carcinogenic risk to humans: some naturally occuring substances: some food, tons and constituents, Heterocyclic Aromatic Amines and Mycotoxins In IARC ed ; Ochratoxin A. IARC Monograph N° 56. Lyon: IARC (sous presse)

- Kanisawa , M. et Suzuki, S. (1978) : Induction of renal and hepatic tumours in mice by ochratoxin A , a mycotoxin. Gann 69, 599-600.

- Krogh P. (1980): Ochratoxins: occurence, biological effects and causal role in disease. In Natural toxins, Eaker D. & Wadstrom, T.eds. Oxford, Pergamon Press pp 673-680.

- Krogh , P. ,Hald, B., Plestina, R. & Ceovic, S. (1977): Balkan endemic nephropathy and food born ochratoxin A : preliminary results of survey of food stuffs. Acta Pathol. Microbiol. Scand. sect B 85, 238-240.

- Malaveille , C., Brun, G. et Bartsch H. (1991): Genotoxicity of ochratoxin A and structurally related compounds in Escherichia coli strains : studies on their mode of action. In : Mycotoxins, Endemic nephropathy and urinary tract tumours IARC Scientific publications N°115 Castegnaro M., Plestina R., Dirheimer G., Chernozemsky I.N. & Bartsch M. eds. Lyon: IARC pp261-266.

- Manolov , G., Manolova, Y. Castegnaro, M., Chernozemsky, I.N. (1991) : Chromosomal alterations in lymphocytes of patients with Balkan endemic nephropathy and healthy individuals in vitro with

ochratoxin A . In: Mycotoxins, Endemic nephropathy and urinary tract tumours. IARC Scientific publications N°115 Castegnaro M., Plestina R., Dirheimer G., Chernozemsky I.N. & Bartsch eds Lyon IARC pp267-272.
- Mayura, K., Edwards, J.F., Maull, E.A. & Phillips , T.D.(1989): the effects of ochratoxin A on postimplantation rat embryos in culture. Arch. Environ. Contam. Toxicol., 18, 411-415.
- Petkova-Bocharova, T., Chernozemski, I.N. & Castegnaro, M. (1988) Ochratoxin A in human blood in relation to Balkan endemic nephropathy and urinary system tumours in Bulgaria. Food Addit. Contam. , 5, 299-301.
- Petkova-Bocharova ,T., Castegnaro, M. (1991): Ochratoxin A in human blood in relation to endemic nephropathy and urinary tract tumours in Bulgaria.In : Mycotoxins, Endemic nephropathy and urinary tract tumours IARC Scientific publications N°115 Castegnaro M., Plestina R., Dirheimer G., Chernozemsky I.N. & Bartsch H. eds. Lyon:IARC pp135-137.
- Pfohl-Leskowicz, A., Chakor, K., Creppy, E.E. & Dirheimer G. (1991) : DNA-adducts formation in mice treated with ochratoxin A . In : Mycotoxins, Endemic nephropathy and urinary tract tumours IARC Scientific publications N°115 Castegnaro M., Plestina R., Dirheimer G., Chernozemsky I.N. & Bartsch H. eds. Lyon:IARC pp245-253.
- Roth , A., Chakor, K., Creppy, E.E., Kane, A., Röschenthaler , R. , Dirheimer , G. (1988): Evidence for an enterohepatic circulation of ochratoxin A in mice . Toxicology, 48, 293-308.
- Tor, L., Steien , K., Stormer , F.C. (1989) : Mechanism of ochratoxin A induced immunosuppression. Mycopathologia, 107, 153-159.

Tableau 1: Fréquence de l'ochratoxicose humaine en Alsace

ORIGINE DES ECHANTILLONS	RURAUX		CITADINS	
	230 echantillons		270 echantillons	
Taux de contamination ng/ml	nb de positif	% positif	nb de positif	% positif
0,1 - 0,2	32	22,1	23	8,5
> 0,2 - 1	18	7,8	6	2,22
> 1 - 2	3	1,3	9	3,3
> 2*	6	2,6	0	0
TOTAL	59	33,6	38	14,02

* maxima détectés chez les ruraux: 11,8 ng/ml; chez les citadins: 1,7 ng/ml

Tableau 2: Fréquence de l'ochratoxicose humaine en Aquitaine.
Echantillon du centre de santé pour les citadins et Mutuelle Agricole pour les ruraux

ORIGINE DES ECHANTILLONS	RURAUX		CITADINS	
	212 echantillons		853 echantillons	
Taux de contamination ng/ml	nb de positif	% positif	nb de positif	% positif
0,1 - 0,2	8	3,8	27	3,2
> 0,2 - 1	6	2,8	49	5,7
> 1 - 2	11	5,2	36	4,2
> 2*	22	10	40	4,7
TOTAL	47	22,2	152	17,8

* maxima détectés chez les ruraux: 130 ng/ml; chez les citadins: 120 ng/ml

Tableau 3: Fréquence de l'ochratoxicose humaine en Aquitaine (echantillon du CTS

ORIGINE DES ECHANTILLONS	RURAUX		CITADINS	
	201 echantillons		789 echantillons	
Taux de contamination ng/ml	nb de positif	% positif	nb de positif	% positif
0,1 - 0,2	4	1,9	24	3
> 0,2 - 1	23	11,4	67	8,5
> 1 - 2	6	2,9	21	2,7
> 2*	4	1,9	35	4,4
TOTAL	37	18,4	149	18,9

* maxima détectés chez les ruraux: 8,4 ng/ml; chez les citadins: 161 ng/ml

Tableau 4: Fréquence de l'ochratoxicose humaien en région Rhône-Alpe.

ORIGINE DES ECHANTILLONS	RURAUX		CITADINS		CITADINS en milieu rural	
	309 echantillons		176 echantillons		30 echantillons	
Taux de contamination ng/ml	nb de positif	% positif	nb de positif	% positif	nb de positif	% positif
0,1 - 0,2	24	7,8	6	3,4	1	3,3
> 0,2 - 1	29	9,4	5	2,9	0	0
> 1 - 2	6	1,9	0	0	3	10
> 2*	1	0,3	0	0	0	0
TOTAL	60	19,4	11	6,3	4	13,3

* maxima détectés chez les ruraux: 4,3 ng/ml; chez les citadins: 1 ng/ml

Summary

The prevalence of human ochratoxicosis has been determinated in three regions of France,Alsace, Aquitaine and Rhône-Alpes from january 1991 to may 1992.
It appears that this prevalence is lower in France as compared to other European countries (less than 20 % in total at the 0.1 ng/ml level). The reason of this low prevalence could be the dryness of the last two years and possibly the species of the toxigenic fungi involved in the contaminations.
This study also reveals that the rural populations are more exposed than the urban ones (19 to 33 % instead of 4 to 18 %) in the Swedish rural areas indicating that the eating habits and prevaling commodities are determinant in the exposure to mycotoxins in general and especially to ochratoxins.

Auteurs

E.E. CREPPY[1], A.M. BETBEDER[1], A. GHARBI[1], M.F. GAURET[1], M. ANDRIEUX[1], J. COUNORD[1]
1. Laboratoire de Toxicologie et d'Hygiène Appliquée. U.F.R. des Sciences Pharmaceutiques, Université Bordeaux II, 3 ter, place de la Victoire, 33076 Bordeaux Cedex

H. BARTSCH[2], M. CASTEGNARO[2], B. FOUILLET[8], P. CHAMBON[8]
2. International Agency for Research on Cancer, Lyon
8. U.F.R. des Sciences Pharmaceutiques, Lyon

Y. GROSSE[3], A. PFOHL-LESZKOWICZ[3], G. DIRHEIMER[3]
3. IBMC du CNRS et U.F.R. des Sciences Pharmaceutiques, 15, rue Descartes, 67084 Strasbourg Cedex

J. MERIAUX[4a], M. GUIOT-GUILLAIN[4a], C. DOUET [4b]
4a. Centre de Bilan de Santé, Caisse d'Assurance Maladie, Bordeaux
4b. Caisse Mutualité Sociale Agricole de la Gironde

C. MANIER[5], P. GIACOMOTTO[5], G. VEZON[5]
5. Centre Régional de Transfusion Sanguine de Bordeaux

P. MONCHARMONT[6]
6. Centre Régional de Transfusion Sanguine de Lyon-Beynost

C. WALLER[7], D. LAUSTRIAT[7], J.P. CAZENAVE[7]
7. Centre Régional de Transfusion Sanguine de Strasbourg

Human ochratoxicosis and its pathologies. Eds E.E. Creppy, M. Castegnaro, G. Dirheimer. Colloque INSERM/ John Libbey Eurotext Ltd. © 1993, Vol. 231, pp. 159-165.

Ochratoxin A in livestock and human sera in Japan quantified by a sensitive ELISA

Osamu Kawamura[(1)], Sanae Maki[(1)], Shoichiro Sato[(2)] and Yoshio Ueno[(1)]

[(1)]Cellular and Molecular Biology, the Research Institute for Biosciences, Chiba and Department of Toxicology and Microbial Chemistry, Faculty of Pharmaceutical Sciences, Science University of Tokyo, Ichigaya, Tokyo 162 and [(2)]Ueda Meat Inspection Station, Ueda, Nagano, Japan

Summary

A sensitive and specific enzyme-linked immunoassay (ELISA) method for ochratoxin A (OTA) has been developed with an anti-OTA monoclonal antibody (mAb) OTA.7 and horseradish peroxidase (POD) as enzyme label. Serum samples were adjusted pH to 3, extracted with chloroform, and the OTA extract dissolved in 10% methanol-phosphate buffer saline was subjected to ELISA analysis with the detection limit of 10 pg/ml serum. OTA in the ELISA-positive extracts was confirmed by HPLC analysis.

All 11 dairy cattle sera sampled in December 1991 were positive of OTA with an average of 116 pg/ml, and one out of 5 cattle sera sampled in January 1992 was contaminated with 10 pg/ml. All 10 swine sera were positive of OTA with an average of 2,059 (min. 880 - max. 4,500) pg/ml. These findings revealed that the sera of swine were contaminated with OTA widely in Nagano areas, and the OTA serum level in swine was significantly higher than that of cattle, in agreements with our previous survey in 1989.

In 20 human sera (15 males and 5 females) sampled in 1992 at Tokyo from healthy volunteers, one female was negative while 19 subjects were positive of OTA with an averages of 103 (males) and 89 (females) pg/ml. It is the first report demonstrating the presence of OTA in human sera in Japan.

INTRODUCTION

Ochratoxin A (OTA), produced by various species of Aspergillus and Penicillium, is one of contaminants in agricultural commodities and feeds, resulting in health hazards in man and livestock. After the findings that OTA is carcinogenic to kidneys and liver in experimental animals (Kanisawa and Suzuki, 1978; Bendele et al., 1985; Boorman, 1988) and this mycotoxin is presumed as a causal factor for urinary system tumors in the endemic area of Balkan nephropathy (Castegnaro et al., 1991), the exposure to OTA is one of great concerns in health authorities. For monitoring the OTA exposure, the detection of OTA in cereals and food is extensively performed, along with the analysis of OTA in serum and edible tissues, by using HPLC and ELISA, as currently reviewed by Scott (1992).

In order to establish the biomonitoring system for OTA exposure, we have developed several monoclonal antibodies (mAb) specific to OTA (Chiba et al., 1985), established the specific ELISA for OTA and applied for livestock sera (Kawamura et al., 1989). The survey performed in 1989 has firstly demostrated the presence of OTA in swine sera in Japan (Kawamura et al., 1990).

In the present experiment, we have further continued the survey of OTA residue by ELISA in livestock sera sampled in 1990-91 from the areas near by those performed before (Kawamura et al., 1990). Furthermore, we have surveyed the residue of OTA in human serum to demonstrate the direct evidence for human exposure to this carcinogenic mycotoxin in Japan.

MATERIALS AND METHODS

Serum samples:

11 and 5 sera from dairy cattle (in total 16 samples) were respectively obtained in Tateshina district, Nagano Prefecture, in December 1990 and in Yamanashi in January 1991. Ten swine sera were sampled in Matsumoto, Nagano Prefecture. The 20 human sera were sampled in 1992 from healthy volunteers in Tokyo. All serum samples were stored at -80℃ until analysis.

Extraction of OTA from serum:

For ELISA, to 6 ml of the serum samples 0.5 ml of acetic acid was added and the samples were left at room temperature for 20 min. The mixtures were then diluted with 6 ml of phosphate buffered saline (PBS), extracted by shaking with 12 ml of chloroform for 1 min, and centrifuged at 3,000 rpm for 10 min. The chloroform layer (3 ml) was evaporated to dryness, and the residue was dissolved in 300 µl of PBS/Tween containing 10% methanol. For HPLC analysis, the serum (12 ml) was mixed with 10 ml of 0.5M H_3PO_4-2M NaCl solution (pH 1.6), shaken at room temperature for 1 hr, and the whole was shaken with 10 ml of chloroform for 30 min, and centrifuged at 3,000 rpm for 10 min. The chloroform layer was evaporated to dryness; the residue was dissolved in 5 ml of chloroform, and extracted with 5 ml of 0.1M $NaHCO_3$ solution by shaken for 1 min. The aqueous layer was adjusted to pH 2 with formic acid, and extracted with 2 ml of chloroform which being evaporated to dryness. The residue was then dissolved in 60 µl of acetonitrile.

Indirect competitive ELISA and HPLC:

The ELISA analysis with the anti-OTA mAb OTA.7, horseradish peroxidase (POD)-labeled second antibody and 3,3,5,5-tetramethylenebenzidine (TMBZ) as substrate was carried out as described previously (Kawamura et al., 1990). The detection limit of OTA was 10 pg/ml. Confirmation of OTA was carried out with HPLC (Shimadzu LC-10AD) fitted with a fluorescence detector (Shimadzu CTO-6A) and ODS column. The mobile phase was composed of acetonitrile-0.1% H_3PO_4 (1:1. v/v), and the fluorescence detector was set to 330 and 460 nm wavelengths for excitation and emission respectively.

RESULTS AND DISCUSSION

OTA contents in the sera of livestocks sampled in 1990 and 1991 are presented in Table 1. The data indicated that 10 swine had OTA positive serum with a mean value estimated to 2,059 pg/ml (minimum, 880 and maximum, 4,500 pg/ml). Our previous survey of OTA in swine demonstrated that most samples were OTA positive with the mean values of 400 (lot A, 19 samples in Ueda district in 1988), 362 (lot B, 124 samples in Iida district in 1985) and 5,201 (lot C, 17 samples in Ueda district in 1988).

Table 1. ELISA analysis of serum OTA in livestock.

	Serum OTA (pg/ml)	
Swine[1)]	Cattle[2)]	Cattle[3)]
1,900	250	ND*
980	87	ND
1,940	130	ND
3,000	108	ND
920	134	ND
2,300	48	10
980	100	
3,100	40	
880	112	
4,500	180	
	92	
(av.) 2,059	116	10

1) Matsumoto district in October 1990
2) Tateshina district in December 1990
3) Yamanashi district in January 1991
* not detected (detection limit; 10 pg/ml)

Therefore, the sera levels of OTA in the swine sampled at Matsumoto district, near by the above districts, were closed to those of lots A and B. It suggested high frequency of OTA residue in swine in different districts in Nagano Prefecture.

As for 11 diary cattles sampled in Tateshina in 1990, the mean value of serum OTA was estimated to 116 pg/ml, with the minimum of 40 and maximum of 250 pg/ml. While, among 5 cattles sampled in Yamanashi in 1991, only one serum was positive for OTA (10 pg/ml), and the other being below the detection limit. We also reported previously that 4 cattles sampled in Ueda district in 1988 were negative for OTA. These findings suggested that the serum contents of OTA in cattles were far less than that in swine, and differed among the areas surveyed. Further surveys of OTA in feeds and edible tissues are underway.

With an aim to investigate the possibility of the presence of

OTA in human serum, sera from 20 healthy volunteers were sampled in 1992 and analyzed by the above mentioned ELISA.

Table 2. OTA in human serum sampled in 1992.

Sex		OTA (pg/ml)				
		ELISA	ELISA	HPLC	ELISA	HPLC
		(July 6)	(Oct. 27)		(Dec. 18)	
Male	1	-	142	323		
	2	101	132	109		
	3	54	278	252		
	4	20	185	132		
	5	-	113	121		
	6	88				
	7	88				
	8	76				
	9	80				
	10	76				
	11	114				
	12	23				
	13	26				
	14	100				
	15	16				
		(av. 98)	(123)	(187)		
Female	1	95	127	165	250	167
	2	92				
	3	98				
	4	18				
	5	ND				
		(av. 87)				

As shown in Table 2, among 5 female sera, only one sample was below the detection limit, and the other 4 were positive for OTA with the average of 87 pg/ml (minimun 18 pg/ml and maximun 98 pg/ml). All 13 sera in the male were positive for OTA with the mean value of 98 (min. 16 - max. 114) pg/ml.

In order to confirm the ELISA data, the HPLC and ELISA analyses were performed on the 6 sera (4 from the same subjects and 2 additionals) sampled 3 months later. The data showed that the values obtained by both the biological and chemical methods are closely associated each other (see Table 2). Additional experiment on the female serum sampled further 2 months later demonstrated the presence of OTA (see Table 2).

The residue of OTA in human serum was already reported in several countries such as Denmark (Hald, 1988), Bulgaria (Pektova-Bocharova et al., 1988) and Canada (Frohlich et al., 1991). Although the level of OTA in human serum in Japan was far less than those reported in the endemic areas of Balkan nephropathy and others, the present trial with a highly sensitive ELISA method was the first to demonstrate the residue of OTA, a carcinogenic mycotoxin, in human serum in Japan. Further survey is underway to confirm the present finding on serum residue of OTA and its origin for human exposure.

REFERENCES

Bendele, S. A., Carlton, W. W., Krogh, P. and Lillehoji, E. B. (1985): Ochratoxin A carcinogenesis in the (C57BL/6YxC3HF)1 mouse. J. Natl. Cancer Inst., 75:733-739.

Boorman G. (1988): Technical report on the toxicology and carcinogenesis studies of ochratoxin A in F334/N rats. National Toxicology Program, NIH Publication No. 88-28/3, Triangle Park, North Crolina: US Department of Health and Human Service.

Castegnaro, M., Plestina, R., Dirheimer, G.,Chernozemsky, I. N. and Bartsch, H. eds (1991): Mycotoxins, Endemic Nephropathy and Urinary Tract Tumours, Lyon, IARC Scientific Publication.

Chiba, J., Kajii, H., Kawamura, O., Ohi, K., Morooka, N. and Ueno, Y. (1985): Production of monoclonal antibodies reactive with ochratoxin A: enzyme-linked immunosorbent assay for detection of ochratoxin A, Proc. Jpn. Assoc. Mycotoxicol., 21: 28-29.

Frohlich, A. A., Marquardt, R. R. and Ominski, K. H. (1991): Ochratoxin A as a contaminant in the human food chain: a Canadian perspective. In Mycotoxins, Endemic Nephropathy and Urinary Tract tumours, eds M. Castegnaro, R. Plestina, G. Dirheimer,I N. Chernozemsky and H. Bartsch, pp. 139-143. Lyon, IARC Scientific Publication.

Hald, B. (1989): Human exposure to ochratoxin A, in Mycotoxins and Phycotoxins '88., eds S. Natori, K. Hashimoto and Y. Ueno, pp. 57-65. Amsterdam: Elsevier.

Kanisawa, M. and Suzuki, S. (1978): Induction of renal and hepatic tumors in mice by ochratoxin A, a mycotoxin. Gann, 69: 599-600.

Kawamura, O., Sato, S., Kajii, H., Nagayama, S. Ohtani, K. Chiba, J. and Ueno, Y. (1989): A sensitive enzyme-linked immunosorbent asssay of ochratoxin A based on monoclonal antibodies, Toxicon, 27: 876-897.

Kawamura, O., Sato, S., Nagura, M., Kishimoto, S., Ueno, I., Sato, S., Uda, T., Ito, Y. and Ueno, Y. (1990): Enzyme-linked immunosorbent assay for detection and survey of ochratoxin A in livestock sera and mixed feeds. Food & Agric. Immunol., 2: 135-143.

Petkova-Bocharova, T., Chernozemsky, I. N. and Castegnaro, M. (1988): Ochratoxin A in human blood in relation to Balkan endemic nephropathy and urinary systems tumours in Bulgaria, Food Addit. Contami. 5: 299-301.

Scott, P. M. (1992): Mycotoxins, J. AOAC Intern., 75: 95-102.

Résumé

Une méthode immunoenzymatique sensible, ELISA, a été développée pour la recherche de l'ochratoxine A (OTA) à l'aide d'un anticorps monoclonal anti-OTA OTA.7 couplé a la péroxyde de raitfort. Les échantillons de sérum ont été ajustés à un pH de 3, extraits avec du chloroforme et le résidu d'OTA redissous dans une solution constituée de 10% de méthanol dans un tampon phosphate pH 7.2 avant l'analyse par ELISA. La limite de détection est de 10 pg/ml de sérum. Les échantillons OTA positifs sont reconfirmés par CLHP.

11 sérums de bovin sur 11 étaient positifs en décembre 1991 avec des taux de 116 pg/ml. En décembre 1992, 1/5 des échantillons des bovins est positif à nouveau, alors que 10/10 des sérums de porcs sont positifs avec un taux minimal de 880 pg/ml et un taux maximal de 4,500 pg/ml.

Ces résultats montrent que le sérum de porc est plus contaminé que celui de bovins, comme nous l'avions trouvé en 1989 dans la région de Nagano.

En ce qui concerne l'ochratoxicose humaine en 1992, 19 échantillons sur 20 étaient contaminés avec des taux de 103 pg/ml pour les hommes de 89 pg/ml pour les femmes. C'est la première fois que l'ochratoxicose humaine est mise en évidence au Japan.

Human ochratoxicosis and its pathologies. Eds E.E. Creppy, M. Castegnaro, G. Dirheimer. Colloque INSERM/ John Libbey Eurotext Ltd. © 1993, Vol. 231, pp. 167-174.

Estimating human exposure to ochratoxin A in Canada

T. Kuiper-Goodman[(1)], K. Ominski[(2)], R.R. Marquardt[(2)], S. Malcolm[(1)], E. McMullen[(1)], G.A. Lombaert[(3)] and T. Morton[(4)]

[(1)]*Health Protection Branch, Health and Welfare Canada, Food Directorate, Ottawa, Ontario, K1A 0L2;* [(2)]*University of Manitoba, Faculty of Agricultural and Food Sciences, Department of Animal Science, Winnipeg, Manitoba, R3T 2N2;* [(3)]*Health Protection Branch, Health and Welfare, Winnipeg, Manitoba, R2J 3Y1;* [(4)]*Health Protection Branch, Health and Welfare, Burnaby, B.C. V5G 4P2, Canada*

Summary

Ochratoxin A (OA) is a mycotoxin which has been found to occur in foods of plant origin, in edible animal tissues, as well as in human blood sera and milk. The ability for OA to move up the food chain is aided by its long half-life in certain edible animal species. The major target for OA toxicity in all mammalian species is the kidney. In addition, OA is teratogenic and immunotoxic. Recent studies have provided 'clear evidence' for the carcinogenicity of OA in two rodent species. In a 2-year carcinogenicity study in rats conducted by the National Toxicology Program, National Institutes of Health in the USA (NTP, 1989; Boorman et al., 1992), OA was found to be a potent renal carcinogen, and on the basis of that study, the estimated tolerable daily intake (TDI) was 1.5 to 5.7 ng/kg bw, depending on the method of extrapolation used (Kuiper-Goodman, 1990; 1991). The Health Protection Branch has recently initiated several surveys of cereal food products, meat products and pig sera. Based on preliminary results of these surveys, worst case estimates of human exposure to OA from the consumption of pork-derived food products, breakfast cereals and bread in Canada were <3.5 ng/kg bw/day. Average exposure was <1.6 ng/kg bw. No association between renal disease and OA levels in human serum were found in a 2-year survey conducted in the Province of Manitoba. In that survey, mean human serum levels of OA were 0.25 ng/ml and 1.3 ng/ml in 1990 and 1991, respectively. The latter value corresponds to an intake of 1.7 ng/kg bw/day. Based on the estimated range of TDI's, these mean exposure estimates are not deemed to represent a health hazard.

Introduction

Careful assessment of human exposure to OA is an important part of the overall risk assessment and risk management of this mycotoxin. In addition such assessments are useful in ecological (population based) studies, designed to investigate a possible association between certain types of renal disease and exposure to OA. In the various surveys, a large proportion of the data has been below the detection limit for OA. In trying to estimate average human exposure from such results, a variety of statistical procedures have been used.

Methods

Human sera
During 1990, human blood samples (n=159) were collected from male and female renal patients and non-renal patients or volunteers at the Health Sciences Centre and at St Boniface Hospital in Winnipeg, Manitoba (Frohlich et al., 1991). No feedback regarding results for OA was provided to the persons sampled, and during 1991, blood was collected from 60 of the same renal and non-renal patients. Samples were stored at -70^{0}C, prior to analysis. Serum was extracted using the method of Hult et al. (1979), and samples were analyzed by reversed phase HPLC, with fluorescence detection. Selected samples were later confirmed using an enzymatic method (Ominski et al., manuscript in preparation). During 1990, the detection limit (DL) was 0.1 ng OA/ml; this increased to 0.5 ng/ml the following year, due to a deterioration in the performance of the detector.

Pig sera
A national survey of OA in pig sera was made during May, June and July of 1990. Blood samples were collected from slaughterhouses by Agriculture Canada, and analyzed by the Health Protection Branch (HPB). Serum was precipitated with methanol, then extracted using the method of Hult et al. (1979). The extract was analyzed by reversed phase HPLC with fluorescence detection (HPB LPFC-157 method with modification). The DL was 0.5 ng/ml. Recovery was about 80%. Confirmations were made by methyl esterification using boron trifluoride methanol.

Organ meats
A national survey of OA in pork kidney, pork liver and chicken liver was conducted by HPB during 1990, 1991, and 1992, with samples collected at the retail level. The method used silica column cleanup (Scott et al., 1991). Recovery was 77 to 112%. The DL was 0.5 ng/g, except for 1990, when the DL was 1.0 ng/ml. Analytical problems were encountered in the analysis of blood sausage and tongue, and such products were deleted from the survey, which has not yet been completed.

Cereal Food Products
A national survey of 70 breakfast cereals (wheat, oat, and corn based products) was conducted by HPB during 1991-1992 and 1992-1993. The survey also included 4 samples of other cereal food products (corn flour, 1; corn meal, 3). Samples were analyzed by the HPB LPFC-157 method (Scott et al., 1991) as for organ meats. The DL was 0.5 ng/g. Recovery was 46-95%. This survey is ongoing.

Intake estimates of OA
Intake estimates of OA, based on HPB/Agriculture Canada pig sera, were calculated, as previously, from similar data for the province of Manitoba (Kuiper-Goodman and Scott, 1989), using the relationships between pork serum data and OA residues in edible pork tissues established from "carry over" studies by Mortensen et al.(1983). Estimates of pork derived food products were made for 1-4 year olds and 12-19 year-old males (eaters only, i.e. those eating the food commodity of concern) based on the Nutrition Canada Survey (Health and Welfare, 1976). Intake estimates for other age groups, relative to body weight, are generally lower. In the same way, food intake estimates for a variety of breakfast cereals (dry weight basis) were made for the same two age groups, and multiplied by the level of OA found in those commodities.

Statistical Methods

In order to estimate the overall mean of OA levels, a non-parametric method was used. The empirical cumulative distribution function (ECDF) of the non-censored data was calculated and a fifth-degree polynomial fit to $-\log_e(1-\text{ECDF})$. The polynomial was fit without an intercept and with the value $-\log_e(1-(nc+\frac{1}{2})/n)$ at the DL, where nc is the number of censored observations. The probability density function and the expected value of the censored observations were estimated from the polynomial. The estimated overall mean was calculated as a weighted average of the estimated expected value of the censored observations and the mean of the non-censored observations. For comparison, extreme means were also calculated, whereby values of 0 or DL were assigned to censored observations. Such values should also be examined in any interpretation of the data. Medians have also been reported as an indicator of average exposure, due to the occurrence of large values which tend to have undue influence on the mean.

Results

Human Sera in Manitoba

OA was detected in sera from both renal and non-renal patients. There was no evidence that the frequencies of detected values differed between the two groups, so that the results were pooled (Table 1). There were significant differences ($p<0.05$, sign test) in ochratoxin levels between 1990 and 1991 based on 60 patients observed in both years. In 32 patients, levels in both years were <0.5. In 23 of the remaining 28 patients, levels in 1991 were higher than levels in 1990. During 1990, one sample had a high value (35.3 ng/ml). Although this value is considered to be a true value, it was decided to also calculate the group mean with it deleted. This same person had a value below the DL in the following year. The overall estimated mean was 0.25 ng/ml serum in 1990, and 1.29 ng/ml in 1991. Using the formula of Breitholz et al. (1991), it was estimated that with a mean serum level of 1.29 ng/ml, the average intake of OA from the consumption of food products would have been 1.72 ng/kg bw/day. During 1991, 11 of 60 persons sampled had serum values of OA above 3.14 ng/ml, which similarly would indicate intakes exceeding 4.2 ng/kg bw/day.

Table 1. Summary of Estimated and Calculated Means (ng/ml) of OA in Human Sera Collected in Manitoba During 1990 and 1991.

Year	n	DL	# >DL	Mean of observations above DL[a]	Estimated mean[b]	Calculated mean[c]	Overall median
1990	159	0.1	63	1.13 (0.16-35)	0.47	0.45-0.51	0.07
1990	158[d]	0.1	62	0.58 (0.16-5.2)	0.25	0.23-0.29	0.07
1990	59[d,e]	0.1	37	0.49 (0.16-3.7)	0.31	0.31-0.34	0.22
1991	60	0.5	23	3.10 (0.82-9.0)	1.29	1.19-1.50	0.3

[a]. Values in brackets are the range of positive values; [b]. Estimated using a non-parametric approach; [c]. calculated using 0 and DL for censored observations; [d]. excluding one sample containing 35.3 ng OA/ml; [e]. sera from these patients were also analyzed in 1991.

HPB and Agriculture Canada Survey of Pig Sera

The mean and median OA levels from a nation-wide survey of pig sera, collected during 1990, are shown in Table 2. The two highest values observed, 24.4 and 76.5 ng OA/ml, occurred in Ontario. The results indicate that exposure to OA from feed ingredients does occur through-

Table 2. Levels of Ochratoxin A in Pig Sera by Region During 1990

Region of Origin	n	No. > 0.5 <10	>10 <20	>20	Levels in ng/ml: Mean of observat-ions above DL	Overall estimated mean[a]	Median
Atlantic	19	14	1	0	3.2	2.5 (2.5-2.6)	0.8
Quebec	60	22	0	0	1.2	0.6 (0.4-0.7)	0.4
Ontario	60	19	1	2	6.6	2.5 (2.4-2.8)	0.3
Central	30	21	2	0	2.8	2.1 (2.1-2.2)	1.2
Western	30	17	2	0	2.9	1.9 (1.9-2.0)	0.8
TOTAL	199	93	6	2	3.3	1.8 (1.7-1.9)	0.5

[a] Based on a non-parametric approach; values in brackets calculated using 0 and DL for censored observations.

out Canada. It will be necessary to repeat the survey over a number of years in order to see possible effects of fluctuations in climatic conditions affecting mold growth and the occurrence of OA in Canada.

HPB Survey of Organ Meats
During 1990, 3/42 pork kidney samples (Table 3), 0/28 pork liver samples, and 0/24 chicken liver samples were found to contain OA at levels greater than the DL (1 ng/g). In a nation-wide survey, conducted during 1991-92 and 1992-93, 20 pork liver and 20 chicken liver samples were negative. However, eight of 44 pork kidneys analyzed (DL=0.5 ng/g) were found to contain between 0.5 and 5.2 ng/g OA. The overall mean and median values are shown in Table 3.

Table 3. Levels of Ochratoxin A in Pork Kidneys in Canada by Region and Year of Survey[a]

Year	Region of origin	n	No. with levels above the DL[b]	Overall Mean in ng/g[c]	Median in ng/g
1990[d]	Atlantic	3	1 (7.2)		
	Quebec	6	0		
	Ontario	6	0		
	Central	27	2 (3.4;3.6)	0.5 (0.3-1.2)	0.20
	Total	42	3	0.5 (0.3-1.3)	0.13
1991	Atlantic	3	0		
	Quebec	6	0		
	Ontario	6	0		
	Central	6	2 (0.8;2.4)	0.6 (0.5-0.9)	0.21
	Western	7	5 (0.8;1.3;1.5;1.5;5.2)	1.5 (1.5-1.6)	1.3
	Total	28	7	0.6 (0.5-0.9)	0.24
1992	Atlantic	3	1 (0.5)		
	Quebec	6	0		
	Central	7	0		
	Total	16	1	<0.5	

[a] Survey not yet completed; [b] values in brackets are actual values; [c] Based on a non-parametric approach; values in brackets calculated using 0 and DL for censored observations, respectively. [d] During 1990 the DL was 1.0 ng/g; the following years the DL was 0.5 ng/g.

HPB Survey of Mainly Breakfast Cereal Food Products
Four of 70 samples analyzed were above the DL (0.5 ng/g) with levels ranging from 0.8-15 ng/g. The overall mean and median values are shown in Table 4. Values above the DL were from 2 whole wheat cereals, 1 mixed cereal, and 1 sample of corn flour.

Estimates of intake of OA in Canada
The estimated intake of OA from pork derived food products, based on pig serum levels, was less than 0.2 ng/kg bw (Table 5). The intake from products containing pork blood as the major ingredient would be higher, but the extent to which these products are consumed in Canada is unknown.

For the intake estimates of pork kidney we have used the value of 1.5 ng OA/g from the western region, since the majority of values here were above the DL. The intake values of OA for the "eaters only" of the 1-4 year olds and 12-19 year-old males are 1.56 and 1.95 ng/kg bw, respectively (Table 5). Using the relationship of Mortensen et al. (1983), estimates were also made for the intake of pork liver and loin based on the higher pork kidney level of 1.5 ng/g found in the western region during 1991 (Table 5). These estimates were about 10-fold higher than those based on the 1990 nation-wide pork serum levels.

Table 4. Levels of Ochratoxin A in Mainly Breakfast Cereal Food Products in Canada by Region and Year of Survey [a]

Year	Region of origin	n	# with levels >0.5 ng/g[b]	Mean in ng/g[c]	Median in ng/g
1991	Atlantic	5	0		
	Quebec	10	1 (1.4)		
	Ontario	20	1 (0.8)		
	Central	5	0		
	Western	10	0		
	Total	50	2	0.13(0.0-0.5)	0.1
1992	Atlantic	5	1 (15.0)		
	Quebec	10	1 (1.6)		
	Central	5	0		
	Total	20	2	0.90(0.8-1.3)	0.1
1991/ 1992	Total	70	4	0.35(0.3-0.7)	0.1

[a.] Survey not yet completed; [b.] values in brackets are actual values.
[c.] Values in brackets are calculated means, using 0 and DL for censored observations, respectively.

For the intake estimates for Canada from mainly breakfast cereals, we have used the mean residue level of OA over the 2-year survey period, since except for one high value, there was not much difference in incidence of OA between the two years. Using a value of 0.35 ng/g as the mean level from the combined 2-year survey data (Table 4), the intake estimates of OA for the two respective age-groups are 0.93 and 0.41 ng/kg bw (Table 5).

The overall "worst case" intake estimates of OA from these foods were made by adding the "eaters only" estimates for the consumption of pork derived food products (western Canada) and mainly breakfast cereals, thus assuming that the same person consumes both food commodities on the same day. In this way, the intake estimates of OA were 2.9 and 2.5 ng/kg bw for 1-4 year-olds and 12-19 year-old males, respectively.

Discussion

The human serum surveys in Manitoba, and the nation-wide surveys of pig serum and certain food commodities indicate and confirm that low-level human exposure to OA occurs in Canada. Indeed, if the relationship of Breitholz et al. (1990) is considered valid, then 18.3% of the human sera analyzed during 1991 indicated exposure that exceeded the TDI of 4.2 ng/kg bw/day, used in Canada. Further random nation-wide surveys of human serum, over a number of years are needed.

Our previous exposure estimate of OA from pork derived food products in Canada was < 1 ng/kg bw/day for both 1-4 year-olds and 12-19 year-old males, and was based on a mean level of 9.2 ng/ml of OA in pork sera collected in slaughter houses in the Province of Manitoba (Kuiper-Goodman and Scott, 1989). In the same paper, an intake of 4.5 and 3.5 ng/kg bw/day was estimated for 1-4 year-olds and 12-19 year-old males (mean of eaters), based on a 3-year national survey (1976-1979) in which 6/315 mainly cereal grains had OA levels above the DL (2.5 ng/g) with a range of 3-8 ng/g. By assigning a value of ½ DL to the censored observations, and presuming no reductions from processing, a mean OA level of 1.47 ng/g for cereal based foods (i.e. bread) was estimated. In 1991, using the same cereal survey data, a lower value was assigned to the censored observations, based on a presumed gamma distribution, giving an estimated mean residue level of 0.27 ng/g and the overall intake estimate from bread was reduced to 0.54 and 0.40 ng/kg bw/day for the same two age groups (Kuiper-

Table 5. Intake Estimates of Ochratoxin A from the Consumption of Pork Derived Food Products in Western Canada and Mainly Breakfast Cereals in Canada.

Food commodity	Age group	Food intake in g/day.[a]	Level of OA in food in ng/g	Intake per person in ng/day	Intake in ng/kg bw/day
Pork serum estimates[b]					
-kidney	12-19 M	70	0.12	8.2	0.15
-liver	12-19 M	188.3	0.05	8.9	0.17
-loin	12-19 M	119	0.06	7.4	0.14
Pork kidney	1-4	15.0	1.5	22.5	1.56
"	12-19 M	70.0	1.5	105.0	1.95
Pork kidney estimates -liver[c]	1-4	17.6	0.60	10.6	0.74
"	12-19 M	188.3	0.60	113.0	2.10
-loin[c]	1-4	38.0	0.75	28.5	1.98
- "	12-19 M	119	0.75	89.3	1.66
Breakfast cereals	1-4	38.4	0.35	13.4	0.93
"	12-19 M	62.8	0.35	22.0	0.41

[a] mean of "eaters only" consumption group; [b] estimates for kidney, liver and loin estimated from pig serum level of 1.8 ng/ml, using relationship of Mortensen et al. 1983 [c] estimates for liver and loin estimated from pork kidney level of 1.5 ng/g, using the relationship of Mortensen et al. 1983

Goodman, 1991). These estimates agree well with the present estimates from breakfast cereals.

The HPB nation-wide survey of pig sera, conducted in 1990 indicated that pig serum levels were generally lower than those observed in Manitoba during 1986 (Marquardt et al.,1988) and 1989/1990 (Frohlich et al., 1991). The intake estimates, calculated on the basis of possible edible tissue residues of OA, were <0.2 ng/kg bw for both 1-4 year-olds and 12-19 year-old males (Table 5). Further nation-wide surveys are needed over several years, to confirm that indeed exposure of pigs to OA in Canada is generally low. Furthermore, the pig sera data may also serve as an indicator of potential human exposure. The levels of OA encountered in our pork kidney survey were low, when compared to similar surveys made in some European countries (Hald, 1991). Intake estimates from pork consumption alone could be up to about 2 ng/kg bw/day.

The results presented here are still preliminary, since the surveys have not yet been completed over a sufficient number of years, and do not adequately cover all susceptible food commodities, such as other wheat products (bread, pastries), beans, coffee, and breast milk. Some of the survey results involved small sample sizes for individual regions and food commodities, which makes it difficult to draw conclusions about year to year or region to region comparisons.

The estimation procedure, for dealing with values below the detection limit, is being developed as an alternative to the widespread use of ½ DL for such observations. This method appeared to work well for the data described here. It is currently being refined and tested using both real and simulated data.

One of the objectives of this Symposium is to compare exposure to OA and occurrence of renal disease. No ecological or other types of epidemiological studies, other than the Manitoba study reported here, have been conducted in Canada to see if exposure to OA is associated with the occurrence of renal disease. In Canada as a whole, the incidence rate of renal cancer (ICD9-189.0,1,2)(age 0-85+), age adjusted to the 1971 Canadian population, has gradually increased from 6.67 (males) and 3.18 (females) per 100,000 in 1970 to 11.39 (males) and 5.73 (females) per 100,000 in 1988 (HPB, Bureau of Chronic Disease Epidemiology, 1993, unpublished tabulations). Similarly, mortality rates, standardized to the 1971 Canadian population, have increased from 3.80 (males) and 1.80 (females) per 100,000 in 1970 to 4.61 (males) and 2.29 (females) per 100,000 in 1990. Similar trends have been observed in other western countries. The increases have been attributed, at least in part, to cigarette smoking. No reliable trends in morbidity rates for other types of renal disease were available, since the available data are based on hospital separations and therefore could have included more than one visit per individual and different types of treatment over the 18 year period. For 1989, the respective morbidity rates for renal disease (age=0-85+) per 100,000, for males and females, were 179.0 and 202.6.

Overall, the "worst case" mean intake estimates of OA for the "eaters only" exposure groups from pork and cereal food products, including bread (based on the 1976-1979 survey), are 3.4 and 2.9 ng/kg bw/day for 1-4 year-olds and 12-19 year-old males, respectively. For comparison, average exposure for the same two age groups based on the "all person" mean intake for all pork and cereal food products was also estimated, and was 1.5 and 1.1 ng/kg bw, respectively. For these latter values an average kidney level of 0.6 ng/g for 1991, and a residue level of 0.35 ng/g in cereal food products was used. Since these estimates fall within the range of estimated TDI's, exposure to OA at these mean levels is not considered to represent a health hazard. Nevertheless, the human sera data indicated higher exposure in some individuals. Therefore, because of the toxic properties and probable long half-life of OA in humans, exposure to this mycotoxin, including peak exposures, should be kept to a minimum.

Acknowledgements

The authors would like to acknowledge and thank Dr. A. Frohlich, Department of Animal Science, University of Manitoba, Dr. F. Madrid, Department of Food Science, University of Manitoba, Dr. J. Manfreda, Health Sciences Centre, Drs. J. McKenzie and A. Fine, St. Boniface Hospital, for their involvement in the human sera project, and G. Nixon, Field Operations Directorate, Health Protection Branch, for coordinating the HPB surveys. The critical review of the manuscript by Dr. P. Scott, L. Vavasour, G. Boulton and J. Salminen is also gratefully acknowledged.

References

Boorman, G. A., McDonald, M. R., Imoto, S., and Persing, R. (1992): Renal lesions induced by ochratoxin A exposure in the F344 rat. *Toxicol. Pathol.* **20**: 236-245.

Breitholtz, A., Olsen, M., Dahlback, A., and Hult, K. (1991): Plasma ochratoxin A levels in three Swedish populations surveyed using an ion-pair HPLC technique. *Food Addit. Contam. 8*: 183-192.

Frohlich, A. A., Marquardt, R. R., and Ominski, K. H. (1991): Ochratoxin A as a contaminant in the human food chain: A Canadian perspective. In *Mycotoxins, Endemic Nephropathy and Urinary Tract Tumours*, eds Castegnaro, M., Plestina, R., Dirheimer, G., Chernozemsky, I. N., Bartsch, H. pp. 139-143. Lyon: International Agency for Research on Cancer,IARC.

Hald, B. (1991): Porcine nephropathy in Europe. In *Mycotoxins, Endemic Nephropathy and Urinary Tract Tumours*, eds Castegnaro, M., Plestina, R., Dirheimer, G., Chernozemsky, I. N., Bartsch, H. pp. 49-56. Lyon: International Agency for Research on Cancer,IARC.

Health and Welfare Canada (1976):Nutrition Canada Food Consumption Patterns Report. Ottawa: Department of Supply and Services.

Hult, K., Hokby, E., Hagglund, U., Gatenbeck, S., Rutqvist, L., and Sellyey, G. (1979): Ochratoxin A in pig blood: Method of analysis and use as a tool for feed studies. *Appl. Environ. Microbiol. 38*:772-776.

Kuiper-Goodman, T. (1991): Risk assessment of ochratoxin A residues in food. In *Mycotoxins, Endemic Nephropathy and Urinary Tract Tumours*, eds Castegnaro, M., Plestina, R., Dirheimer, G., Chernozemsky, I. N., Bartsch, H. pp. 307-320. Lyon: International Agency for Research on Cancer,IARC.

Kuiper-Goodman, T. (1990): Uncertainties in the risk assessment of three mycotoxins: aflatoxin, ochratoxin A, and zearalenone. *Can. J. Physiol. Pharmacol. 68*: 1017-1024.

Kuiper-Goodman, T., and Scott, P M. (1989): Risk assessment of the mycotoxin ochratoxin A. *Biomed. Environ. Sci. 2*: 179-248.

Marquardt, R. R., Frohlich, A. A., Sreemannarayana, O., Abramson, D., and Bernatsky, A. (1988): Ochratoxin A in blood from slaughter pigs in western Canada. *Can. J. Vet. Res. 52*:186-190.

Mortensen, H. P., Hald, B., and Madsen, A. (1983): Feeding experiments with ochratoxin A contaminated barley for bacon pigs. 5. Ochratoxin A in pig blood. *Acta Agric. Scand. 33*:235-239.

NTP (National Toxicology Program) (1989): NTP Technical Report on the Toxicology and Carcinogenesis studies of ochratoxin A (CAS No. 303-47-9) in F344/N rats (gavage studies). ed Boorman, G.A, pp. 1-141. Research Triangle Park, NC: US Department of Health and Human Services, National Institutes of Health.

Scott, P. M., Kanhere, S. R., Canela, R., Lombaert, G. A., and Bacler, S. (1991): Determination of ochratoxin A in meat by liquid chromatography. *Prehrambeno-tehnol. Biotehnol. Rev. 29*: 61-64.

III. Ochratoxicosis and associated pathologies in human and animal

III. Ochratoxicose et pathologies associées chez l'homme et l'animal

Human ochratoxicosis and its pathologies. Eds E.E. Creppy, M. Castegnaro, G. Dirheimer. Colloque INSERM/ John Libbey Eurotext Ltd. © 1993, Vol. 231, pp. 177-187.

Is the oxidative pathway implicated in the genotoxicity of ochratoxin A ?

A. Pfohl-Leszkowicz[1], Y. Grosse[1], A. Kane[1, 3], A. Gharbi[2], I. Baudrimont[2], S. Obrecht[1], E.E. Creppy[2] and G. Dirheimer[1]

[1]Institut de Biologie Moléculaire et Cellulaire du Centre National de la Recherche Scientifique and Université Louis Pasteur, 15 rue Descartes, 67084 Strasbourg, France; [2]Laboratoire de Toxicologie et d'Hygiène Appliquée, Université Bordeaux II, 3 ter, place de la Victoire, 33076 Bordeaux, France; [3]Institut de Technologie Alimentaire, BP 2765, Dakar, Sénégal

Summary:

In order to check if OTA lipid peroxidation ability is linked to its genotoxic effect, expressed by DNA adducts formation in mice on kidney and testicle DNA, we tested the oxidative pathways in OTA metabolism and the possibilities of glutathione conjugaison.

SOD and catalase given prior OTA administration induced a decrease of the total DNA-adducts in kidney and testicles, by 90% and 30% respectively. Thus these adducts were probably produced by peroxidative pathway. Phoron given to mice depleted the glutathione level by 66% and 11% respectively in kidney and testicles after 4h. This substance given prior to OTA administration led to the decrease by 70% at 8h and 70% at 24h of total DNA-adducts level in kidney and testicles respectively. N-acetylcysteine which also depleted glutathione level by 34% at 8h in kidney and by 14% at 24h in testicles, led also to decrease of DNA-adducts. These results indicate that glutathione plays a prominent role in the genotoxicity of OTA through conjugation of OTA (or its metabolites) with glutathione or through the pathway of the oxidoreduction properties of glutathione. In addition, incubation of homologue DNA with OTA and kidney microsomes, which contained high amounts of peroxidases, led to DNA-adducts formation. This last result reinforced the importance of the peroxidasic pathway to be responsible for formation of the majority of DNA-adducts induced by OTA in kidney and testicles.

Introduction:

Ochratoxin A (OTA) has been shown to have a number of toxic effects, the most prominant being nephrotoxicity. It induces in animals, a tubulo intersticial nephropathy similar to Balkan endemic nephropathy (BEN) in human (Krogh *et al*, 1977, 1979; Petkova-Bocharova *et al*, 1991). Dietary feeding of ochratoxin A has been shown to induce also renal adenomas and hepatocellular carcinomas in mice (Kanizawa and Suzuki, 1978; Kanizawa, 1984; Bendele *et al*, 1985) and in rats (Boorman, 1989). The carcinogenicity of OTA was also suspected in human because of the high incidence of kidney, pelvis, ureter and urinary bladder carcinomas among patients suffering from Balkan endemic nephropathy where OTA is implicated (Castegnaro *et al*, 1987; Ceovic *et al*, 1992; Vukelic *et al*, 1992). The covalent binding of chemicals or their reactive metabolites to DNA is generally believed to be a key step in the initiation of chemically induced carcinogenesis (Miller and Miller, 1969). It has been shown that OTA induces DNA single-strand breaks (Creppy *et al*, 1985, Kane *et al*,

1986) and many DNA-adducts in several animal tissues. They were completely repaired in a time depending manner except in kidney where several adducts were still detected after 16 days. (Pfohl-Leszkowicz *et al*, 1991, 1993a,b). The metabolites of OTA implicated in the genotoxicity are not known.

Rahimtula et al. have recently shown that OTA enhances NADPH dependent lipid peroxidation in kidney and liver *in vitro* and *in vivo*. (Rahimtula *et al*, 1988; Omar *et al*, 1990; Omar *et al*, 1991). The peroxidation of polyunsaturated fatty acids present in the membrane lipids has been proposed as a mechanism by which a number of foreign compounds produce structural tissue injury (Halliwell and Gutteridge, 1988; Slater, 1984). Transition metal are involved in lipid peroxidation (Halliwell and Aruoma, 1991). Recently, Omar *et al* (1991) have shown that OTA induces lipid peroxidation by chelating Fe^{3+} ions. Furthermore, cytochrome P450 stimulates the OTA induced lipid peroxidation. Superoxide dismutase (SOD) removes reactive oxygen by converting it into hydrogen peroxide; this enzyme works in conjonction with two enzymes, catalase and glutathione peroxidase (GSH-Px). This last enzyme is the most important in removing H_2O_2, by oxidizing the reduced glutathione (GSH) into oxidized glutathione (GSSG). It has been shown that SOD and catalase inhibit nephrotoxicity of cyclosporin A in rats. (Wolf *et al*, 1992). We therefore studied the implication of the above quoted oxydative pathways in the OTA induced genotoxicity. Another OTA metabolism pathway which could lead to genotoxic derivatives is the conjugation of the parent compound or one of its metabolites to glutathione. We therefore tested the effect of phorone (depletor of glutathione) and N-acetylcysteine (precursor of glutathione) on the formation of DNA adducts by OTA.

Materials and Methods:

Chemicals:

Ochratoxin A, proteinase K, ribonucleases A and T1, N-acetylcysteine, superoxide dismutase and catalase were obtained from Sigma (St.Louis, MO, USA), T4 polynucleotide kinase from PL Biochemicals (Milwaukee, WI, USA), micrococcal nuclease and spleen phosphodiesterase from Worthington Biochemicals (Freehold, NJ, USA), nuclease P1 from Boehringer (Mannheim, Germany), [γ^{32}P] ATP, 5000 Ci/mmole from Amersham (Buckinghamshire, England), phoron from Aldrich, Steinheim, Germany.

Animals treatment :

Swiss male mice (Sexal, Vigneul sous Montmédy, France) weighing 25 ± 2g, aged seven weeks, were given OTA (in 0.1 M $NaHCO_3$ pH 7.4) 0.2 to 2.5 mg/kg body weight by gastric intubation. SOD and catalase in 0.2ml distilled water were administered (20 mg/kg body weight each) by subcutaneous injection, one hour before OTA administration. Phoron and N-acetylcysteine dissolved in olive oil were given *per os* 1h before OTA administration and every 12h (250 and 500 mg/kg respectively). Mice were killed by decapitation after 48h. Kidney and testicles were excised and frozen at -80°C until further processing.

Glutathione and OTA contents determination:

Glutathione levels were determined as described by Ecobichon (1984). OTA content was determined as described by Creppy *et al*, 1991.

*Preparation of microsomes:*Microsomes from kidney, liver and testicles were obtained as described by Monod *et al* (1988). Briefly, organs were homogenized with a potter-Elvehjem in a buffered solution (0.15 M KCl, 50 mM phosphate buffer pH 7.4). The microsomal fraction was obtained by differential centrifugations and all microsomal pellets were resuspended in a solution of 50 mM phosphate buffer pH 7.4 containing 0.15 M KCl, 1 mM EDTA, 1 mM DTT and 20% glycerol. These fractions were stored at -80°C until use.

^{32}P-postlabelling of DNA adducts:

The DNAs were extracted and purified as described previously (Pfohl-Leszkowicz *et al*,

1991). The method used for ^{32}P postlabelling was that previously described by Reddy and Randerath (1986) with minor modifications. Briefly, DNA (6 µg) was digested at 37°C for 4h with micrococcal nuclease (183 mU) and spleen phosphodiesterase (12 mU) in a reaction mixture (total volume 10 µl) containing 20 mM sodium succinate and 10 mM $CaCl_2$, pH 6.

Digested DNA was treated with nuclease P1 (6 µg) at 37°C for 45 min. before ^{32}P-postlabelling. Normal nucleotides and excess of ATP were removed by chromatography on polyethyleneimine-cellulose plates in 2.3 M NaH_2PO_4 pH 5.7 (D1) overnight. Origin areas containing labelled adducted nucleotides were cut out and transferred on to another polyethyleneimine-cellulose plate, run in 4.77 M lithium formate and 7.65 M urea pH3.5 (D2). Two further migrations (D3 and D4) were performed perpendicularly to D2. The solvent for D3 was 0.6 M NaH_2PO_4 and 5.95 M urea pH 6, and the solvent for D4 was 1.7M NaH_2PO_4 pH 6. Autoradiography was carried out at -80°C for 24 or 60h exposure in the presence of an intensifying screen (Cronex). Spots were scrapped off and their radioactivity was counted by the Cerenkov technique.

Results and Discussion:

In order to test oxidative pathways in OTA metabolism and possibilities of conjugation with glutathione, the following compounds were given to mice before and after OTA treatment : superoxide dismutase and catalase (which are known to be oxidative radicals scavenger), N-acetylcysteine (NAC) (which is a precursor of glutathione synthesis *in vivo*) and phoron (which is a depletor of glutathione).

Kidney and testicles which seem to be the main target of OTA in mice (Kane *et al*, 1986; Gharbi *et al*, 1993) were excised at different times for determination of OTA and glutathione contents, and DNA-adducts.

Figure 1 and Table I show that SOD and catalase given together to mice 1h before OTA reduced total DNA-adducts levels in kidney and testicles.

Table 1: Total DNA-adduct level

	OTA	OTA + SOD + cat.	OTA + phor	OTA + NAC
Kidney				
8h	5.4[a]	Nd	1.1	1.7
24h	40.6	Nd	*	*
48h	103.1	10	0	*
testicles				
8h	2.4	Nd	2.5	2.4
24h	8.3	Nd	2.3	2.54
48h	14.7	11.7	0.3	2.3

[a]*Results are expressed as modified nucleotides per* 10^9 *nucleotides. * limit of detection*

In kidney, this reduction is of about 90% whereas in the testicles it is only of 30%. In fact, most of the previous DNA-adducts completely disappeared in testicles, while new adducts appeared (mentionned by arrows on the picture, fig.1H) in mice treated by SOD and catalase. This result is indicative of modified metabolic pathways for the production of the ultimate OTA reactive entities in presence of SOD and catalase. The DNA adducts which disappeared upon the action of SOD and catalase were probably produced by pathways implicating peroxidation or lipoperoxidation.

In order to check the action of glutathione on OTA genotoxicity, we first studied the ability of phoron and N-acetylcysteine to modify the glutathione level in the different organs.

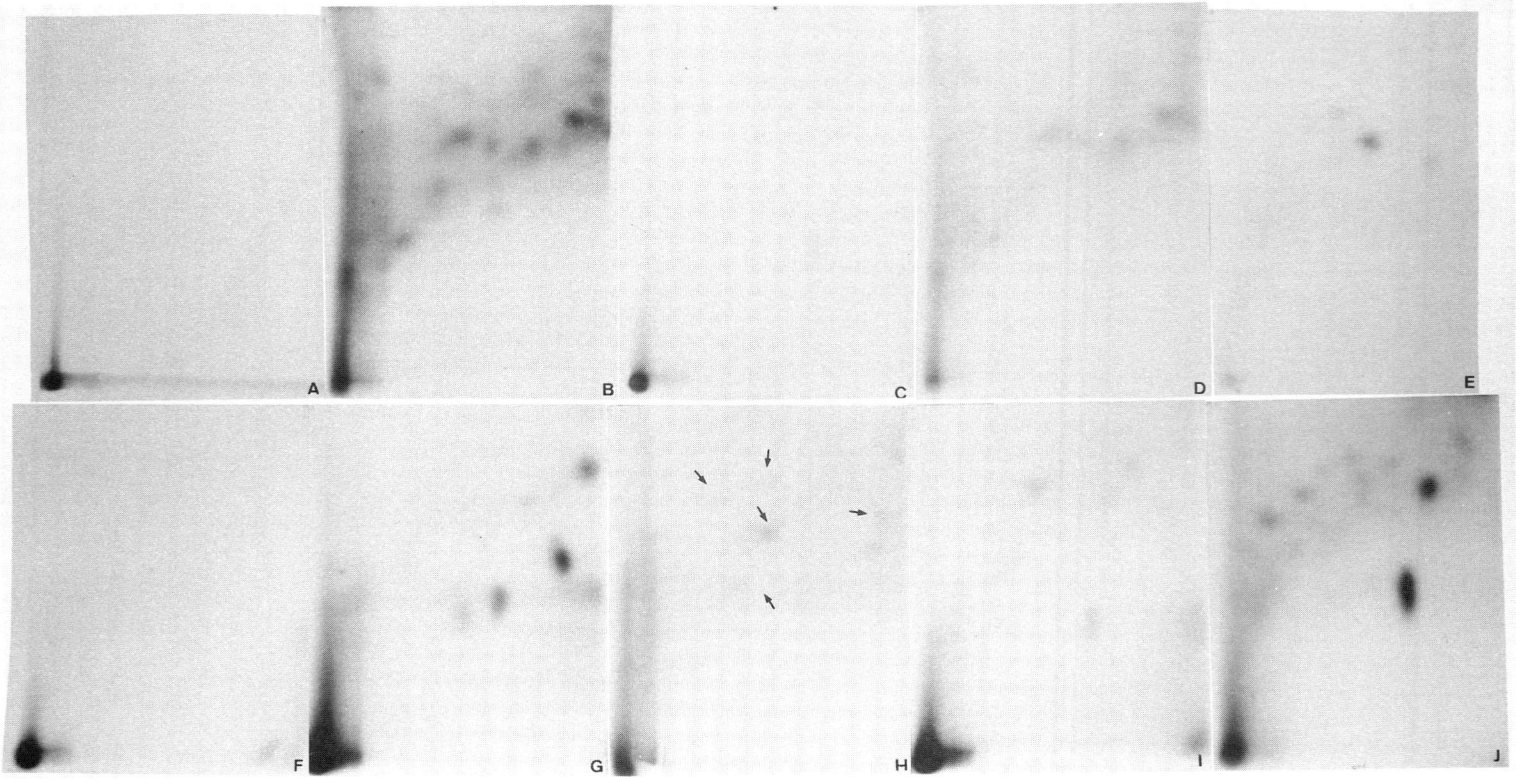

Fig. 1: Autoradiography of DNA-adducts in mice kidney and testicle after OTA treatment. A to E correspond to kidney DNA-adducts and F to J to testicle DNA-adducts. Non-treated mice (A, F); mice treated with 8ppm of OTA (B, G); mice pretreated one hour before OTA administration : by SOD and catalase (C, H); by phorone (D, I) and by NAC (E, J).

Table 2: Glutathione level in mice kidney, testicles and liver after administration of phoron and N-acetylcysteine.

	phoron i.p.	phoron per os	phoron per os	OTA + phor.	NAC per os	NAC per os	OTA + NAC
Kidney		*one dose*	*every 12h*		*one dose*	*every 12h*	
0h	1240 ± 50	1240 ± 50	1240 ± 50	1240 ± 50	1240 ± 50	1240 ± 50	1240 ± 50
1	625 ± 50	400 ± 50	Nd	400 ± 50	1250 ± 50	Nd	1250 ± 50
4	370 ± 20	425 ± 50	Nd	425 ± 50	975 ± 10	Nd	975 ± 10
8	1110 ± 100	1075 ± 50	1020 ± 50	700 ± 50	880 ± 50	850 ± 50	825 ± 50
24	Nd	Nd	850 ± 50	750 ± 10	Nd	800 ± 30	750 ± 10
48	Nd	Nd	800 ± 20	750*	Nd	820 ± 30	700*
testicles							
0	925 ± 100	925 ± 100	925 ± 100	925 ± 100	925 ± 100	925 + 100	925 ± 100
1	875 ± 100	1000 ± 20	Nd	1000 ± 20	1050 ± 50	Nd	1050 ± 50
4	638 ± 20	825 ± 50	Nd	825 ± 20	1080 ± 50	Nd	1080 ± 50
8	606 ± 50	810 ± 50	810 ± 50	650 ± 70	900 ± 50	900 ± 50	800 ± 10
24	Nd	Nd	850 ± 50	700 ± 10	Nd	650 ± 70	725 ± 70
48	Nd	Nd	850 ± 50	700*	Nd	700 ± 10	750*
Liver							
0	1425 ± 25	1425 ± 25	1425 ± 25	1425 ± 25	1425 ± 25	1425 ± 25	1425 ± 25
1	480 ± 50	500 ± 50	Nd	500 ± 50	2213 ± 200	Nd	2213 ± 200
4	525 ± 10	525 ± 50	Nd	525 ± 50	2750 ± 300	Nd	2750 + 300
8	1062 ± 60	625 ± 50	625 ± 50	580 ± 50	1280 ± 200	1280 ± 100	1575 ± 200
24	Nd	Nd	925 ± 70	800 ± 50	Nd	1700 ± 100	2113 ± 250
48	Nd	Nd	1300 ± 100	900*	Nd	1625 ± 100	1825*

*Results are expressed as µg glutathione/g organ. * One animal died.*

Table 2 shows that in liver and in kidney, phoron given *per os* decreased significantly the glutathione level. The decrease after one hour was of 70% and 65% respectively in kidney and liver. This decrease was maintained in kidney until 4h and in liver until 8h. The glutathione level increased again from 4h in kidney. In the testicles, no decrease was observed after 1h, the glutathione level in this organ decreased by 13% only after 4h and 20% after 8h. For maintaining a reduced glutathione level, particularly in kidney and testicles, phoron was given every 12 h (i.e. at 12h, 24h and 36h). Table 2 also shows that the oral administration led to a lower level of glutathione in liver than the i.p. administration. We therefore preferred oral route of administration. In addition, OTA administration did not significantly affect the glutathione content of the three organs of the phoron treated animals (Table 2). Pretreatment of mice by phoron one hour before OTA poisoning led to the decrease of DNA-adducts in the two organs analyzed (kidney and testicles) (fig.1 and table 1). The decrease was of 70% in kidney after 8h, whereas in testicles the total DNA-adducts decreased significantly only after 24h by about 70%. The shift in time for the reduction of DNA-adducts level in testicles is probably due to the lack of depletion of glutathione by phoron when OTA was given to mice, so in this tissue, the decrease of DNA-adducts occured only after the depletion of glutathione. As phoron decreased OTA-DNA-adducts formation, it must be that glutathione plays a prominent role in genotoxicity of this toxin. This role could be achieved through conjugation of GSH with OTA or its metabolites or by the pathway of the oxido-reduction properties of glutathione. (see below)

NAC when given *per os* to mice increased the glutathione level in liver until 4 hours after injection. Unexpectively, NAC decreased the glutathione level in kidney by 22%, 34% and 40% respectively after 4h, 8h and 24h as compared to control (Table 2). NAC also decreased the total DNA-adducts level in kidney by 80% at 8h and completely afterwards. However, in the testicles, the glutathione level did not vary significantly within 1 to 8 hours. In addition, no modification of DNA-adduct level could be seen in testicles at 8h, confirming the correlation between glutathione level and DNA-adduct formation. Afterwards, since NAC was given every 12h, a decrease of glutathione (25% at 24h) and a simultaneous decrease of DNA-adducts in this tissue occured (Table 2). Thus there was a clear correlation between glutathione level and DNA-adduct formation which is in favor of the hypothesis of genotoxicity of OTA-glutathione or OTA metabolites-glutathione conjugates *in vivo*.

In general, conjugation with glutathione is regarded as a detoxication process of xenobiotics (Jakoby and Habig, 1980). However, conjugation with GSH has been implicated in the activation of xenobiotics to mutagenic and carcinogenic electrophilic compound (Van Bladeren *et al*, 1980;1981; Vadi *et al*, 1985, Dekant and Vamvakas, 1989, Lafleur and Retèl, 1993) and evidence is accumulating that GSH conjugates of a variety of compounds are nephrotoxic (for a review see Monks and Lau, 1987; Dekant *et al*, 1988). Two major classes of toxic GSH-conjugates are known : those that are toxic without further metabolism (liver S9 independent acting) and those that must be metabolized further to express toxicity. GSH-conjugates undergo renal metabolism to the corresponding cysteine S-conjugates by the action of γ-glutamyltransferase and dipeptidase activities on the renal brush border membrane or in the peritubular capillaries. These conjugates are transported into the renal cells. (Ishikawa, 1992) Inside the cells, cysteine S-conjugates have three alternatives fates: (i) they may be secreted unchanged into the plasma (ii) they may be N-acetylated to form the corresponding mercapturic acids (Jakoby and Habig, 1980) or (iii) they may undergo β-lyase catalyzed reaction to yield potentially toxic thiols (for reviews see, Monks and Lau, 1987; Dekant *et al*, 1989). Another explanation of the decrease of adducts in kidney and testicles DNA after the treatment of NAC could be its antioxidant activity towards superoxide anions (De Flora *et al*, 1984, 1985, 1986, 1991; Halliwell, 1991; Halliwell and Gutteridge , 1984).

The importance of the oxidative pathway in DNA adduct formation by OTA was reinforced by *in vitro* experiments: OTA (20µM-100 µM) was incubated with microsomes from kidney and testicles for 1h. Microsomes from renal tissue produced several DNA-adducts which were OTA dose dependant (fig.2) whereas no DNA-adducts were found with testicles microsomes (not shown). Peroxidases were found in large amount in kidney microsomes, but not in testicles microsomes. All together these results pointed at the peroxidasic pathway to be responsible for the formation of the majority of DNA-adducts. This was confirmed by the fact that no adducts were found after liver microsomes incubation with OTA and DNA even if these microsomes contained high amounts of cytochromes P450 (results not shown).

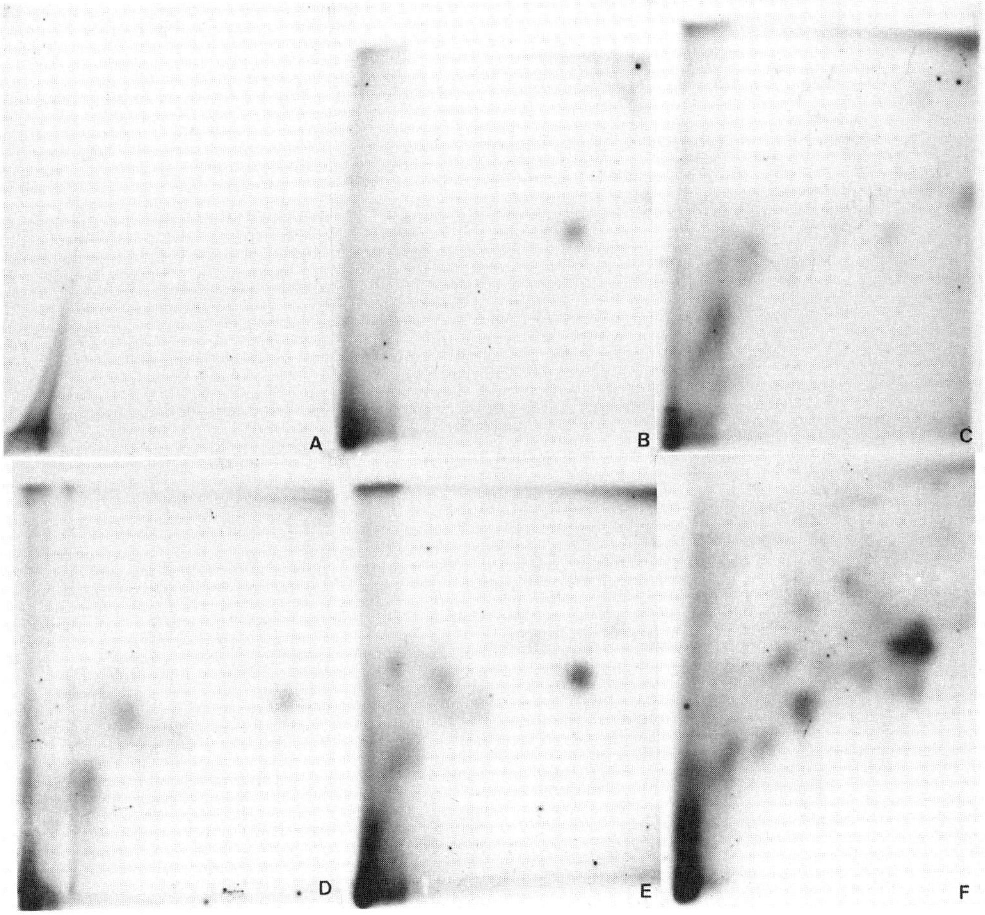

Fig. 2: DNA-adducts formation after incubation of ochratoxin A in presence of renal microsomes. DNA was incubated with renal microsomes in absence (A) or in presence of increasing amount of OTA : (B) 20µM; (C) 40µM; (D) 60µM; (E) 80µM; (F) 100µM.

OTA content of the tissues was determined in order to correlate them eventually with the damage. This seemed necessary because of inter-individual variation in the distribution of OTA and also because of the effect of the different treatments, SOD and catalase, NAC, phoron, which could lower the OTA concentration in the tissues. In fact, the tissue distribution was not significantly modified by these substances (results not shown). For example with a single dose of 4 ppm of OTA (289 µg/kg) in the feed, SOD and catalase treatment increased the OTA concentration in urine from 1.2 µg/24h to 2.3 µg/24h. But in fact this elimination was very low as compared to the dose administrated. Thus, the increase of elimination of OTA cannot by itself explain completely the decrease of genotoxicity, unless one considers that only the free form of OTA (which corresponds to less than 1%) is bioavailable for genotoxicity.

Acknowledgement:
This research was supported by grants from the Ligue Nationale contre le Cancer, Comités Départementaux de la Gironde et du Haut-Rhin, the Ministère de la Recherche et de la Technologie (Action Toxicologie), the Fondation pour la Recherche Médicale and the Région Aquitaine, University of Bordeaux II.

References:

Bendele, A.M., Carlton, W.W., Krogh, P. and Lillehoj, E.B. (1985) Ochratoxin A carcinogenesis in the (C57BL/6J x C3H)F1 mouse. *J.Natl Cancer Inst.*, 75, 733-742.

Boorman, G;, ed. (1989). NTP *Technical Report on the Toxicology and Carcinogenesis Studies of OchratoxinA (CAS NO. 303-47-9) in F344/N rats (Gavage studies)* (NIH Publication No 89-2813.) Research Triangle Park, NC, US Departement of Health and Human Services, National Institutes of Health.

Castegnaro, M, Bartsch, H., Chernozemsky, I.N. (1987) Endemic nephropathy and urinary tract tumours in the Balkans. *Cancer Res.*, 47, 3608-3609.

Ceovic, S., Hrabar, A. and Saric, M. (1992) Epidemiology of Balkan endemic nephropathy. *Fd Chem. Toxic.*, 3, 183-188.

Creppy, E.E., Kane, A., Dirheimer, G., Lafarge-Frayssinet, C., Mousset, S. and Frayssinet, C. (1985). Genotoxicity of ochratoxin A in mice: DNA single-strand break evaluation in spleen, liver and kidney. *Toxicol. Lett.*, 28, 29-35.

Creppy, E.E., Betbeder, A.M., Gharbi, A., Counord, J., Castegnaro, M., Bartsch, H., Moncharmont, P., Fouillet, B., Chambon, P. and Dirheimer, G. (1991) Human ochratoxicosis in France. In: Castegnaro, M., Plestina, R., Dirheimer, G., Chernozemsky, I.N. and Bartsch, H., eds, *Mycotoxin, endemic nephropathy and urinary tract tumours. (IARC Scientific publications N°115)*, Lyon; International Agency for Research on Cancer, pp 145-151.

De Flora, S., Bennicelli, C., Zanacchi, P, Camoirano, A., Morelli, A. and De Flora, A. (1984). In vitro effects of N-acetylcysteine on the mutagenicity of direct-acting compounds and procarcinogens. *Carcinogenesis*, 5, 505-510.

De Flora, S., Bennicelli, C., and Camoirano, A. (1985) In vivo effects of N-acetylcysteine on glutathione metabolism and on the biotransformation of carcinogenic and /or mutagenic compounds. *Carcinogenesis*, 6,1735-1745.

De Flora, S., Astengo, M., Serra, D. and Bennicelli, C.(1986) Inhibition of urethan-induced lung tumors in mice by dietary N-acetylcysteine. *Cancer Lett.*, 32, 235-241.

De Flora, S., Izzotti, F., D'Agostini, F. and Cesarone, C.F. (1991). Antioxidant activity and other mechanisms of thiols involved in chemoprevention of mutation and cancer. *The*

American Journal of Medicine, 91 (suppl 3C), 122-130.
Dekant W., Lash, L.H. and Anders, M.W. (1988) Fate of glutathione conjugates and bioactivation of cysteine S-conjugates by cysteine conjugate β-Lyase. In: Sies, H. and Ketterer, B. eds, *Glutathione conjugaison, mechanisms and biological significance*. Academic Press, London pp415-443.
Dekant, W. and Vamvakas, S. (1989) Bioactivation of nephrotoxic haloalkenes by glutathione conjugation: formation of toxic and mutagenic intermediates by cysteine cinjugate β-lyase. *Drug Metabolism Reviews*, 20, 43-83.
Ecobichon, D.J. (1984) Glutathion depletion and resynthesis in laboratory animals. *Drug and Chemical Toxicology*, 7, 345-355.
Gharbi, A., Trillon, O., Betbeder, A.M., Counord, J., Gauret, A., Pfohl-leszkowicz, A., Dirheimer, G. and Creppy, E.E. (1993) Some effects of ochratoxin A, a mycotoxin contaminating feeds and food on rat testis. Toxicology (in press)
Halliwell, B.. (1991) Reactive oxygen species in living systems: source, biochemistry and role in human disease. *The American Journal of Medicine*, 91 (suppl3C), 14-22.
Halliwell, B. and Gutteridge, J.M.C. (1988) Free radicals and antioxidant protection: mechanisms and significance in toxicology and disease. *Hum. Toxicol.*, 7, 7-13.
Halliwell, B. and Aruoma, O.I. (1991) DNA damage by oxygen-derived species. Its mechanism and measurement in mammalian systems. FEBS Lett, 281, 9-10.
Ishikawa, T. (1992). The ATP-dependent glutathione S-conjugate export pump. *TIBS*, 17, 463-469.
Jakoby, W.B. and Habig, W.H. (1980) Glutathione transferases. In: Jakoby, W.B., ed, *"Enzymatic Basis of Detoxification"*, vol.2,pp.63-94, Academic Press, Orlando, FL.
Kane, A., Creppy, E.E., Roth, A., Röschenthaler, R. and Dirheimer , G. (1986). Distribution of the [3H]-label from low doses of radioactive ochratoxin A ingested by rats, and evidence for DNA single-strand breaks caused in liver and kidneys. *Arch. Toxicol.*, 58, 219-224.
Kanizawa, M. (1984) Synergistic effect of citrinin on hepatorenal carcinogenesis of ochratoxin A in mice. In: *Toxigenic fungi, their toxins and health hazard*,(H. Kurata and Y. Ueno, eds) Amsterdam, Oxford, New York, Elsevier Science Publishers, pp 245-254 (Developments in Food Science N° 7)
Kanizawa, M. and Suzuki, S. (1978) Induction of renal and hepatic tumors in mice by ochratoxin A, a mycotoxin. *Gann*, 69, 599-600.
Krogh, P., Hald, B., Plestina, R. and Ceovic, S. (1977) Balkan (endemic) nephropathy and foodborn ochratoxin A: preliminary results of survey of foodstuffs. *Acta Pathol. Microbiol. Scand.* Sect.A, 246, (Suppl.) 1-21.
Krogh, P., Elling, F., Friss, C.H.R., Hald, B., Larsen, A.E., Lillehoj, E.B., Madsen, A., Mortensen, H.P., Rasmussen, F. and Ravoskov, V. (1979) Porcine nephropathy induced by long term ingestion of ochratoxin A. *Vet. Pathol.*, 16, 466-475.
Lafleur, M.V.M. and Retèl, J. (1993) Contrasting effects of SH-compounds on oxidative DNA damage: repair and increase of damage. *Mutat.Research*, 295, 1-10.
Miller, J.A. and Miller, E.C. (1969). The metabolic activation of carcinogenic aromatic amines and amides. *Prog.exp.Tumor Res.*, 11, 273-275.
Monks, T.J. and Lau, S.S. (1987) Commentary: renal transport process and glutathione conjugate-mediated nephrotoxicity. *Drug Metabolism and disposition*, 15, 437-441.
Monod, G., Devaux, A. and Rivière, J-L. (1988) Effect of chemical pollution on the activities of hepatic xenobiotic metabolizing enzymes in fish from the river Rhône. *Sci.Tot.Environ.*, 73, 189-201.

Omar, R.F., Hasinoff, B.B., Mejilla, F. and Rahimtula, A.D. (1990) Mechanism of ochratoxin A stimulated lipid peroxidation. *Biochem.Pharmacol.*, 40, 1183-1191.

Omar, R.F., Rahimtula, A.D. and Bartsch, H. (1991) Role of cytochrome P-450 in ochratoxin A-stimulated lipid peroxidation. *J.Biochem.Toxicology*, 6, 203-209

Petkova-Bocharova, T., Chernozemsky, I.N. and Castegnaro,M. (1991) Ochratoxin A in human blood in relation to Balkan endemic nephropathy and urinary tract tumours in Bulgaria. *Food Addit.Contam*, 5, 299-301.

Pfohl-Leszkowicz, A., Chakor, K, Creppy, E.E. and Dirheimer, G. (1991). DNA adduct formation in mice treated with ochratoxin A . In : Castegnaro, M., Plestina, R., Dirheimer, G., Chernozemsky, I.N. and Bartsch, H., eds, *Mycotoxin, endemic nephropathy and urinary tract tumours. (IARC Scientific publications N°115)*, Lyon, International Agency for Research on Cancer, pp 245-253.

Pfohl-Leszkowicz, A., Grosse, Y., Kane, A., Creppy, E. and Dirheimer, G. (1993a) Genotoxicity and DNA binding of Ochratoxin A, a ubiquitous mycotoxin found in food and feed. In: Dengler, H.J. and Mutschler, E. eds, *Metabolism of xenobiotics and clinical pharmocology* Gustav Fischer Verlag. (in press)

Pfohl-Leszkowicz, A., Grosse, Y., Kane, A., Creppy, E.E. and Dirheimer, G. (1993b) Differential DNA adduct formation and disappearance in three mice tissues after treatment by the mycotoxin ochratoxin A. *Mutation Res.* (accepted for publication)

Rahimtula, A.D., Béréziat, J.-C., Bussacchini-Griot, V. and Bartsch, H. (1988) Lipid peroxidation as a possible cause of ochratoxin A toxicity. *Biochemical Pharmacology*, 37, 4469-4477.

Reddy, M.V. and Randerath,K.(1986) Nuclease P1 mediated enhancement of sensitivity of ^{32}P-postlabeling test for structurally diverse DNA adducts. *Carcinogenesis*, 7 , 1543-1551.

Slater, T.F. (1984). Free radical mechanisms in tissue injury. *Biochem.J.*, 222, 1-15.

Vadi, H.V., Schasteen, C.S. and Reed, D.J. (1985) Interaction of S-(2-haloethyl)-mercapturic acid analogs with plasmid DNA. *Toxicol. Appl.Pharmacol.*, 80, 386-396.

Van Bladeren, P.J., Breimer, D.D., Rotteveel-Smijs, G. M.T., Jong, R.A., Buijs, W., Van der Gen, A. and Mohn, G.R. (1980). The role of glutathione conjugation in the mutagenicity of 1,2-dibromoethane. *Biochem.Pharmacol.*, 29, 2975-2982.

Van Bladeren, P.J., Breimer, D.D., Rotteveel-Smijs, G. M.T., De Kniff, P., Mohn, G.R., Van Meeteren-Walchi, B., Buijs, W. and Van der Gen, A. (1981). The relation between the structure of vicinal dihalogen compounds and their mutagenic activation via conjugation to glutathione. *Carcinogenesis,* 2, 499-505.

Vukelic, M., Sostaric, B. and Belicza, M. (1992) Pathomorphology of Balkan endemic nephropathy. *Fd.Chem.Toxic.*, 30, 193-200.

Wolf, A., Tschopp, M; and Ryffel, B. (1992) Inhibition of Sandimhun (cyclosporin A)-induced adverse effects in rat kidneys by superoxide dismutase and catalase. Evidence for the involvement of free reactive oxygen species in the pathomechanism of the drug. *Toxicology Lett.,* suppl. 1992, 97.

Résumé

Dans le but de savoir si la voie oxydative de métabolisation de l'ochratoxine A (OTA) et / ou si sa conjugaison au glutathion, ou celle d'un de ces métabolites, étaient en relation avec sa génotoxicité, nous avons testé la formation d'adduits à l'ADN dans le rein et le testicule de souris en faisant varier ces voies de métabolisation. L'administration simultanée de superoxyde dismutase et de catalase avant le gavage des souris par l'OTA provoque une diminution des adduits totaux dans l'ADN de rein et de testicule de 90% et 30% respectivement. Ces adduits sont donc probablement produits par voie oxydative. L'administration de phorone au souris abaisse le taux de glutathion de 66% et de 11% respectivement dans le rein et le testicule après 4h. Cette substance administrée avant l'OTA baisse de 70% le taux d'adduits à 8h dans le rein et à 24h dans le testicule. La N-acétylcystéine qui abaisse également le taux de glutathion de 34% à 8h dans le rein et de 14% dans le testicule, induit aussi une diminution des adduits dans ces tissus. Ces résultats indiquent que le glutathion joue un rôle primordial dans la génotoxicité de l'OTA, soit par la formation de dérivés conjugués génotoxiques, soit par ces propriétés oxydoréductrices. De plus, l'incubation d'un ADN homologue avec des microsomes de rein en présence d'OTA conduit à la formation d'adduits. Ces microsomes de reins sont riches en peroxydases. En conclusion, tous ces résultats indiquent clairement que la voie oxydative de métabolisation est responsable de la formation de la majorité des adduits à l'ADN induits dans le rein et le testicule par l'OTA.

Human ochratoxicosis and its pathologies. Eds E.E. Creppy, M. Castegnaro, G. Dirheimer. Colloque INSERM/ John Libbey Eurotext Ltd. © 1993, Vol. 231, pp. 189-198.

Influence de la superoxyde dismutase associée à la catalyse sur la néphrotoxicité induite par l'ochratoxine A chez le rat

I. Baudrimont[(1)], A.M. Betbeder[(1)], A. Gharbi[(1)], A. Pfohl-Leszkowicz[(2)], G. Dirheimer[(2)] et E.E. Creppy[(1)]

[(1)]Laboratoire de Toxicologie et d'Hygiène Appliquée, Université de Bordeaux II, 3 ter, place de la Victoire, Bordeaux; [(2)]Institut de Biologie Moléculaire et Cellulaire du CNRS, 15, rue René Descartes, Strasbourg, France

Mots clés : Ochratoxine A - Prévention - Néphrotoxicité - Superoxyde dismutase - Catalase.

Résumé

L'ochratoxine A (OTA), mycotoxine élaborée par des moisissures des genres *Aspergillus* et ***Penicillium***, est un contaminant alimentaire que l'on trouve essentiellement dans les céréales, les oléagineux, les fruits secs, ainsi que dans le sang et les tissus d'animaux ayant consommé des aliments contaminés. Elle est néphrotoxique (agent causal principal de la Néphropathie Endémique des Balkans - BEN -), immunosuppressive, génotoxique, tératogène et cancérogène. L'OTA inhibe la synthèse des protéines en entrant en compétition avec la phénylalanine dans la réaction d'aminoacylation du t - RNA phénylalanine ; elle provoque, également, des dommages oxydatifs (elle augmente la peroxydation lipidique). Il est, alors, apparu intéressant d'étudier les effets de la superoxyde dismutase associée à la catalase sur la néphrotoxicité induite par l'OTA, chez le rat. Ces deux enzymes sont administrées (à raison de 20 mg / kg chacune) par injection sous - cutanée, toutes les 48 heures, pendant trois semaines, une heure avant chaque administration d'OTA (289 µg / kg / 48 h). La superoxyde dismutase et la catalase préviennent , en grande partie, la néphrotoxicité induite par l'OTA (enzymurie, protéinurie, créatininémie) et favorisent l'élimination urinaire de la toxine. Ces résultats indiquent (i) que les radicaux superoxydes et le peroxyde d'hydrogène sont, vraisemblablement, impliqués dans les lésions induites par l'OTA ; (ii) que la superoxyde dismutase et la catalase pourraient être utilisées pour prévenir les lésions rénales dans les cas d'ochratoxicose humaine.

Introduction

L'ochratoxine A (OTA) est une mycotoxine élaborée par diverses moisissures, en particulier ***Aspergillus ochraceus*** et ***Penicillium viridicatum*** (Van der Merwe et coll., 1965). Elle a été identifiée dans de nombreux produits végétaux (céréales, oléagineux, fruits secs...) ainsi que dans le sang et les tissus d'animaux ou de personnes ayant consommé des aliments contaminés (notamment dans les pays des Balkans, en Scandinavie, en Allemagne, en Angleterre et en France) (Krogh et coll., 1987 ; Petkova - Bocharova et coll., 1988 ; Bauer et coll., 1987 ; Breitholtz et coll., 1991 ; Creppy et coll., 1991 ; Kuiper - Goodman et coll., 1989).

L'OTA est un analogue structural de la phénylalanine constitué d'une partie dihydroisocoumarinique chlorée liée par une liaison peptidique à une molécule de L - phénylalanine (Van der Merwe et coll., 1965). L'OTA inhibe la synthèse des protéines en entrant en compétition avec la phénylalanine dans la réaction d'aminoacylation du t - RNA phénylalanine (Bunge et coll., 1978 ; Creppy et coll., 1979 ; Creppy et coll., 1983). L'OTA provoque également des dommages oxydatifs (elle augmente la peroxydation lipidique, in vitro et in vivo) (Omar et coll., 1990 ; Rahimtula et coll., 1988 ; Dirheimer et Creppy, 1991). L'OTA est considérée comme l'agent causal principal de la Néphropathie Endémique des Balkans (BEN). Elle provoque, in vivo, une néphropathie tubulo - interstitielle (Krogh et coll., 1974 ; Kane et coll., 1986 ; Petkova - Bocharova et coll., 1988). En outre, elle perturbe la coagulation sanguine (Galtier et coll., 1979 ; Gupta et coll., 1979), le métabolisme glucidique (Pitout et coll., 1968) et est immunosupressive (Haubeck et coll., 1981 ; Creppy et coll., 1983 ; Dwivedi et coll., 1985 ; Tor et coll., 1989), tératogène (Aora et coll., 1981 ; Mayura et coll., 1984), génotoxique (Creppy et coll., 1985 ; Kane et coll., 1986 ; Pfol - Leszkowicz et coll., 1991) et cancérogène (Kanisawa et Suzuki, 1978 Bendele et coll., 1985 ; Markovic et coll., 1985 ; NTP, 1989).

Etant donné l'importance de l'OTA sur le plan de la santé publique et l'impossibilité d'empêcher la prolifération des moisissures toxinogènes, il est nécessaire de mettre au point des moyens permettant d'éliminer l'OTA de l'organisme et de réduire, voire d'éliminer, ses effets toxiques.

L'objectif de ce travail était d'étudier l'influence de la superoxyde dismutase (SOD) associée à la catalase (enzymes intervenant dans le processus de détoxication des radicaux libres) sur la néphrotoxicité induite par l'OTA.

Materiels et méthodes

Traitements des animaux

Des rats mâles de souche Wistar (160 ± 10 g) provenant de SEXAL (France) sont distribués, de manière aléatoire, en trois lots de six animaux.

Traitement par l'Ochratoxine A

Les animaux reçoivent, par intubation oesophagienne, toutes les 48 heures, pendant trois semaines, une quantité d'OTA (Sigma) correspondant à 289 µg / kg de poids corporel ; la toxine est mise en solution dans 0,3 ml de $NaHCO_3$ 0,1 M, pH 7,4.

Traitement par la SOD et la catalase

La SOD (Sigma) et la catalase (Sigma) solubilisées dans 0,1 ml d'eau (20 mg / kg chacune) sont administrées, par injection sous - cutanée, une heure avant chaque administration d'OTA.

Témoins

Les animaux reçoivent uniquement 0,3 ml de $NaHCO_3$ 0,1 M, pH 7,3.

Prélèvement des échantillons biologiques

Prélèvement des urines

Après trois semaines de traitement, les rats sont placés, individuellement, dans des cages métaboliques pendant 24 heures. Les urines sont centrifugées à 3000 x g pendant 10 minutes à 5°C. Les surnagents sont aliquotés et conservés à - 20°C jusqu'au moment du dosage.

Prélèvement du sang

Au moment du sacrifice, le sang est prélevé au niveau du sinus rétro - orbital puis centrifugé à 2000 x g pendant 5 minutes à 5°C. Le sérum est ensuite conservé à - 20°C jusqu'au moment du dosage.

Méthodes de dosage et traitement statistique des données

La créatinine est dosée photométriquement par la réaction de Jaffé (Merck, 3385). Le dosage des protéines urinaires est réalisé selon la méthode de Bradford (1976) (Sigma, 610 A). La lactate déshydrogénase (LDH) et la leucine aminopeptidase (LAP) sont déterminées photométriquement, selon la méthode décrite par la "Deutsche Gesellschaft für Klinische Chimie" (1972), en utilisant des kits enzymatiques (respectivement Merck 3399 et Merck 3359). Le dosage de la γGT est réalisé selon la méthode décrite par Szasz et coll. (1974), (Merck 14302). Le glucose urinaire est dosé photométriquement en utilisant la méthode de la glucose déshydrogénase (Merck 1219).

Tous les dosages effectués sont réalisés à l'aide d'un spectrophotomètre Coulter Asa Junior relié à une imprimante Oki 182.

L'OTA urinaire et sérique est dosée, après extraction (fig. 1), par CLHP et détection fluorimétrique dans les conditions suivantes :

Appareillage : Pompes Bishoff Model 2200, injecteur automatique Alcott Model 738 Autosampler, précolonne Lichrosorb 10 µ, C 18, colonne Sphérisorb ODS 10 µ (30 x 0,25 cm), détecteur de fluorescence Jasco 821 - FP.

Conditions chromatographiques : Phase mobile : méthanol : acétonitrile : solution d'acétate de sodium 5 mM : acide acétique (300 / 300 / 400 / 14 ; v / v / v / v) ; débit : 1ml/min ; temps de rétention : 8 min ; longueurs d'onde d'exitation : 340 nm et d'émission : 465 nm.

La quantification se fait par rapport à des standards (solution méthanolique d'OTA à 4 ng / ml et 10 ng / ml) à l'aide d'un logiciel informatique ICS ("Pic 3") qui intègre les surfaces sous pics et calcule la concentration relative par rapport aux standards.
Les résultats sont analysés, par groupes, en utilisant le test, non - paramétrique, de Wilcoxon (Wilcoxon Rank - Sum Test).

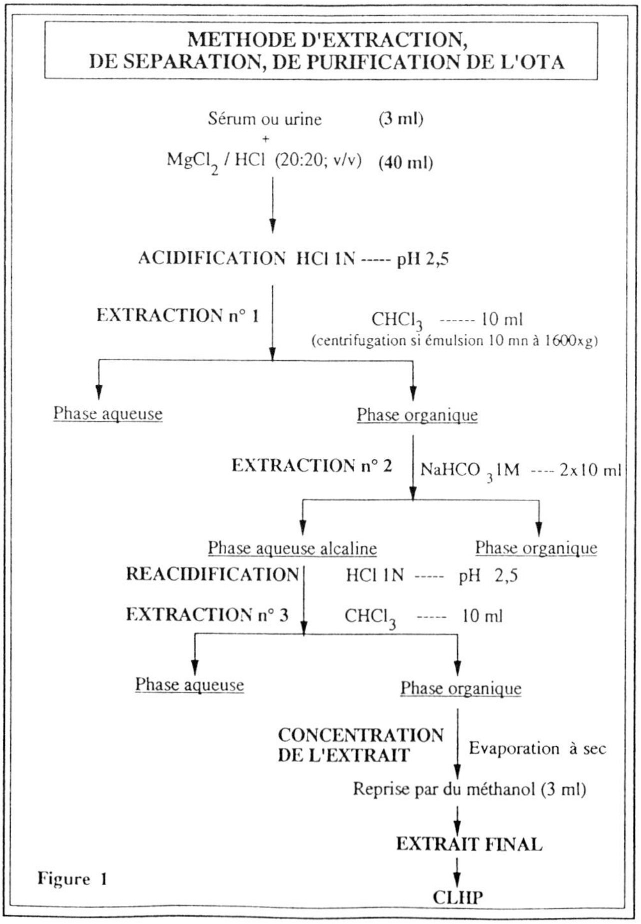

Figure 1

Résultats et discussion

Enzymurie et protéinurie

Le tableau 1 et la figure 2 montrent, qu'après trois semaines de traitement par de faibles quantités d'OTA (4 ppm), la protéinurie et l'enzymurie sont significativement augmentées comparativement aux valeurs des témoins. Ceci confirme les résultats obtenus par Kane et coll. (1986).

Le taux de protéines dans les urines augmente de 55% par rapport à celui des rats témoins. Cette protéinurie (environ 0,5 g de protéines / 1 d'urine) est probablement la conséquence d'une atteinte glomérulaire.

En ce qui concerne les enzymes urinaires, on observe une augmentation moyenne d'environ 90% pour la LDH et de 45 à 50% pour la LAP et la γGT. Nous avons exprimé les résultats en rapportant les activités enzymatiques au volume urinaire (U / L) et non au taux de créatinine, car la concentration de celle ci dans les urines varie. Cependant, lorsque l'on exprime les résultats en unité par rapport au poids de créatinine, le sens de variation reste le même.

L'augmentation des activités enzymatiques dans les urines témoigne de lésions tubulaires proximales et distales.

Tableau 1 : Variations du taux de protéines dans les urines de rats traités pendant trois semaines, toutes les 48 h avec l'OTA (effets de la SOD + catalase)

TRAITEMENTS	Taux de protéines mg / 24 h
Témoins ($NaHCO_3$)	4.68 ± 0.37 *
OTA	7.14 ± 1.47 * (a)
OTA + SOD + catalase	4.93 ± 0.56 *(b)

* déviation standard pour n = 6
a significativement différents des témoins à p = 0.005
b non significativement différents des témoins à p = 0.005

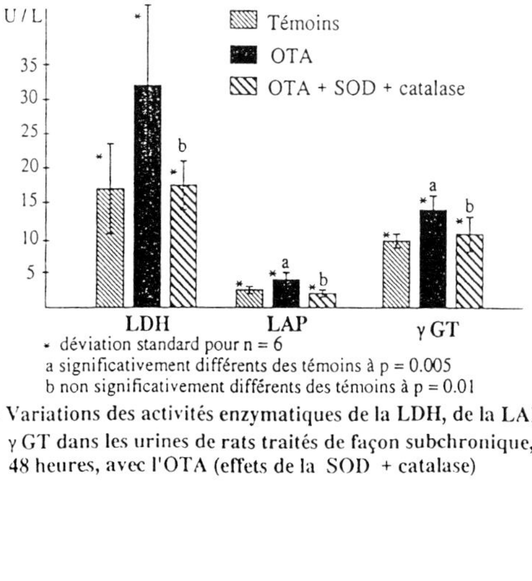

* déviation standard pour n = 6
a significativement différents des témoins à p = 0.005
b non significativement différents des témoins à p = 0.01

Figure 2 : Variations des activités enzymatiques de la LDH, de la LAP et de la γ GT dans les urines de rats traités de façon subchronique, toutes les 48 heures, avec l'OTA (effets de la SOD + catalase)

La SOD et la catalase atténuent l'augmentation de l'enzymurie et de la protéinurie provoquée par l'OTA (valeurs se rapprochant de celles des témoins) (tableau 1 et fig. 2). Ces deux enzymes semblent donc prévenir, en grande partie, les lésions oxydatives induites par l'OTA.

Fonctions rénales

Les résultats des dosages urinaires et sériques de la créatinine montrent qu'il y a, simultanément, une élévation de la créatininémie et une diminution de la créatininurie chez les rats traités par l'OTA (fig. 3). Des résultats similaires ont été rapportés par Kane et coll. (1986) après six semaines de traitement par l'OTA chez le rat.

En présence d'OTA, nous constatons une légère augmentation de la glucosurie (0,34 ± 0,09 g / l contre 0,32 ± 0,06 g / l chez les témoins). Compte tenu de la diurèse qui augmente, suite au traitement par l'OTA, la glucosurie passerait de 3,2 ± 0,6 g / 24 h à 5,3 ± 1,4 g / 24 h. La glucosurie observée témoigne d'un défaut de réabsorption tubulaire, anomalie qui n'interviendrait, apparemment, qu'au stade terminal de la néphropathie. Une expérimentation à long terme serait nécessaire pour confirmer ce résultat.

Les traitements par la SOD et la catalase atténuent significativement l'augmentation de la créatininémie observée chez les rats traités par l'OTA, et augmentent, par contre, la créatininurie qui se rapproche de la valeur des témoins (fig. 3). La clairance de la créatinine diminue significativement après traitement par la SOD + catalase (tableau 2).

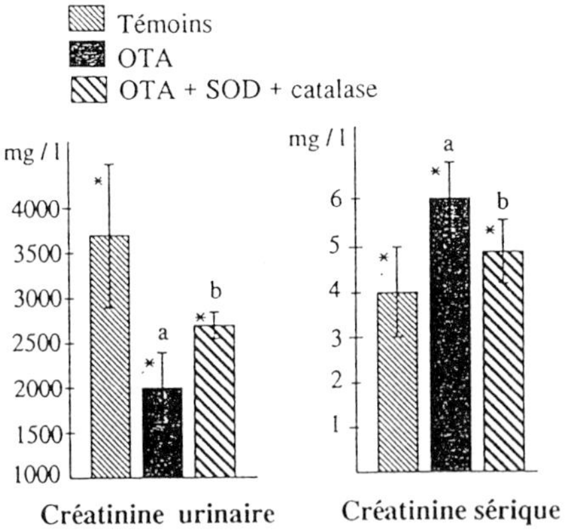

⁎ déviation standard pour n = 6
a significativement différents des témoins à p = 0.005
b significativement différents des résultats obtenus avec les rats traités avec l'OTA à p = 0.005

Figure 3 : Variations de la créatinine urinaire et sérique chez des rats traités, par gavage, toutes les 48 heures, pendant trois semaines, avec l' OTA (effets de la SOD + catalase)

Tableau 2 : Variations de la clairance de la créatinine che[z] rats traités pendant trois semaines, toutes les [48 h] avec l'OTA (effets de la SOD + catalase)

TRAITEMENTS	Clairance de la créatinine ml / min
Témoins (NaHCO3)	6.53 ± 1 *
OTA	3.01 ± 0.43 * (a)
OTA + SOD + catalase	4.43 ± 0.75 *(b)

* déviation standard pour n = 6
a significativement différent des témoins à p = 0.01
b significativement différent des résultats obtenus avec les rats traités avec l'OTA à p = 0.01

Dosages urinaire et sérique de l'OTA

L'OTA éliminée dans les urines des rats intoxiqués est filtrée au niveau du glomérule. Elle est sécrétée et réabsorbée au niveau du tubule proximal. Tous ces systèmes de transport dépendent de l'intégrité membranaire.

- La SOD associée à la catalase favorise l'élimination urinaire de l'OTA (78%) (tableau 3).

- **L'action de l'OTA** au niveau de la cellule tubulaire pourrait s'expliquer par un mécanisme de peroxydation membranaire qui aurait pour effet d'entraver l'excrétion tubulaire de la toxine (diminution de la fluidité membranaire) .

- **La SOD et la catalase** agiraient, probablement, en inhibant ce processus oxydatif (diminution de la peroxydation lipidique) ce qui aurait pour conséquence d'augmenter l'excrétion et l'élimination de l'OTA, comme en témoingnent les résultats (tableau 3).

Tableau 3 : Variations des taux d'OTA sérique et urinaire chez des rats traités pendant trois semaines, toutes les 48 heures (effets de la SOD + catalase)

TRAITEMENTS	Sérum µg / ml	Urine ng / ml
Témoins (NaHCO3)	-	-
OTA	1.85	126
OTA + SOD + catalase	1.77	224

En conclusion, ces molécules semblent donc prévenir, en grande partie, la néphropathie induite par l'OTA. Elles pourraient, par conséquent, être utilisées en tant qu'agents protecteurs contre les effets de l'ochratoxicose humaine.

Il serait nécessaire, à la faveur de travaux ultérieurs, de préciser les modalités d'utilisation, de déterminer les doses et les voies d'administration, de manière à utiliser ces molécules dans des conditions de parfaite innocuité.

BIBLIOGRAPHIE

ARORA, R. G. and FROLEN, H. (1981): Interference of mycotoxins with prenatal development of the mouse. II. Ochratoxin A induced teratogenic effects in relation to the dose and stage of gestation. Acta Vet. Scand., 22, 535 - 552.

BAUER, J., GAREIS, M. (1987): Ochratoxin A in der Nahrungsmittelkette.Z. Veterinärmed. B. 34, 613 - 627.

BENDELE, A. M., CARLTON, W. W., KROGH, P., LILLEHOJ, E. B. (1985): Ochratoxin A carcinogenesis in the (C 57BL / 6J X C3H)F1 mouse. J. Natl. Cancer Inst., 75, 733 - 742.

BRADFORD, M. M. (1976): A rapid and sensitive method for the quantification of microgram quantities of protein utilizing the principle of protein dye. Anal. Biochem., 72, 248 - 254.

BREITHOLTZ , A., OLSEN, M., DAHLBACK, A., HULT, K. (1991): Plasma Ochratoxin A levels in three Swedish populations surveyed using and ion - pair HPLC technique. Fd Addit and Cont.

BUNGE, I., DIRHEIMER, G. and RÖSCHENTHALER, R. (1978): In vivo and in vitro inhibition of protein synthesis in *Bacillus stearothermophilus* by ochratoxin A. Biochem. Biophys. Res. Commun., 83, 398 - 405.

CREPPY, E. E., LUGNIER, A. A. J., FASIOLO, F., HELLER, K., RÖSCHENTHALER, R., DIRHEIMER, G. (1979): In vitro inhibition of yeast phenylalanine - tRNA synthetase by ochratoxin A. Chem. Biol. Interact., 24, 257 - 262.

CREPPY, E. E., KERN, D., STEYN, P. S., VLEGAAR, R., RÖSCHENTHALER, R., DIRHEIMER, G. (1983a): Comparative study of the effect of ochratoxin analogues on yeast aminoacyl - tRNA synthetases and on the growth and protein synthesis of hepatoma cells. Toxicol. Lett. 19, 217 - 224.

CREPPY, E. E. STØRMER, F. C., RÖSCHENTHALER, R., DIRHEIMER, G. (1983b): Effects of two metabolites of ochratoxin A, (4 R) - 4 - hydroxy ochratoxin A and ochratoxin α, on the immune response in mice. Infect. Immun., 39, 1015 - 1018.

CREPPY, E. E., KANE, A., DIRHEIMER, G., LAFARGE - FRAYSSINET, C., MOUSSET, S. (1985): Genotoxicity of ochratoxin A in mice : DNA single - strand breaks. Evaluation in spleen, liver and kidney. Toxicol. Lett., 28, 29 - 35.

CREPPY, E. E., BETBEDER, A. M. , GHARBI, A., COUNORD, J., CASTEGNARO, M., BARTSCH, H., MONCHARMONT, P., FOUILLET, B., CHAMBON, P., DIRHEIMER, G. (1991): Human ochratoxicosis in France. In mycotoxins, nephropathy and urinary tract tumours. I A R C Lyon Ed., 145 - 151.

Deutsche Gesellschaft für Klinische Chemie, (1972): Z. Kin. Chem. Biochem., 10, 182.

DIRHEIMER, G., CREPPY, E. E. (1991): Mechanism of action of ochratoxin A. IARC Lyon Ed., 171 - 176.

DWIVEDI, P., BURNS, R. B. (1985): Immunosuppressive effect of ochratoxin A in young turkeys. Avian Pathol., 14, 213 - 225.

GALTIER, P., BONEN, B., CHARPENTEAU, J. L., BODIN, G., ALVINERIE, M., MORE, J. (1979): Physiopathology of hemorrhagic syndrom related to ochratoxin A intoxication in rats. Food Cosmet. Toxicol., 17, 49 - 53.

GUPTA, M., BANDYOPADHYAY, S., PAUL, B., MAJUMDER, S. K. (1979): Hematological changes produced in mice by ochratoxin A. Toxicology, 14, 95 - 98.

HAUBECK, H. D.,LORKOWSKI, G., KÖLSCH, E., RÖSCHENTHALER, R. (1981): Immunosuppression by ochratoxin A and its prevention by phenylalanine. Appl. Environ. Microbiol., 41, 1040 - 1042.

KANE, A. (1986): Intoxication subchronique par l'ochratoxine A, mycotoxine contaminant les aliments : effets néphrotoxiques et génotoxiques. Thèse Université Louis Pasteur, Strasbourg.

KANE, A., CREPPY, E. E., RÖSCHENTHALER, R., DIRHEIMER, G. (1986a): Changes in urinary and renal tubular enzymes caused by subchronic administration of ochratoxin A in rats. Toxicology, 42, 233 - 243.

KANE, A., CREPPY, E. E., ROTH, A., RÖSCHENTHALER, R., DIRHEIMER, G. (1986b): Distribution of the (3H) label from low doses of radioactive ochratoxin A ingested by rats, and evidence for DNA single - strand breaks caused in liver and kidneys. Arch. Toxicol., 58, 219 - 224.

KANISAWA, M., SUSUKI, S. 1978): Induction of renal ad hepatic tumors in mice by ochratoxin A, a mycotoxin. Gann, 69, 599 - 600.

KROGH, P., AXELSEN, N. H., ELLING, F., GYRD - HANSEN, N., HALD, B., HYLDGAARD - JENSEN, J., LARSEN, A. E., MADSEN, A., MORTENSEN, H. P., MÖLLER, T., PETERSEN, O. K., RAVNSKOV, U., ROSTGAARD, M., AALUND, O. (1974): Experimental porcine nephropathy : changes of renal function and structure induced by ochratoxin A - contamined feed. Acta Path. and Microbiol. Scand., section A, Suppl., 246, 1 - 21.

KROGH, P. (1987): Ochratoxin in food. In : Mycotoxins in food. Ed. P. KROGH (Academic Press, New york), 97 - 121.

KUIPER - GOODMAN, SCOOT, P. M. (1989): Risk assessment of the mycotoxin ochratoxin A. Biomed Environ.Sci., 2, 179 - 248.

MARKOVIC, B. (1985): Nephropathie des Balkans et carcinomes à cellules transitionnelles. Journal d' Urologie, 91, 215 - 220.

MAYURA, K., PARKER, R., BERNDT, W. O., PHILLIPS, T. D. (1984): Ochratoxin A induced teratogenesis in rats : partial protection by phenylalanine. Appl. Environm. Microbiol., 48, 1186 - 1188.

NTP. (1989): Technical Report on the Toxicology and Carcinogenesis Studies of Ochratoxin A (C A S N0 - 303 - 47 - 9) in F 344 / N rats (gavage studies).

OMAR, R. F., HASINOFF, B. B., MEJILLA, F., RAHIMTULA, A. D. (1990): Mechanism of ochratoxin A stimulated lipid peroxydation. Biochem. Pharmacol., 40, 1183 - 1191.

PETKOVA - BOCHAROVA, T., CHERNOZEMSKY, I. N., CASTEGNARO, M. (1988): Ochratoxin A in human blood in relation to Balkan endemic nephropathy and urinary system tumours in Bulgaria. Food Addit. Contam., 5, 299 - 301.

PFOHL-LESZKOWICZ, A., CHAKOR, K., CREPPY, E. E., DIRHEIMER, G. (1991): DNA adduct formation in mice treated with ochratoxin A. IARC Lyon Ed., 245 - 253.

PITOUT, M. J. (1968): The effect of ochratoxin A on glycogen storage in rat liver. Toxicol. Appl. Pharmacol., 13, 299 - 306.

RAHIMTULA, A. D., BEREZIAT, J. C., BUSSACCHINI-GRIOT, V., BARTSCH, H. (1988): Lipid peroxydation as possible cause of ochratoxin toxicity. Biochem., Pharmacol., 37, 4469 - 4477.

SZASZ, G., WENMAN, G., STAHLER, F., WHALEFELD, A. W., PERSIJN, J. P. (1974): New substrates for measuring γ-glutamyl-transpeptidase activity. Z. Klin. Chem. Klin -Biochem., 12, 228.

TOR, L., STEIEN, K., STØRMER, F. C. (1989): Mechanism of ochratoxin A induced immunosuppression. Mycopathologia, 107, 153 - 159.

VAN DER MERWE, K. J., STEYN, P. S. FOURRIE, L., SCOTT, D. B., THERON, J. J. (1965): Ochratoxin A, a toxic metabolite produced by *Aspergillus ochraceus* Wilh. Nature, 205, 1112- 1113.

Summary

Ochratoxin A (OTA), a mycotoxin produced by fungi of *Aspergillus* and *Penicillium* genera, is found, essentially, in cereals, dried fruits, seeds and, also, in blood of animals and human, after consumption of contaminated feeds and food. OTA is nephrotoxic and suspected to be the main causal agent of Balkan Endemic Nephropathy. It is, also, immunosuppressive, teratogenic and carcinogenic. OTA inhibits protein synthesis by competition with phenylalanine in the t - RNA specific of phenylalanine aminoacylation reaction. Recently, lipid peroxidation induced by OTA has been reported indicating that the lesions induced by this mycotoxin could, also, be related to oxidative pathways. It was, then, interesting to study the effects of superoxide dismutase (SOD) and catalase on the nephrotoxicity induced by OTA. The two enzymes (20mg/kg each) were given one hour before gavage by OTA (289µg/kg every 48H) for 3 weeks. SOD and catalase prevented most of the toxic effects induced by ochratocxin A (enzymuria, proteinuria, creatinemia), and increased urinary excretion of OTA. These results indicated (i) that lipoperoxydation is probably one of the damaging process of OTA in vivo (ii) that SOD and catalase could be used for prevention of renal lesions in cases of ochratoxicosis.

Human ochratoxicosis and its pathologies. Eds E.E. Creppy, M. Castegnaro, G. Dirheimer. Colloque INSERM/ John Libbey Eurotext Ltd. © 1993, Vol. 231, pp. 199-207.

Preponderance of DNA-adducts in kidney after ochratoxin A exposure

A. Pfohl-Leszkowicz[1], Y. Grosse[1], S. Obrecht[1], A. Kane[1, 2], M. Castegnaro[3], E.E. Creppy[4] and G. Dirheimer[1]

[1]*Institut de Biologie Moléculaire et Cellulaire du CNRS, 15, rue Descartes, 67084 Strasbourg, France;* [2]*Institut de Technologie Alimentaire, BP 2765, Dakar, Sénégal;* [3]*International Agency for Research on Cancer, 150, cours Albert-Thomas, 69008 Lyon, France;* [4]*Laboratoire de Toxicologie et d'Hygiène Appliquée, Université Bordeaux II, 3 ter, place de la Victoire, 33076 Bordeaux, France*

Summary:

Ochratoxin A (OTA) is a mycotoxin which has been implicated in Balkan endemic nephropathy (BEN), a disease characterized by tubulonephritis, and may be involved in the high incidence of urinary tract tumors associated to BEN. It induces DNA single-strand breaks and has been shown to be carcinogenic in two rodent species. For a better understanding of the OTA genotoxic effects, OTA-DNA adducts formation and disappearance has been measured using the ^{32}P postlabelling method after oral administration of 2.5 mg/kg of OTA to mice. In kidney, liver and spleen, several modified nucleotides were clearly detected in DNA, 24h after administration of OTA, but their levels in the various tissues varied significantly in time dependent manner over a 16 day period. Total DNA-adducts reached a maximum at 48 h when 103, 42 and 2.2 adducts per 10^9 nucleotides were found respectively in kidney, liver and spleen, indicating that kidney is the main target for the genotoxicity and likely carcinogenicity of OTA. The main adduct was different in kidney and in liver. All adducts disappeared in liver and spleen 5 days after OTA administration, whereas some adducts persisted up to 16 days in the kidney. Some adducts were organ and species specific. DNA-adducts could also be found in liver, spleen and kidney of rat. The finding that the adducts are not quantitatively and qualitatively the same in organs and species tested is likely due to differences of metabolism in these organs and species, leading to different ultimate carcinogens and may also result from differences in the efficiency of repair processes. Incubations of DNA with kidney or liver microsomes in presence of OTA lead to DNA-adducts formation only with kidney microsomes. In order to provide further evidence that OTA is involved in the etiology of human urinary tract cancer, DNA from urinary tract tumor tissues was collected from Bulgarian subjects who lived in an high risk area for BEN and thus may have been exposed to OTA. Several characteristic DNA-adducts were detected in these DNA. By comparison and co-migration with DNA-adducts identified in OTA-treated mice kidney, some of these DNA-adducts were tentatively related to OTA exposure with the conclusion that the tumor causal agent is likely ochratoxin A.

Introduction:

The covalent binding of chemicals or their reactive metabolites to DNA is generally believed to be a key step in the initiation of chemically induced carcinogenesis (Miller and Miller, 1969). Several studies have demonstrated a strong relationship between extend of

DNA-adduct formation and tumorigenicity (Brookes and Lawley, 1964, Lutz, 1979, Phillips *et al.*, 1979, Taningher *et al.*, 1990).
Ochratoxin A is an ubiquitous mycotoxin contaminating food and feed. Dietary feeding of ochratoxin A has been shown to induce renal adenomas in mice and rats and hepatocellular carcinomas in mice (Kanizawa and Suzuki, 1978; Kanizawa, 1984; Bendele *et al.*, 1985a, Boorman, 1989). The carcinogenicity of OTA was also suspected in human because of the high incidence of kidney, pelvis, ureter and urinary bladder carcinomas among patients suffering from Balkan endemic nephropathy where OTA is implicated (Castegnaro *et al.*, 1987; Ceovic *et al.*, 1992; Vukelic *et al.*, 1992). In the areas of Balkan endemic nephropathy, high levels of ochratoxin A were found in human blood (Petkova-Bocharova and Castegnaro, 1991) but the mycotoxin has also been found in other European countries like Czechoslovakia, Germany, Poland, Danmark and Sweden (for a review see Hald, 1991) and more recently in France (Creppy *et al.*, 1991).
The genotoxicity of OTA was controversal as almost all microbial and mammalian assays were negative (Engel and von Milczewski, 1976; Ueno and Kubota, 1976; Umeda *et al.*, 1977; Kuczuk *et al.*, 1978; Wehner *et al.*, 1978; Bartsch *et al.*, 1980). A weakly positive response for induction of unscheduled DNA synthesis in primary hepatocytes from mice ACI C3H strain and rats ACI strain was found by Mori *et al.* (1984) but not by Bendele *et al.* (1985 b). Contradictory results were also found for sister chromatid exchange in CHO cells (for a review, see Kuiper-Goodman and Scott, 1989). Recently, Manolova *et al.* (1990) demonstrated chromosomal aberrations, particularly on X chromosomes, produced in human lymphocytes in culture by ochratoxin A at a concentration of 15 nM. Moreover, Creppy *et al.* (1985) detected DNA damage reflected by single-strand breaks in kidney, liver and spleen of male Balb/c mice injected intraperitoneally with OTA. The DNA damage to splenic cells was confirmed *in vitro*. DNA single-strand breaks have also been reported in renal and hepatic tissues of rats administred subchronically with small amounts of OTA by gavage (Kane *et al.*, 1986). Recently, Malaveille *et al.* (1991) have shown that OTA induces SOS-DNA repair directly in *E.coli* PQ37 strain. In addition, a rat hepatocyte mediated mutagenic response was demonstrated in *Salmonella typhimurium* TA 1535, 1538 and 100 strains by Hennig *et al.* (1991).

To better understand the mechanism of OTA carcinogenesis, we searched for OTA-DNA-adducts using the nuclease P1 enrichment version of the ^{32}P postlabelling method (Reddy and Randerath, 1986).

Materials and Methods:

Ochratoxin A, proteinase K and ribonucleases A and T1 were obtained from Sigma (St.Louis, MO, USA), T4 polynucleotide kinase from PL Biochemicals (Milwaukee, WI, USA), micrococcal nuclease and spleen phosphodiesterase from Worthington Biochemicals (Freehold, NJ, USA), nuclease P1 from Boehringer (Mannheim, Germany) and [$\gamma^{32}P$] ATP, (5000 Ci/ mmol) from Amersham (Buckinghamshire, England).

Animals treatment :

Swiss male mice and Wistar male rats (Sexal, Vigneul sous Montmédy, France) weighing 25 $\pm$ 2g and 160 $\pm$ 10g respectively, aged seven weeks, were given OTA (in 0.3 ml of 0.1 M $NaHCO_3$ pH 7.4) 0.04 to 2.5 mg/kg body weight by gastric intubation. The control animals received only 0.3 ml of 0.1M $NaHCO_3$, pH 7.4.

*Preparation of microsomes:*Microsomes from kidney, liver and testicles were obtained as described by Monod *et al.* (1988). Briefly, organs were homogenized with a potter-Elvehjem in a buffered solution (0.15 M KCl, 50 mM phosphate buffer pH 7.4). The microsomal fraction was obtained by differential centrifugations and all microsomal pellets were resuspended in 50 mM phosphate buffer pH 7.4 containing 0.15 M KCl, 1 mM EDTA, 1 mM DTT and 20% glycerol. These fractions were stored at -80°C until use.

Results and discussion:

Our results concern several species including mice, rats and human. DNA from kidney, liver and spleen in animals; kidney and bladder in human were analysed. In this study, the intention was to correlate the administration of OTA or its putative presence in BEN suffering people with the amount of total DNA-adducts in different organs.

In mice, the treatment by OTA (2.5 mg/kg single dose given orally) induced from day one, several adducts in kidney, liver and spleen. Total DNA-adducts represented at 48h : 103, 42 and 2.2 adducts per 10^9 nucleotides respectively in kidney, liver and spleen (Fig. 1). They seem to be completely repaired in liver and spleen after 8 days, whereas some of them persisted more than 16 days (seven adducts representing 7.7 adducts per 10^9 nucleotides) in kidney.

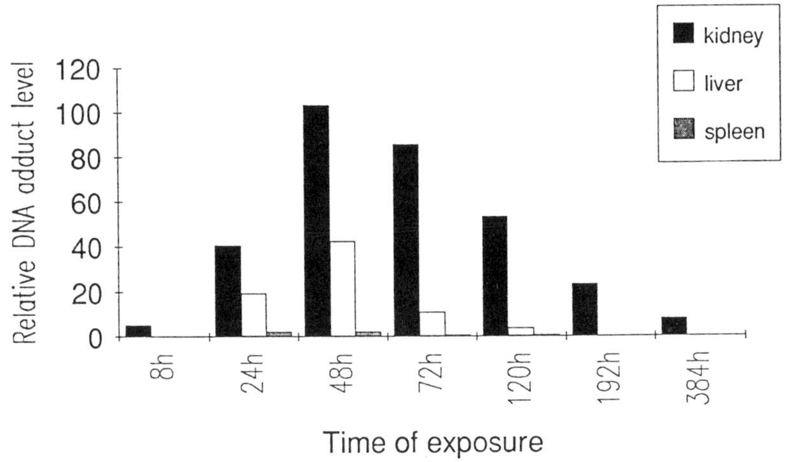

Fig. 1: *Total DNA adduct levels in kidney, liver and spleen after treatment with 2.5 mg/kg of ochratoxin A. Results are expressed as number of adduct per 10^9 nucleotides (relative adduct level)*

In another experiment, the OTA doses were lowered (0.04 to 1.2 mg/kg). In these conditions, after 48h and for a single dose, there were no DNA-adducts in liver under 0.6 mg/kg, whereas some DNA-adducts were detected even for 0.04 mg/kg in the kidney. It was interesting to compare the genotoxicity of OTA in mice and rat (Fig.2). For this purpose, male Wistar rats have received low doses of OTA : a single dose ranging from 0.04 to 1.2 mg/kg or 289 µg/kg of OTA every 48h for 8 weeks. This last treatment corresponds to a contamination of 2 ppm/day in the feeds. Some DNA-adducts in rat tissues were also detected. The DNA-adduct patterns were not exactly the same in rats and mice. Nevertheless, some of the DNA-adducts detected in rat kidney seemed to be identical to those found in mice tissues. Some other were specific to the rats. After subchronic intoxication, 4 main DNA-adducts of equal intensity were detected in rat kidney. Three of these 4 adducts corresponded to DNA-adducts persisting up to 16 days in mice kidney (Pfohl-Leszkowicz *et al.*1993b). In liver, DNA-adduct pattern was very different in rat as compared to mice.

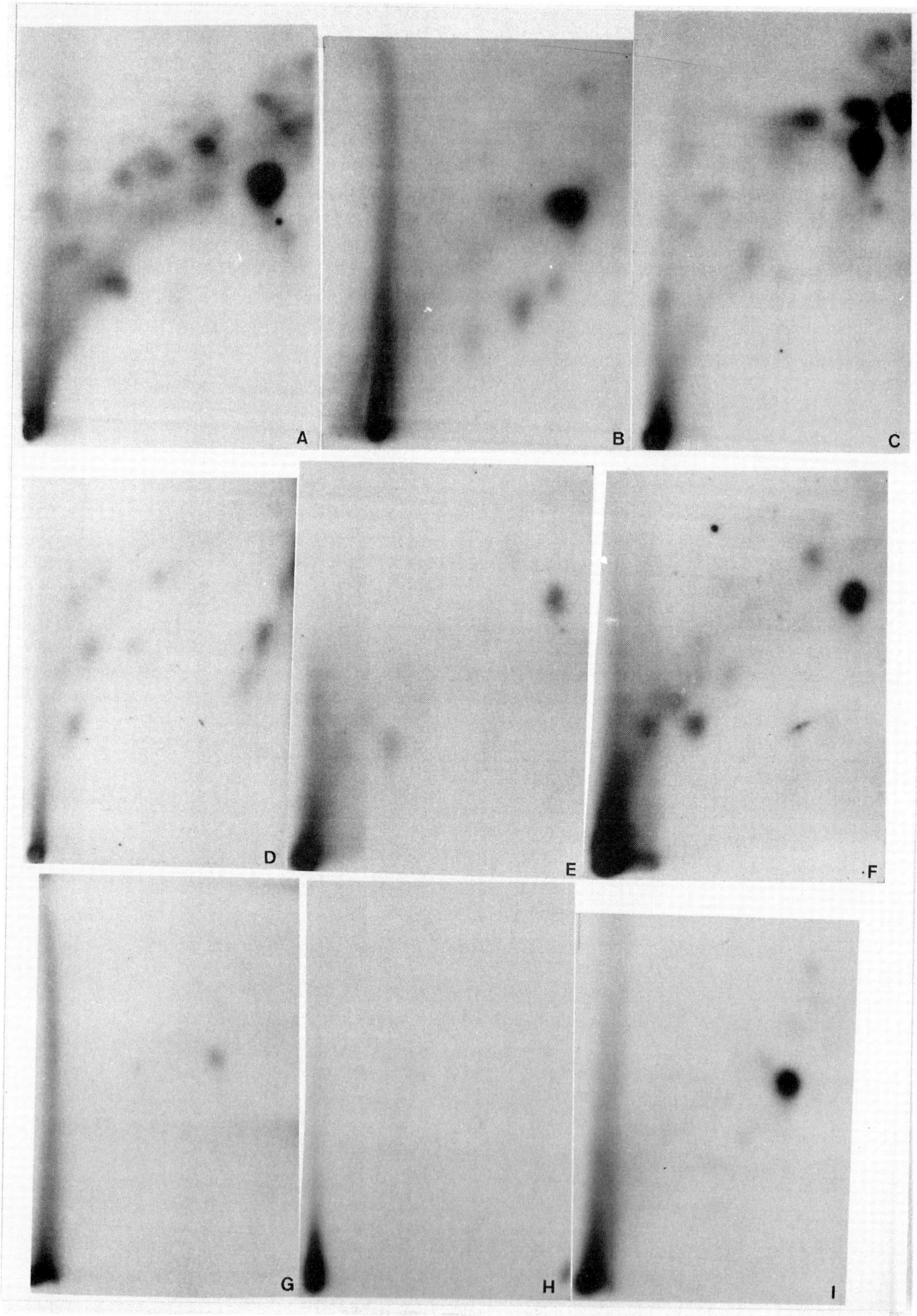

Fig. 2: Autoradiography of DNA-adducts : (A,D,G) in mice, 24h after a single dose of OTA 2.5 mg/kg; (B,E,H) in rats, 48h after a single dose of OTA 1.2 mg/kg; (C,F,I) after 8 weeks treatment with 4ppm of OTA every 48h. A,B,C : in kidney ; D,E,F : in liver ; G,H,I : in spleen.

It was thus important to search for the formation of OTA DNA-adducts *in vitro* after incubation with microsomes from the two tissues (kidney and liver). After 1h of incubation with OTA (20µM-100µM) microsomes from rat renal tissue produced several DNA-adducts whereas no DNA-adduct was found with liver microsomes. Higher amounts of cytochromes P450 were found in liver as compared to kidney, whereas peroxidases were found in high amount only in kidney. These results seem to indicate that oxidative pathways could induce the formation of reactive entities leading to DNA-adducts. (Pfohl-Leszkowicz *et al.*, 1993c this volume)
In order to provide further evidence that OTA is involved in the etiology of human urinary tract tumors, tumorous tissues were collected from patients with putative past OTA exposure, and DNA-adducts were analysed by ^{32}P-postlabelling and a comparison was made with DNA adducts observed in OTA treated mice (Fig.3).

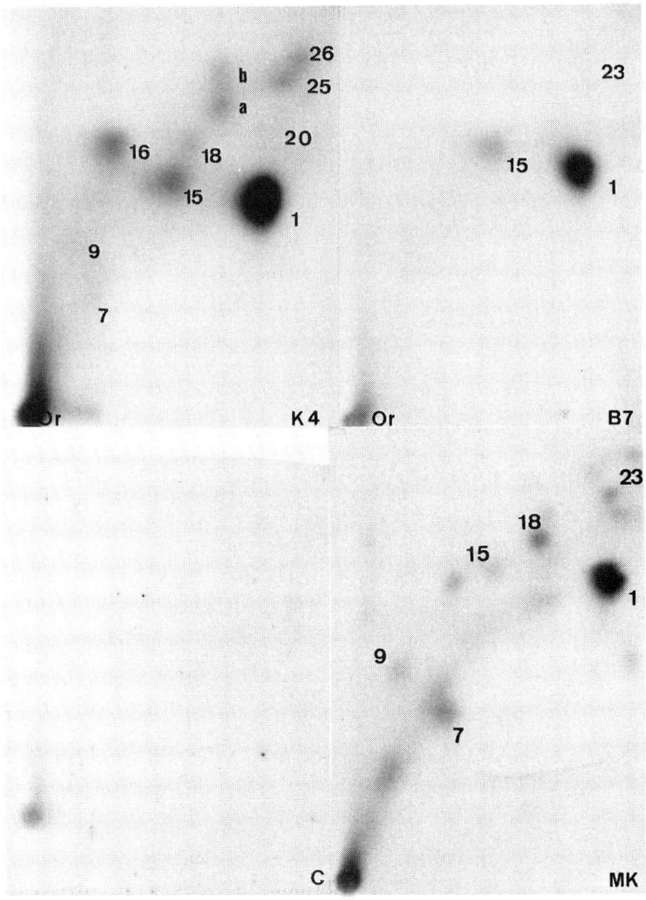

Fig. 3 : Patterns of DNA-adducts from human and mice kidney. K4 : human kidney tumors DNA; B7 : human bladder tumors DNA; C : untreated mice kidney DNA; MK : mice kidney DNA five days after OTA treatment (single dose of 2.5 mg/kg).

Several characteristic DNA-adducts were detected in urinary tract tumors from Bulgarian subjects that lived in an high risk BEN area and thus may have been exposed to OTA. As compared to kidney tumors, the human bladder tumors contained fewer adducts and at lower levels. Up to date, only cancer of renal pelvis and ureter (but not of the bladder) is known to be associated with BEN. (Chernozemsky *et al.*, 1977; Ceovic *et al.*, 1992). By comparison with DNA-adducts from OTA-treated mice kidney, some of the DNA-adducts from human origin were tentatively related to OTA exposure (Pfohl-Leszkowicz *et al.*, 1993 a).

In conclusion, the genotoxic effect of OTA was clearly established. The finding that DNA-adducts were not quantitatively and qualitatively the same in all organs and species tested might be due to different metabolisms, leading to different ultimate carcinogens. The different kinetics of disappearance of the adducts is likely due to differences in repair processes. Thus, the detected DNA-adducts correspond to those which are not rapidly repaired. This may mean that the lower level of DNA adducts in liver is due to more efficient repair, whereas in kidney this repair is less rapid. Moreover, the *in vitro* results suggest that special metabolic pathways exist in the kidney explaining the higher amount of adducts in this organ. The higher DNA-adduct formation joined to their less efficient repair processes may explain why kidney is the target organ for carcinogenicity.

Acknowledgement:

This research was supported by grants from the Ligue Nationale contre le Cancer, Comités Départementaux de la Gironde et du Haut-Rhin, the Ministère de la Recherche et de la Technologie (Action Toxicologie), the Fondation pour la Recherche Médicale and the Région Aquitaine, University of Bordeaux II. We thank Dr Chernozemsky for the human samples.

References:

Bartsch, H., Malaveille, C., Camus, A.M., Martel-Planche, G., Brun, G., Hautsabadie, N., Hauterfeuille, A., Sabadie, N., Barbin, A., Kuroki, T;, Drevon, C;, Piccoli, C. and Montesano, R. (1980) Validation and comparative studies on 180 chemicals with *Salmonella typhimurium* strains and V79 Chinese hamster cells in the presence of various metabolizing systems. *Mutat. Res.*, 76, 1-50.

Bendele, A.M., Carlton, W.W., Krogh, P. and Lillehoj, E.B. (1985a) Ochratoxin A carcinogenesis in the (C57BL/6J x C3H)F1 mouse. *J.Natl Cancer Inst.*, 75, 733-742.

Bendele, A.M., Neal, S.B., Oberly, T.J., Thompson, C.Z., Bewsey, B.J., Hill, L.E., Rexroat, M.A., Carlton, W.W. and Probst, G.S. (1985b) Evaluation of ochratoxin A for mutagenicity in a battery of bacterial and mammalian cell assays. *Fd Chem.Toxicol.*, 23, 911-918.

Boorman, G;, ed. (1989). NTP *Technical Report on the Toxicology and Carcinogenesis Studies of OchratoxinA (CAS NO. 303-47-9) in F344/N rats (Gavage studies)* (NIH Publication No 89-2813.) Research Triangle Park, NC, US Departement of Health and Human Services, National Institutes of Health.

Brookes, P. and Lawley, P.D. (1964). Evidence for the binding of polynuclear aromatic hydrocarbons to the nucleic acids of mouse skin: relation between carcinogenic power of hydrocarbons and their binding to DNA. *Nature*, 202, 781-784.

Castegnaro, M, Bartsch, H., Chernozemsky, I. (1987) Endemic nephropathy and urinary tract tumours in the Balkans. *Cancer Res.*, 47, 3608-3609.

Ceovic, S., Hrabar, A. and Saric, M. (1992) Epidemiology of Balkan endemic nephropathy. *Fd Chem. Toxic.*, 3, 183-188.

Chernozemsky, I.N., Stoyanov, I.S., Petkova-Bocharova, T.K., Nicolov, I.G, Draganov, I.V., Stoichev, I.I., Tanchev, Y., Naidenov, D. and Kalcheva, N.D. (1977). Geographic correlation between the occurrence of endemic nephropathy and urinary tract tumours in Vratza district, Bulgaria. *Int. J.Cancer*, 19, 1-11.

Creppy, E.E., Kane, A., Dirheimer, G., Lafarge-Frayssinet, C., Mousset, S. and Frayssinet, C. (1985). Genotoxicity of ochratoxin A in mice: DNA single-strand break evaluation in spleen, liver and kidney. *Toxicol. Lett.*, 28, 29-35.

Creppy, E.E., Betbeder, A.M., Gharbi, A., Counord, J., Castegnaro, M., Bartsch, H., Montcharmont, P., Fouillet, B., Chambon, P. and Dirheimer, G. (1991) Human ochratoxicosis in France. In: Castegnaro, M., Plestina, R., Dirheimer, G., Chernozemsky, I.N. and Bartsch, H., eds, *Mycotoxin, endemic nephropathy and urinary tract tumours. (IARC Scientific publications N°115)*, Lyon, International Agency for Research on Cancer, pp 145-151.

Engel, G. and von Milczewski, K.E. (1976) Zum Nachweis von Mykotoxin nach Aktivierung mit Rattenleber-homogenaten mittels Histidin-mangel-mutanten von Salmonella typhimurium. Kiel Milchwirtsch. *Forschungsger.*, 28, 359-366.

Hald, B. (1991) Ochratoxin A in human blood in European countries. In: Castegnaro, M., Plestina, R., Dirheimer, G., Chernozemsky, I.N. and Bartsch, H., eds, *Mycotoxin, endemic nephropathy and urinary tract tumours. (IARC Scientific publications N°115)*, Lyon, International Agency for Research on Cancer, pp 159-164.

Hennig, A., Fink-Gremmels, J. and Leistner, L. (1991) Mutagenicity and effects of ochratoxin A on the frequency of sister chromatid exchange after metabolic activation. In: Castegnaro, M., Plestina, R., Dirheimer, G., Chernozemsky, I.N. and Bartsch, H., eds, *Mycotoxin, endemic nephropathy and urinary tract tumours. (IARC Scientific publications N°115)*, Lyon, International Agency for Research on Cancer, pp 255-260.

Kane, A., Creppy, E.E., Roth, A., Röschenthaler, R. and Dirheimer , G. (1986). Distribution of the [3H]-label from low doses of radioactive ochratoxin A ingested by rats, and evidence for DNA single-strand breaks caused in liver and kidneys. *Arch. Toxicol.*, 58, 219-224.

Kanizawa, M. (1984) Synergistic effect of citrinin on hepatorenal carcinogenesis of ochratoxin A in mice. In: *Toxigenic fungi, their toxins and health hazard*,(H. Kurata and Y. Ueno, eds) Amsterdam, Oxford, New York, Elsevier Science Publishers, pp 245-254 (Developments in Food Science N° 7)

Kanizawa, M. and Suzuki, S. (1978) Induction of renal and hepatic tumors in mice by ochratoxin A, a mycotoxin. *Gann*, 69, 599-600.

Kuczuk, M.H., Benson, P.M., Heath, H. and Hayes, W. (1978). Evaluation of the mutagenic potential of mycotoxins using Salmonella typhimurium and Saccharomyces cerevisiae. *Mutat. Res.*, 53, 11-20.

Kuiper-Goodman, T. and Scott, P.M. (1989) Risk assessment of the mycotoxin ochratoxin A. *Biomed. Environ.Sci.*, 2, 179-248.

Lutz, W.K. (1979). In vivo covalent binding of organic chemicals to DNA as a quantitative indicator in the process of chemical carcinogenesis. *Mutat. Res.*, 65, 289-356.

Mälaveille, C., Brun, G. and Bartsch, H. (1991) Genotoxicity of ochratoxin A and structurally related compounds in *Escherichia coli* strains: studies on their mode of action. In: Castegnaro, M., Plestina, R., Dirheimer, G., Chernozemsky, I.N. and Bartsch, H., eds, *Mycotoxin, endemic nephropathy and urinary tract tumours. (IARC Scientific publications N°115)*, Lyon, International Agency for Research on Cancer, pp 261-266.

Manolova, Y., Manolov, G., Parvanova, L., Petkova-Bocharova, T., Castegnaro, M. and Chernozemsky,I.N. (1990); Induction of characteristic chromosomal aberrations, particularly X-trisomy, in cultured human lymphocytes treated by ochratoxin A, a mycotoxin implicated in Balkan endemic nephropathy. *Mutat.Res.*, 231, 143-149.

Miller, J.A. and Miller, E.C. (1969). The metabolic activation of carcinogenic aromatic amines and amides. *Prog.exp.Tumor Res.*, 11, 273-275.

Monod,G., Devaux, A. and Rivière J-L. (1988) Effect of chemical pollution on the activities of hepatic xenobiotic metabolizing enzymes in fish from the river Rhône. *Sci. Tot. Environ.*, 73, 189-201.

Mori, H., Kawai, K., Ohbayashi, F., Kuniyasu, T., Yamazaki, M., Hamasaki, T. and Williams, G.M. (1984). Genotoxicity of a variety of mycotoxins in the hepatocyte primary culture/DNA repair test using rat and mouse hepatocytes. *Cancer Res.*, 44, 2918-2923.

Petkova-Bocharova, T. and Castegnaro,M. (1991) Ochratoxin A in human blood in relation to balkan endemic nephropathy and urinary tract tumours in Bulgaria. In: Castegnaro, M., Plestina, R., Dirheimer, G., Chernozemsky, I.N. and Bartsch, H., eds, *Mycotoxin, endemic nephropathy and urinary tract tumours. (IARC Scientific publications N°115)*, Lyon, International Agency for Research on Cancer, pp 135-137.

Pfohl-Leszkowicz, A., Chakor, K, Creppy, E.E. and Dirheimer, G. (1991). DNA adduct formation in mice treated with ochratoxin A . In : Castegnaro, M., Plestina, R., Dirheimer, G., Chernozemsky, I.N. and Bartsch, H., eds, *Mycotoxin, endemic nephropathy and urinary tract tumours. (IARC Scientific publications N°115)*, Lyon, International Agency for Research on Cancer, pp 245-253.

Pfohl-Leszkowicz, A., Grosse, Y., Castegnaro, M., Petkova-Bocharova, T., Nicolov, I.G., Chernozemsky, I.N., Bartsch, H., Betbeder, A.M., Creppy, E.E. and Dirheimer, G. (1993a) Ochratoxin A related DNA adducts in urinary tract tumours of Bulgarian subjects. In: *"Postlabelling methods for the detection of DNA adducts"(IARC Scientific publications N°124)*, Lyon, International Agency for Research on Cancer, pp 115-122.

Pfohl-Leszkowicz, A., Grosse, Y., Kane, A., Creppy, E. and Dirheimer, G. (1993b) Genotoxicity and DNA binding of Ochratoxin A, a ubiquitous mycotoxin found in food and feed. In: Dengler, H.J. and Mutschler, E. eds, *Metabolism of xenobiotics and clinical pharmocology* Gustav Fischer Verlag. (in press)

Pfohl-Leszkowicz, A., Grosse, Y., Kane, A., Gharbi, A., Baudrimont, I., Obrecht, S., Creppy, E.E. and Dirheimer, G. (1993 c) Is the oxidative pathway implicated in the genotoxicity of ochratoxin ? *this volume.*

Philips, D.H., Grover, P.L. and Sims, P. (1979). A quantitative determination of the covalent binding of a series of polycyclic hydrocarbons to DNA in mouse skin. *Int.J.Cancer*, 23, 201-208.

Reddy, M.V. and Randerath,K.(1986) Nuclease P1 mediated enhancement of sensitivity of ^{32}P-postlabeling test for structurally diverse DNA adducts, *Carcinogenesis*, 7 , 1543-1551.

Umeda, T.M., Tsutsui, T., Ithoh, M. and Saito, M. (1977) Mutagenicity and inducibility of DNA single-strand breaks and chromosome aberrations by various mycotoxins. *Gann*, 68, 619-625.

Taningher, M., Scaccomanno, G., Santi, L., Grilli, S., Parodi, S. (1990). Quantitative predictability of carcinogenicity of the covalent binding index of chemicals to DNA: comparison of the in vivo and in vitro assays. *Environ. Health Perspect.*, 84, 183-192.

Ueno, Y. and Kubota, K. (1976) DNA-attacking ability of c arcinogenic mycotoxins in recombination-deficient mutant cells of *Bacillus subtilis*. *Cancer Res.*, 36, 445-451.

Vukelic, M., Sostaric, B. and Belicza, M. (1992) Pathomorphology of Balkan endemic nephropathy. *Fd.Chem.Toxic.*, 30, 193-200.

Wehner, F.C., Thiel, p.G., Van Rensburg, S.J. and Demasius, I.P.PC. (1978) Mutagenicity to Salmonella typhimurium of some Aspergillus and Penicilium toxins. *Mutat. Res.*, 58, 193-203.

Résumé

L'ochratoxine A est une mycotoxine impliquée dans la néphropathie endémique des Balkans, maladie caractérisée par une tubulonéphrite associée à des tumeurs du tractus urinaire. Elle provoque des cassures simple brin de l'ADN et induit des cancers chez les rongeurs. Afin de mieux comprendre la génotoxicité de l'OTA, la formation d'adduits à l'ADN, suite à l'administration de 2,5 mg/kg d'OTA à des souris, et leur disparition ont été testées par la méthode de post-marquage au phosphore 32. Dans l'ADN de rein, de foie et de rate, plusieurs adduits ont été observés 24h après une administration unique d'OTA. Cependant leurs taux varient significativement suivant l'organe et en fonction du temps. Le taux maximum d'adduits est obtenu après 48h et correspond à 103, 42 et 2,2 adduits par 10^9 nucléotides respectivement dans le rein, le foie et la rate, indiquant que le rein est l'organe cible pour la génotoxicité comme il l'est pour la carcinogénicité. Les adduits majeurs sont différents dans le rein et dans le foie. Tous les adduits disparaissent au bout de cinq jours dans le foie et la rate alors que quelques adduits persistent dans le rein au delà de seize jours après l'administration d'OTA. Certains adduits sont spécifiques de l'organe et de l'espèce. Des adduits à l'ADN peuvent aussi être détectés chez le rat. Le fait que les adduits ne soient pas les mêmes qualitativement et quantitativement d'une espèce à l'autre et d'un organe à l'autre est dû à une différence dans le métabolisme au niveau des organes et des espèces, conduisant à la formation de métabolites ultimes différents ainsi qu'à une différence dans l'efficacité des systèmes de réparation. L'incubation d'ADN avec de l'OTA en présence de microsomes de rein ou de foie conduit à la formation d'adduits uniquement avec les microsomes de rein. Dans le but de mettre en évidence le rôle de l'OTA dans l'étiologie de cancers humains de rein et de la vessie, l'ADN de personnes souffrant de néphropathie endémique des Balkans et ayant été sans doute exposées à l'OTA, a été analysé. Plusieurs adduits ont été détectés. Par comparaison et comigration, certains des adduits trouvés dans les tumeurs humaines ont pu être assimilés à des adduits produits par l'OTA ou un de ses métabolites chez la souris, indiquant que l'agent causal pourrait être identique.

Human ochratoxicosis and its pathologies. Eds E.E. Creppy, M. Castegnaro, G. Dirheimer. Colloque INSERM/ John Libbey Eurotext Ltd. © 1993, Vol. 231, pp. 209-215.

Distribution in Denmark of porcine nephropathy and chronic disorders of the urinary tract in humans

J.H. Olsen[1], B. Hald[2], I. Thorup[3] and B. Carstensen[1]

[1]*Division for Cancer Epidemiology, Danish Cancer Society, Copenhagen;* [2]*Department of Veterinary Microbiology, Royal Veterinary and Agricultural University, Copenhagen, and* [3]*Institute of Toxicology, National Food Agency, Copenhagen, Denmark*

Epidemiological studies on the possible association between exposure of humans to ochratoxin A (OTA) and the risk of urinary tract cancer and nephropathy have provided no clear answer. Recently, OTA was evaluated by a working group at the International Agency for Research on Cancer which found sufficient evidence for carcinogenicity in experimental animals, but inadequate evidence in humans. The overall evaluation resulted in the categorization of OTA in Group 2B (possibly carcinogenic to humans) (IARC, 1993). The contamination of crops with OTA is relatively high in Denmark (Kuiper-Goodman et al., 1989) and may constitute a public health problem.

Material:
A country-wide correlational study was performed in Denmark of the geographic distribution of porcine nephropathy during the late 1970s, and the distribution in the human population of urinary tract cancers and chronic kidney disorders during the 1970s and 1980s. The correlational analysis was performed on the level of the municipality of which there exists 276 in Denmark.

Occurrence of human cancer. All cases of cancer of the kidney parenchyma, renal pelvis, urinary bladder and lung were extracted from the files of the Danish Cancer Registry when diagnosed during the period 1972-87. Benign papillomas of the bladder were included under this term. The Cancer Registry, which was started in 1942, covers all inhabitants of the country, and is regarded to be almost complete (Jensen et al., 1985). Besides detailed diagnostic information on cancer, the Registry contains the municipality of residence of the patient at the date of diagnosis. Incidence rates for each of the selected tumour entities were calculated for each municipality after due adjustment for size and age-composition of the local population over the study period. Finally, we calculated per municipality the ratio between the standardized incidence rates (SMR ratio) of each of the urinary tract cancers (kidney parenchyma, pelvis and bladder) and the incidence rates of lung cancer. The latter calculation was performed in order to establish a local measure on the municipality level of the load on the population of urinary tract carcinogens after an approximate adjustment for smoking habits. Thus we assumed that smoking habits on the municipality level is closely reflected by the incidence rate of lung cancer.

Occurrence of other chronic kidney disorders. Since 1977, regional inpatient registration files in Denmark have been converted into a standardized format and transferred to the Central Hospital Discharge Register (Olsen et al., 1992). All discharged records for 1977 through 1987 that included a diagnosis of a chronic kidney disorder (nephrotic syndrome, chronic glomerulonephritis, nephropathia and similar conditions) were abstracted from this Register, together with the municipality of residence for each patient at the date of hospital discharge. Discharge frequencies of these disorders were calculated for each of the 276 municipalities of the country.

Occurrence of porcine nephropathy. Since the late 1970s hog slaughtering in Denmark has been controlled in regard to cases of nephropathy among the pigs. This supervision has been performed country-wide and any case of nephropathy has been reported to the local veterinary authorities by the veterinarians associated with the individual slaughterhouses. Based on these reports and information on local productivity it was possible to establish regional incidence rates of porcine nephropathy in Denmark.

Methods and results:

The geographic correlation between the incidence rates of urinary tract cancers (UTC) and cancer ratios (UTC/lung cancer) on the one hand, and chronic kidney disorders in humans and porcine nephropathy on the other hand was estimated. Figures 1a-c shows among women the geographic variation in the ratio of the incidence rate of each of the three urinary tract cancers and that of the lung. Women has been chosen in this example as they consume less tobacco than men and are supposed to be exposed to OTA to the same extend as men. The correlational analysis is currently in progress and will be presented and discussed at the meeting.

Porcine nephropathy has been linked strongly to the contamination of feedstuff with OTA. As the occurrence of porcine nephropathy has shown a considerable geographic variation in Denmark as has the OTA contamination of crops used for human and animal consumption, any indication of a correlation between porcine nephropathy and chronic urinary tract disorders in humans will lend support to the hypothesis of a causal association between these human disease entities and OTA exposures.

Acknowledgement:

We thank Mr. Castegnaro for kind advise in the preparation of this study. The research was supported by grants from the Environmental Medical Research Center of the Danish Board of Health.

Figures 1a-c (pages 212-214). Geographic variation among women in the ratio of the incidence of kidney cancer (1a), renal pelvis cancer (1b) and bladder cancer, and the incidence of lung cancer, Denmark, 1972-1987

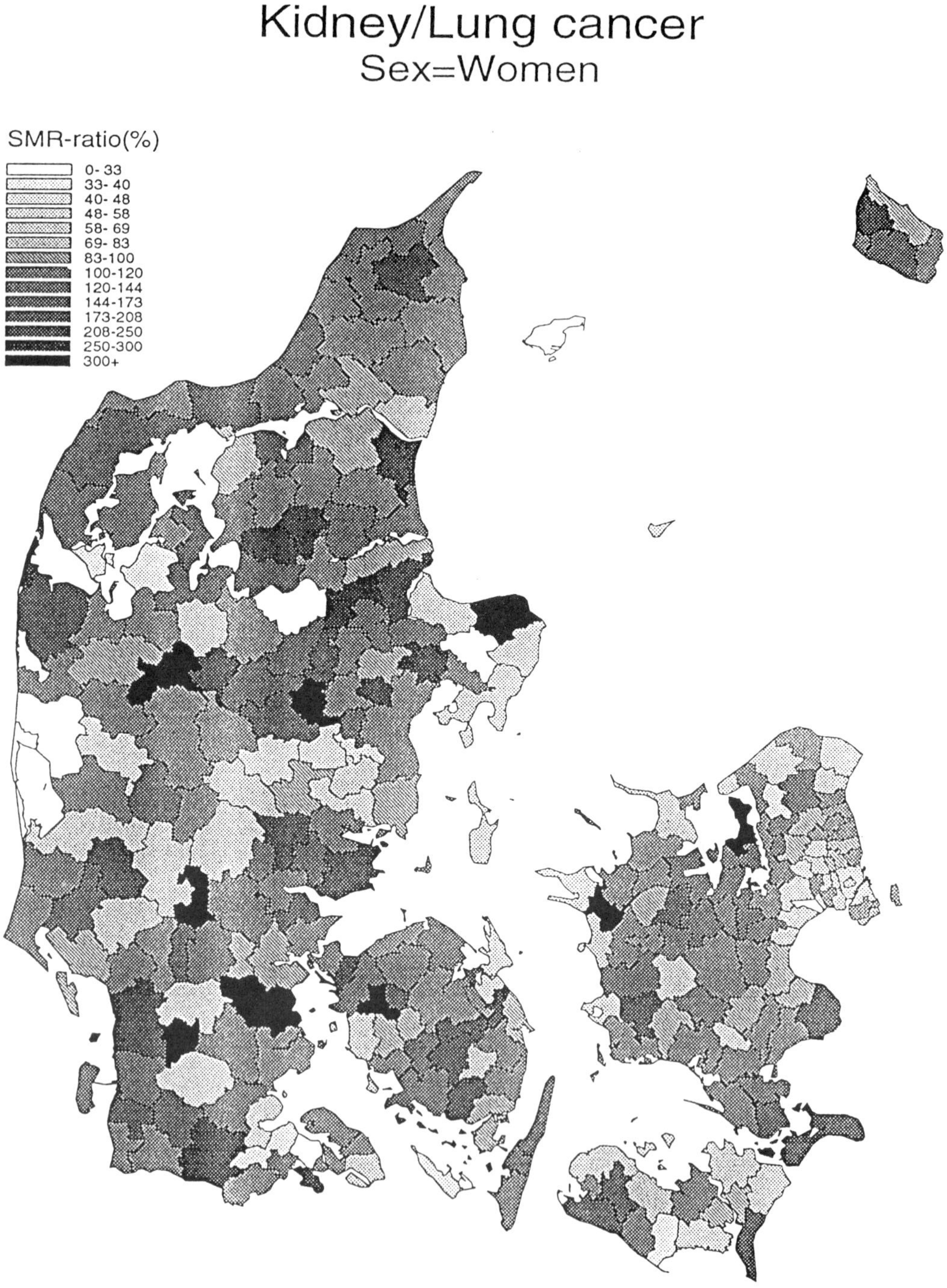

Figure 1a. Patients diagnosed 1968-87, all ages

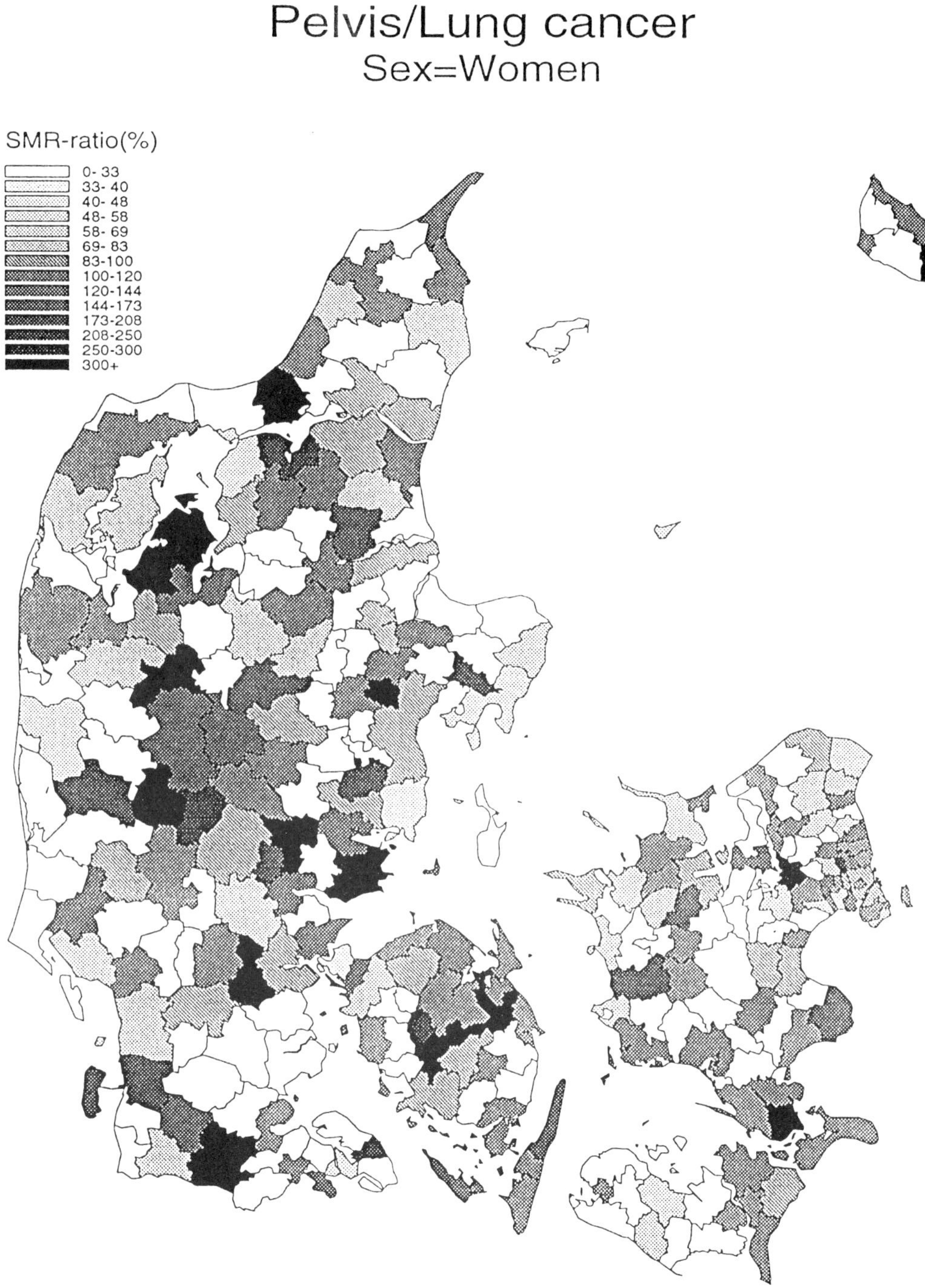

Figure 1b. Patients diagnosed 1968-87, all ages

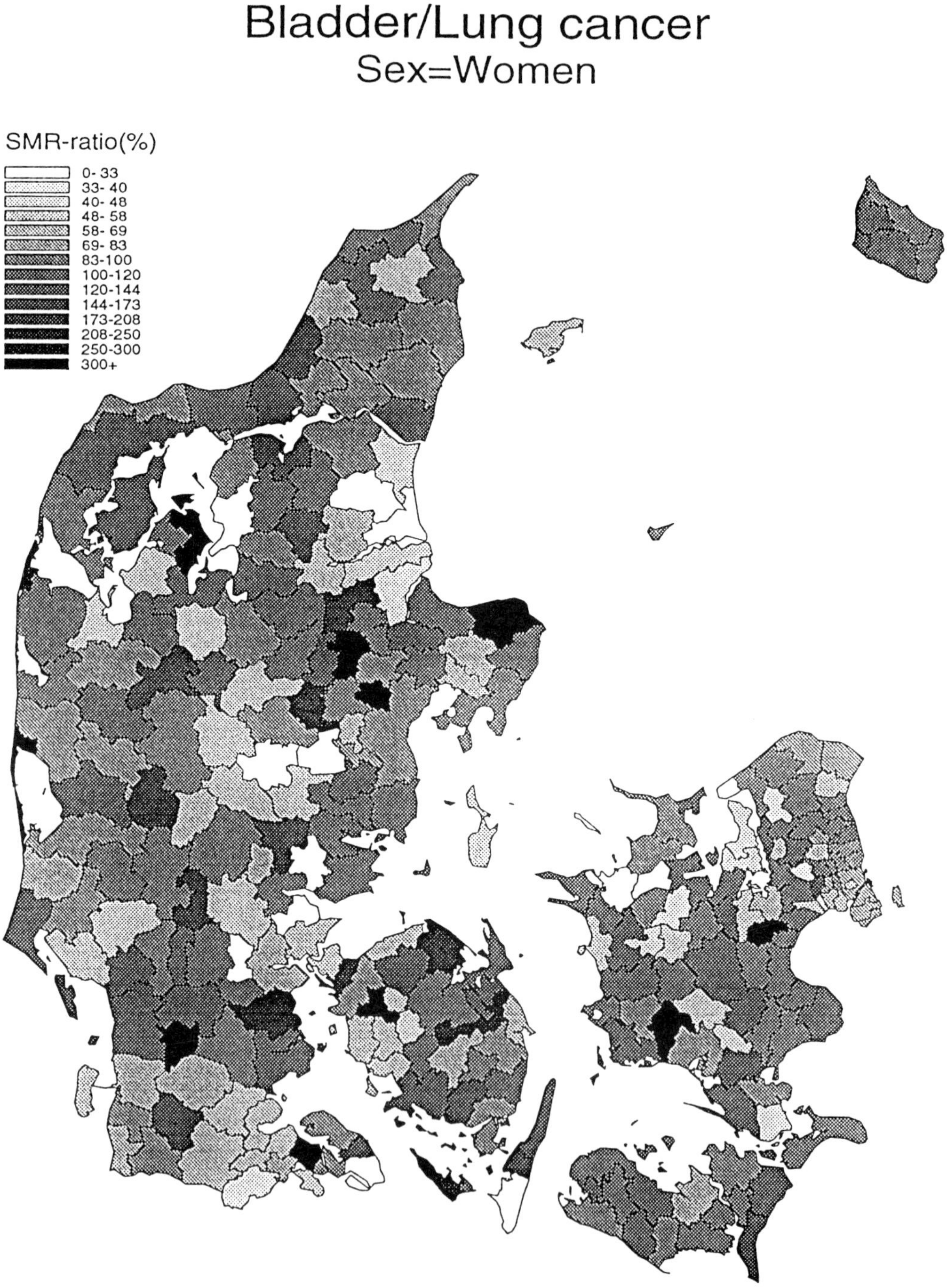

Figure 1c. Patients diagnosed 1968-87, all ages

References:

International Agency for Research on Cancer. Some naturally occurring substances: Some food items and constituents, spices, pyrolysis products and mycotoxins. Lyon 1993. IARC monographs on the evaluation of carcinogenic risks to humans, volume 56.

Kuiper-Goodman T, Scott PM. Risk assessment of the mycotoxin ochratoxin A. Biomed Environ Sci 1989; 2: 179-248.

Jensen OM, Storm HH, Jensen HS. Cancer registration in Denmark and the study of multiple primary cancers, 1943-80. Natl Cancer Inst Monogr 1985; 68: 245-51.

Olsen JH, Møller H, Frentz G. Malignant tumors in patients with psoriasis. J Am Acad Dermatol 1992; 27: 716-22.

Résumé:

L'ochratoxine A est néphrotoxique pour toutes les espèces animales étudiées à ce jour. L'ochratoxicose porcine a été découverte au Danemark dans les années 70.
La similitude des lésions rénales causées par l'ochratoxine chez l'animal avec celles observées chez l' Homme exposé naturellement par son alimentation à cette même mycotoxine est la raison majeure de son implication comme agent causal principal dans la Néphropathie endémique des Balkans. Or chez les personnes souffrant de cette maladie une fréquence très élevée de tumeurs du tractus urinaire est observée.
Cette publication montre la répartition géographique de la fréquence des néphropathies porcines et des troubles du tractus urinaire chez l' Homme au Danemark.

Summary

In the seventies porcin nephropathies have been reported in Denmark due to ochratoxin A contamination of feed and food . This is another source of human exposure to ochratoxin A. In the Balkans, ochratoxin A tentatively related nephropathy in B.E.N. areas is very frequently associated to urinary tract and bladder tumours.
The present paper shows geographical distribution of porcine nephropathy and chronic disorders of the urinary tract in human in Denmark.

Human ochratoxicosis and its pathologies. Eds E.E. Creppy, M. Castegnaro, G. Dirheimer. Colloque INSERM/ John Libbey Eurotext Ltd. © 1993, Vol. 231, pp. 217-225.

Markers for mycotoxin nephrotoxicity in domestic animals and man. Why are there no selective or specific ways of assessing the lesion ?

Peter H. Bach

Drug Development and Chemical Safety Research Unit, Faculty of Science, University of East London, Romford Road, London E15 4LZ, England

Introduction

Renal disease is classified by clinical manifestations, pathologic changes or aetiological agent. Nephrotoxicity (any renal dysfunction caused by chemicals) has been considered by Bach *et al.*, (1989, 1991); Bach and Lock, (1985, 1987, 1989); WHO, (1991) in terms of both structural and functional changes. The European Commission has established that end stage renal disease is of major health significance and costs the Community at least 3,555 million ECU (1 ECU approximately 1 US$) per year in direct health care costs. At least 20% of this health care burden is directly attributable to chemicals, environmental pollutants, natural toxicants and the therapeutic use and the abuse of medicines. *This represents at least 700 million ECU per year.* About 50% of end stage renal disease does not have an identifiable aetiology, but it is likely that chemicals play a significant role in causing or exacerbating such conditions (Bach *et al.*, 1989).

Mycotoxin nephrotoxicity

Citrinin produces an acute S_1 proximal tubular necrosis in several species (Berndt and Hayes, 1977; Phillips *et al.*, 1979; Phillips *et al.*, 1980a,b; Lockard *et al.*, 1980). Porcine nephropathy was associated with fungally contaminated feed (Elling and Moller, 1973; Rutqvist *et al.*, 1978; Sandor *et al.*, 1982) and both *A. ochraceus* cultures or pure ochratoxin A (OA) cause impairment of proximal tubular function such as polydipsia, polyuria, low urinary specific gravity, proteinuria, glucosuria, enzymuria and increased blood urea (Krogh *et al.*, 1974, 1976a). Early structural changes include condensation of cellular material, abnormal mitochondria, disappearance of membranes and excessive lipid droplets in the proximal convoluted tubules and the continuous desquamation of epithelial cells, tubular degeneration and necrosis, interstitial fibrosis and atrophy, and many dilated tubules. There were also atrophied and sclerotic glomerular tufts and glomerular basement membrane thickening (Elling *et al.*, 1985; Szczech *et al.*, 1973a,b). Almost identicle structural and functional changes have been described in avian species (Elling *et al.*, 1975; Krogh *et al.*, 1976b; Svendsen and Skadhauge, 1976; Dwivedi and Burns, 1984), the rat (Maxwell *et al.*, 1987; Purchase and Theron, 1968; Berndt and Hayes, 1979), dog (Szczech *et al.*, 1973a,b) and pig (Szczech *et al.*, 1973c). Long-term administration of the toxin induced progressive nephropathy, but not end stage renal failure (Krogh, 1979; Krogh *et al.*, 1979).

Balkan endemic nephropathy (BEN) is a fatal condition that represents a cascade or progression from an initial focal insult to chronic degenerative change. The clinical picture of BEN is that of a slowly progressive renal degeneration, including proteinuria and uraemia, shrunken kidneys with atrophied tubules, interstitial nephritis and marked glomerular basement membrane thickening and hyalinization, leading to end stage renal disease.

The aetiology is uncertain, but mycotoxins, particularly ochratoxin-A (Hall and Dammin, 1978; Hall, 1982; WHO, 1979), have been implicated as they have been identified in food (Krogh *et al.*, 1977; Pepeljnjak and Cvetnic, 1985; Petkova-Bocharova and Castegnaro, 1985; Cvetnic and Pepeljnjak, 1990) and human blood (Hult and Fuchs, 1986) in areas where the nephropathy is endemic (see Delacruz and Bach, 1990). Evidence for the involvement of OA in BEN is also based on the similarities between the pathological and functional features of porcine nephropathy, (WHO, 1979) and the human disease, but upper urothelial carcinoma has never been described in pigs or other animals. Thus, the animal and human data appear to be adequate to suggest that OA can be considered as a risk factor for toxic nephropathies and in the aetiology of human nephropathy. The causal role of OA in BEN awaits further evidence before it can be confirmed as the only aetiological factor.

The role of multiple mycotoxins in the expression of mycotoxicoses has been outlined elsewhere (McLean and Bach, 1993; Bach and McLean, 1993). OA and aflatoxin B1 enhanced kidney injury (Brown *et al.*, 1986; Brownie and Brownie, 1988; Campbell *et al.*, 1983; Harvey *et al.*, 1989; Huff and Doerr, 1981; Huff *et al.*, 1984, 1988; Micco *et al.*, 1988; Pier *et al.*, 1980; Rati *et al.*, 1981) and in the case of OA and penicillic acid the renal proximal tubular necrosis is affected synergistically (Kubena *et al.* 1984, 1988, 1989; Sansing *et al.*, 1976; Shepherd *et al.*, 1981).

Focal lesions and their consequences

The kidney consists of about 20 different cell types many of which work together in concert. Nephrotoxic agents generally target for a single renal cell type (such as the glomerulus, proximal or distal tubule or medulla) which undergoes degenerative changes for one of several mechanistic reasons. If multiple anatomical targets are affected toxicity becomes manifest in more than one clinical syndrome. The response to injury is dynamic, and adapts to maintain homeostasis during a cascade of repair and recovery (Bach, 1989). Depending on the type, frequency and region of the damaged tissue the kidney undergoes progressive degenerative change, has a reduced functional reserve or the organ recovers fully (WHO, 1991).

Functional reserve.

Renal functional reserve buffers short-lived or protracted demands on function. This is achieved by using only a limited number of nephrons or part of their cellular functions at any one time. Part of this functional reserve is used to meet the response to perturbation of the homeostatic system by water or electrolyte loading. Most of the understanding of renal functional reserve relate to changes in glomerular filtration rate (GFR) and renal blood flow (Friedlander *et al.*, 1989), but other types of functional reserve also occur . A reduced functional reserve may sensitize the kidney to subsequent renal injury, and an initiated degenerative cascade may either stablise or progress to acute or chronic renal failure. It is not possible to differentiate between total recovery, a kidney with a reduced functional reserve and an organ with early progressive degenerative change except in experimental studies where function and morphology can be assessed simultaneously (WHO, 1991).

Clinical investigations.

Nephropathies in man are generally of unknown origin, and clinical diagnosis relies on history, symptoms and/or laboratory abnormalities using retrospective data about exposure (WHO, 1991).

Markers of nephrotoxicity

Markers to screen population groups should be noninvasive, specific and sensitive to early renal functional changes and predictive of risk for renal insufficiency. They should not be affected by physiological conditions (e.g. physical workload, posture, pharmacological effects of exogenous substances or meat meals). Analytical methods must be easy, reproducible and applicable to numbers of appropriately stored samples. To overcome the methodological problems of timed samples, there is an increasing tendency to use spot samples (WHO, 1991).

Glomerular filtration

GFR and renal blood flow can be assessed by the clearance of labelled EDTA, DTPA or hippuran, but these clearance procedures are seldom adopted when screening population groups, since only single blood and spot urine samples are available (WHO, 1991).

Tests designed to assess selective dysfunction.

Serious glomerular and proximal tubular renal diseases may occur in the absence of haemodynamic changes and can be revealed by single plasma proteins in urine. High molecular weight proteins reflects glomerular dysfunction whereas urinary excretion of low molecular weight protein indicate tubular disfunction (WHO, 1991).

Tests designed to assess tissue damage.

Tissue constituents may be shed into urine following toxic damage to specific structures. The localised distribution of a range of enzymes to discrete regions of the nephron have been presented as a way of assessing target selective injury, although it is not very effective. Freshly voided urine is a hostile environment where acceptable stability and analytical precision do not exist for most enzymes. Renal tissue constituents can be detected by immunochemical methods but these also suffer from the same problems as outlined for enzymes. All of these recently developed tests need further validation in well designed longitudinal studies, since their prognostic value is currently unknown (WHO, 1991).

Clinical risk factors.

Risk factors that have been defined for nephrotoxicity and factors that predispose patients to renal failure (Porter, 1989) include:-

1. the age related decline in GFR (Davies and Shock, 1950; Porter, 1989; Porter and Bennett, 1989) and renal function reserve (anderson and Brenner, 1986), and diminished repair capacity (Laurent *et al.*, 1988) due to lack of responce to hypertrophic growth factors (Avendano and Lopez-Novoa, 1987).
2. pre-existing renal disease and loss of renal parenchyma predisposes to abnormal accumulation of many nephrotoxins.
3. high protein diet, multiple myeloma and other conditions where there is an added level of protein excretion where the kidney is under an additional work load.
4. male rodents are considerably more susceptible to nephrotoxicity and carcinogenicity from many environmental toxins (NTP 1983, 1986, 1987, 1988).

The ability to estimate risks that arise from multiple exposures is limited.

Extrapolation of animal data to man.

Risk assessment has been best defined for therapeutic agents. Despite marked species, strain, dietary and sex related renal structural and functional differences (Stolte and Alt 1980, 1982; Mudge 1985) there are many examples where data from common laboratory species can be sensibly extrapolated to man (Bach, 1990).

Mechanisms of renal injury

An understanding of the basic mechanisms of nephrotoxicity (Bach *et al.*, 1989, 1991; Bach and Lock, 1985, 1987, 1989; WHO 1991) have been key to the rational assessment of risk assessment where diagnostic techniqes have not facilitated the identification of renal degeneration. Glomerular lesions may be immunologically induced or direct toxicity to the glomerular apparatus or by materials that may be deposited in basement membrane (Caufield *et al.*, 1976; Burkholder, 1982). Damage to the glomeruli may also occur as a result of fibrin deposition which may *per se* damage glomeruli in several ways including occlusion of glomerular capillaries, involvement in an inflammatory reaction, or by direct toxicity to glomerular mesangial cells (Kanfer, 1989). The other mechanisms that are central to renal cell death include impaired lysosmal function, membrane changes, oxidative stress, altered Ca^{2+}-homeostasis (Recknagel, 1983; Pounds and Rosen, 1988; Moore *et al.*, 1986; Pounds, 1984; Olorunsogo *et al.*, 1985) by disruption of plasma membrane, cytoskeleton, endoplasmic reticulum and mitochondria and initial DNA damage, or by receptor mediated programmed cell death described as apoptosis, oxygen free radical injury to renal mitochondria (Malis and Bonventre, 1986). Lipid peroxidation, free radicals, and thereby result in membrane damage, altered permeability, or fluidity. Proximal renal tubular cells are particularly vulnerable to the toxic action of chemicals due to their high energy demand (such as reabsorptive and secretory functions). Redox-active agents may cause extensive oxidation of GSH to GSSG and cause oxidation of cellular enzymes, depletion of cellular ATP and loss of mitochondrial function (Trump *et al.*, 1989). Reactive, electrophilic metabolites bind covalently to tissue proteins, and it has been suggested that cell injury and death are a consequence of the interaction of such reactive intermediates with critical cellular molecules. The early mechanism of OA action (Dirheimer and Creppy, 1991) in the case of renal proximal tubules appears to be mitochondrial dysfunction, iron-mediated lipid peroxidation does not appear to contribute to death (Aleo *et al.*, 1991).

Application to nephromycotoxicoses

The complexity of renal function and the diversity of nephrotoxicty makes the diagnosis of chemically induced nepropathy difficult using the current invasive on non-invasive criteria unless renal function is severely and possibly irreversible compromised. There is adequate data to show that several mycotoxins are nephrotoxic, but much the data accrued from investigative studies have focused on causing lesions rather than identifying new sensitive and selective methods for their assessment. The better assessment of the role of mycotoxins in human nephropathy should be based on:-

1. identifying and study risk populations and their cohorts from knowing the prevalence of mycotoxins,
2. understanding the molecular and cellular mechanisms by which each mycotoxin causes renal lesions as the only rational approach to develop sensitive - specific markers
3. applying this multi-disciplinary knowledge to assess renal changes in animal models
4. adapting new methods of assessing nephrotoxicity as they are published.

Acknowledgements

Dr. A.T. Evans and Stephen Brant provided useful critical comment. The authors' research was supported by the Wellcome Trust, The Cancer Research Campaign, The International Agency for Research on Cancer, The Kidney Research Fund of Great Britain, Humane Research Trust and European Commission.

References

Aleo, M.D., Wyatt, R.D., and Schnellmann, R.G. (1991): Mitochondrial dysfunction is an early event in ochratoxin A but not oosporein toxicity to rat renal proximal tubules. *Toxicol. Appl. Pharmacol.*, **107**: 73-80

Anderson, S., and Brenner, B.M. (1986): Effects of aging on the renal glomerulus. *Am. J. Med.*, **80**: 435-442.

Avendano, L.H. and Lopez-Novoa, J.M. (1987): Glomerular filtration and renal blood flow in the aged. In: *Renal Function and Diseases in the Elderly*. Macias-Nunez, I.F. and Cameron, J.S. [Editors]. London, Boston, Toronto, Butterworth. pp. 27-48.

Bach, P.H. (1989): The detection of chemically induced renal injury, the cascade of degenerative morphological and functional changes that follow the primary nephrotoxic insult and the evaluation of these changes by in virto methods. *Toxicol. Lett.*, **46**: 237-250.

Bach, P.H. (1990): Interpretation of animal nephrotoxicity data as it relates to man: Our present level of understanding as the basis for risk assessment. In: *Acute Renal Failure*. Solez. K., Racusen, L.C. and Williams, G.M. [Editors]. Chapter 12, pp. 187-210. Marcel Dekker, New York.

Bach, P.H. and Lock, E.A., [Editors]. (1985): *Renal Heterogeneity and Target Cell Toxicity*. Chichester, Wiley.

Bach, P.H. and Lock, E.A., [Editors]. (1987): *Nephrotoxicity in the Experimental and the Clinical Siuation*. Dordrecht, Nijhoff.

Bach, P.H. and McLean, M. (1993): Research Priorities for Assessing the Risk of Multiple Mycotoxin Exposure to Domestic Animals and Man: What we know and what we need to know! In: *Human Ochratoxicosis and Associated Pathologies in Africa and Developing Countries* Creppy, E., Dirheimer, G. [Editors]. John Libbey and Co., Montrouge. This volume.

Bach, P.H., Berlin, A., Heseltine, E., Krug, E., Lauwerys, R., Smith, E. and Van Der Venne, M.-T., [Editors]. (1989): Proceedings of the International Workshop on the Health Significance of Nephrotoxicity. *Toxicol. Lett.*, **46**: 1-306.

Bach, P.H., Gregg, N.J., Wilks, M.W. and Delacruz, L. (1991): [Editors]. *Nephrotoxicity: Mechanisms, Early Diagnosis and Therapeutic Management.* Marcel Dekker, New York.

Berndt, W.O. and Hayes, A.W. (1977): Effects of citrinin on renal transport processes. *J. Environ. Path. Tox.*, **1**: 93-103.

Berndt, W.O. and Hayes, A.W. (1979): *In vivo* and *in vitro* changes in renal function caused by ochratoxin A in the rat. Toxicol., **12**: 5-17.

Brown T.P., Manning R.O., Fletcher O.J. and Wyatt R.D. (1986): The individual and combined effects of citrinin and ochratoxin A on renal ultrastructure in layer chicks. *Avian Dis.*, **30**: 191-198.

Brownie, C.F. and Brownie, C. (1988): Preliminary study on serum enzyme changes in Long Evans rats given parenteral ochratoxin A, aflatoxin B_1 and their combination. *Vet. Human Toxicol.*, 30, 211-214.

Burkholder, P.M. (1982): Functions and pathophysiology of the glomerular mesangium (editorial). *Lab. Invest.*, **46**: 239-241.

Campbell M.L., May J.D., Huff W.E. and Doerr J.A. (1983): Evaluation of immunity of young broiler chickens during simultaneous aflatoxicosis and ochratoxicosis. *Poult. Sci.*, **62**: 2138-2144.

Caulfield, J.P., Reid, J.J. and Farqubar, M.G. (1976): Alterations of the glomerular epithelium in aminonucleoside nephrosis. *Lab. Invest.*, **34**: 43-59.

Cvetnic, Z., and Pepeljnjak, S. (1990): Ochratoxinogenicity of Aspergillus ochraceus strains from nephropathic and non-nephropathic areas in Yugoslavia. *Mycopathologia*, **110**: 93-9

Davies, D.F. and Shock, N.W. (1950): Age changes in glomerular filtration rate, effective renal plasma flow and tubular excretory capacity in adult males. *J. Clin. Invest.*, **29**: 496-507.

Delacruz, L. and Bach, P.H. (1990): The role of ochratoxin A metabolism and biochemistry in animal and human nephrotoxicity. *J. Biopharm. Sci.*, **1**: 277-304.

Dirheimer, G., and Creppy, E.E. (1991): Mechanism of action of ochratoxin A. *IARC. Sci. Publ.*, 171-86

Dwivedi, P. and Burns, R.B. (1984): Pathology of ochratoxicosis A in young broiler chicks. *Res. Vet. Sci.*, **36**: 92-103.

Elling, F., Nielsen, J.P., Lillehoj, E.B., Thomassen, M.S. and Stormer, F.C. (1985): Ochratoxin A-induced porcine nephropathy: Enzyme and ultrastucture changes after short-term exposure. *Toxicon.*, **23**: 247-254.

Elling, F. and Moller, T. (1973): Mycotoxic nephropathy in pigs. *Bull. Wld. Hlth. Org.*, **49**: 411-418.

Elling, F., Hald, B., Jacobsen, Chr. and Krogh, P. (1975): Spontaneous cases of toxic nephropathy in poultry associated with ochratoxin A. *Acta Path. Microbiol. Scand. Sect. A*, **83**: 739-741.

Friedlander, G., Blanchet, F. and Amiel, C. (1989): Renal functional reserve. *Toxicol. Lett.*, **46**: 227-235.

Hall III, P.W. and Dammin, G.J. (1978): Balkan Nephropathy, *Nephron*, **22**: 281-300.

Hall III, P.W. (1982): Endemic Balkan Nephropathy, In: *Nephrotoxic Mechanisms and Environmental Toxins*. G. Porter, [Editor]. New York, Plenum. pp. 227- 240.

Harvey R.B., Huff W.E., Kubena L.F. and Phillips T.D. (1989): Evaluation of diets cocontaminated with aflatoxin and ochratoxin fed to growing pigs. *Am. J. Vet. Res.*, **50**: 1400-1405.

Huff W.E. and Doerr J.A. (1981): Synergism between aflatoxin and ochratoxin A in broiler chickens. *Poult. Sci.*, **60**: 550-555.

Huff W.E., Doerr J.A., Wabeck C.J., Chaloupka G.W., May J.D. and Merkley J.W. (1984): The individual and combined effects of aflatoxin and ochratoxin A on various processing parameters of broiler chickens. *Poult. Sci.*, **63**: 2153-2161.

Huff W.E., Kubena L.F., Harvey R.B. and Doerr J.A. (1988): Mycotoxin interactions in poultry and swine. *J. Animal Sci.*, **66**: 2351-2355.

Hult, K. and Fuchs, R. (1986): Analysis and dynamics of ochratoxin A in biological systems. In: *Mycotoxins and Phycotoxins. Sixth International IUPAC Symposium Mycotoxins and Phycotoxins.* Steyn, P.S. and Vleggaar, R. [Editors]. Amsterdam, Elsevier Science Publishers B.V. pp. 365-376.

Kanfer, A. (1989): The role of coagulation in glomerular injury. *Toxicol. Lett.*, **46**: 83-92.

Krogh, P. (1979): Environmental Ochratoxin A and Balkan (Endemic) Nephropathy: Evidence for support of causal relationship. In: *Endemic Balkan Nephropathy.* Strahinjic, S. and Stafanovic, V., [Editors]. Proceedings of the 4th symposium on Endemic (Balkan) Nephropathy, NIS, pp 35-43.

Krogh, P., Azelsen, N.H., Elling, F., Gyrd-Hansen, N., Hald, B., Hyldgaard-Jensen, J., Larsen, A.E., Madsen, A., Mortensen, H.P., Moller, T., Petersen, O.K., Ravnskov, U., Rostgaard, M. and Aalund, O. (1974): Experimental porcine nephropathy. Changes of renal function and structure induced by ochratoxin A-contaminated feed. *Acta Pathol. Microbiol. Scand. Suppl.*, **246**: 1-21.

Krogh, P., Elling, F., Gyrd-Hansen, N., Hald, B., Larsen, A.E., Lillehoj, E.B., Madsen, A., Mortensen, H.P. and Ravnskov, U. (1976): Experimental porcine nephropathy: changes of renal function and structure perorally induced by chrystalline ochratoxin A. *Acta Path. Microbiol. Scand.*, **84**: 429-434.

Krogh, P., Elling, F., Hald, B., Jylling, B., Petersen, V.E., Skadhauge, E. and Svendsen, C.K. (1976): Experimental avian nephropathy: changes of renal function and structure induced by ochratoxin A-contaminated feed. *Acta Pathol. Microbiol. Scand., Sect. A*, **84**: 429-434.

Krogh, P., Hald, B., Plestina, R. and Ceovic, S. (1977): Balkan (endemic) nephropathy and foodborne ochratoxin A: Preliminary results of a survey of foodstuffs. *Acta. Pathol. Microbiol. Scand. Sect. B*, **85**: 238-240.

Krogh, P., Elling, F., Friis, Chr., Hald, B., Larsen, A. E., Lillehoj, E. B., Madsen, A., Mortensen, P., Rasmussen, F. and Ravnskov, U. (1979): Porcine nephropathy induced by long-term ingestion of ochratoxin A. *Vet. Path.*, **16**: 466-475.

Kubena L.F., Phillips T.D., Witzel D.A. and Heidelbaugh N.D. (1984): Toxicity of ochratoxin A and penicillic acid to chicks. *Bull. Environ. Contam. Toxicol.*, **32**: 717-723.

Kubena L.F., Huff W.E., Harvey R.B., Corrier D.E., Phillips T.D. and Creger C.R. (1988): Influence of ochratoxin A and deoxynivalenol on growing broiler chicks. *Poult. Sci.*, **67**: 253-260.

Kubena L.F., Harvey R.B., Huff W.E., Corrier D.E., Phillips T.D. and Rottinghaus G.E. (1989): Influence of ochratoxin A and T-2 toxin singly and in combination on broiler chickens. *Poult. Sci.*, **68**: 867-872.

Laurent, G., Toubeau, G., Heuson-Steinnon, J.A., Tulkens, P. and Maldauge, P. (1988): Kidney tissue repair after nephrotoxic injury: biochemical and morphologic characterization. *CRC Critical Rev. Toxicol.*, **19**: 147-183.

Lockard, W.V., Phillips, R.D., Hayes, A.W., Berndt, W.O. and O'Neal, R.M. (1980): Citrinin nephrotoxicity in rats: A light electron microscopic study. *Exp. Molec. Path.*, **32**: 266-340.

Malis, C.D. and Bonventre, J.V. (1986): Mechanism of calcium potentiation of oxygen free radical injury to renal mitochondria. *J. Biol. Chem.*, **261**: 14201-14208.

Maxwell, M.H., Burns, R.B. and Dwivedi, P. (1987): Ultrastructural study of ochratoxicosis in quail (*Coturnix coturnix japonica*). *Res. Vet. Sci.*, **42**: 228-231.

McLean, M. and Bach, P.H. (1993): Multiple mycotoxin exposure: Risk to domestic animals and man. *Food Chem. Toxicol.*, **Submitted**

Micco C., Miraglia M., Benelli L., Onori R., Ioppolo A. and Mantovani A.L. (1988): Long term administration of low doses of mycotoxins in poultry. 2. Residues of ochratoxin A and aflatoxin in broilers and laying hens after combined administration of ochratoxin A and aflatoxin B1. *Food Add. Contam.*, **5**: 309-314.

Moore, M.A., Nakamura, T., Shirai, T. and Ito, N. (1986): Immunohistochemical demonstration of increased glucose-6-phosphate dehydrogenase in preneoplastic and neoplastic lesions induced by propylnitrosamines in F 344 rats and Syrian hamsters. *Gann*, **77**: 131-138.

Mudge, G.H. (1985): Pathogenesis of nephrotoxicity: Pharmacological principles. In: *Renal Heterogeneity and Target Cell Toxicity*. Bach, P.H. and Lock, E.A. [Editors]. Chichester, Wiley. pp. 1-12.

NTP (1983): Technical Report on the Carcinogenesis Bioassay of Pentachloroethane (CAS No. 76-01-7) in F344/N Rats and B6c3F1 Mice (Gavage Studies). NIH Publication No. 83-1788, Department of Health and Human Services, Research Triangle Park, NC.

NTP (1986): Technical Report on the Toxicology and Carcinogenesis Studies of 1,4-Dichlorobenzene (CAS No. 106-46-7) in F344/N Rats and B6c3F1 Mice (Gavage Studies). NIH Publication No. 86-2575, Department of Health and Human Services, Research Triangle Park, NC.

NTP (1987): Toxicology and carcinogenesis studies of 1,4-dichlorobenzene in F344/N rats and B6C3F1 mice. NTP TR 319. U.S. Department of Health and Human Services, Public Health Service, National Institutes of Health.

NTP (1988): NTP Draft technical report on the toxicology and carcinogenesis studies of ochratoxin A (CAS No. 303-47-9) in F344/N rats (gavage studies). NIH Publication No. 88-2813 (G. Boorman, Ed.), US Department of Health and Human Services National, Institutes of Health, Research Triangle Park, N.C.

Olorunsogo, O.O., Malomo, S.O. and Bababunmi, E.A. (1985): Protonophoric properties of fluorinated arylalkylsulphonamides: Observations with perfluidone. *Biochem. Pharmacol.*, **34**: 2945-2952.

Pepeljnjak, S. and Cvetnic, Z. (1985): The mycotoxicological chain and contamination of food by ochratoxin A in the nephropathic and non-nephropathic areas in Yugoslavia. *Mycopathologia*, **90**: 147-153.

Petkova-Bocharova, T. and Castegnaro, M. (1985): Ochratoxin A contamination of cereals in an area of high incidence of Balkan endemic nephropathy in Bulgaria. *Food Addit. Contam.*, **2**: 267-270.

Phillips, R., Hayes, A.W. and Berndt W.O. (1979): Disposition of ^{14}C-citrinin in the rat. *Toxicology*, **12**: 285-298.

Phillips, R.D., Hayes, A.W. and Berndt, W.O. (1980a): High pressure liquid chromatographic analysis of citrinin and its application to biological systems. *J. Chromatog.*, **190**: 419-425.

Phillips, R.D., Hayes, A.W., Berndt, W.O. and Williams. W.L. (1980b): Effects of citrinin on renal function and structure. *Toxicology*, **16**: 123-137.

Pier A.C., Richard J.L. and Cysewski S.J. (1980): Implications of mycotoxins in animal diseases. *J. Am. Vet. Med. Assoc.*, **176**: 719-724.

Porter, G. A. (1989): Risk factors for toxic nephropathies. *Toxicol. Lett.*, **46**: 269-279.

Porter, G.A. and Bennet, W.M. (1989): Drug-induced renal effects of cyclosporine, aminoglycoside antibiotics and lithium: extrapolation of animal data to man. In: *Nephrotoxicity: Extrapolation from In Vitro to In Vivo, and Animals to Man*. Bach, P. H. and Lock, E. A. [Editors]. New York and London, Plenum Press. pp. 147- 170.

Pounds, J.G. (1984): Effect of lead intoxication on calcium-mediated cell function: A review. *Neurotoxicology*, **5**: 295-332.

Pounds, J.G. and Rosen, J.F. (1988): Cellular Ca^{2+} Homeostasis and Ca^{2+}-mediated cell processes as critical targets for toxicant action: Conceptual and methodological pitfalls. *Toxicol. Appl. Pharmacol.*, **94**: 331-341.

Purchase, I.F.H. and Theron, J.J. (1968): The acute toxicity of ochratoxin A to rats. *Fd. Cosmet. Toxicol.*, **6**: 479-483.

Rati E.R., Basappa S.C., Sreenivasa Murthy V., Ramesh H.P. and Ramesh B.S. (1981): The synergistic effect of aflatoxin B1 and ochratoxin A in rats. *J. Food Sci. Technol.*, **18**: 176-179.

Recknagel, R.O. (1983): Carbon tetrachloride hepatotoxicity: *status quo* and future prospects. *Trends Pharmacol. Sci.*, 129-131.

Rutqvist, L., Bjurklund, N.E., Hult, K., Hukby, E. and Carlsson, B. (1978): Ochratoxin A as the cause of spontaneous nephropathy in fattening pigs. *Appl. Environ. Microbiol.*, **36**: 920-925.

Sandor, G., Glavits, R., Vajda, L., Vanyi, A. and Krogh, P. (1982): Epidemiological study of ochratoxin A-associated porcine nephropathy in Hungary. In: *Proceedings of the V International IUPAC Symposium on Mycotoxins and Phycotoxins*, Vienna. 349-352.

Sansing G.A., Lillehoj J.B., Detroy R.N. and Miller M.A. (1976): Synergistic toxic effects of citrinin, ochratoxin A and penicillic acid in mice. *Toxicon*, 14, 213-220.

Shepherd E.C., Phillips T.D., Joiner G.N., Kubena L.F. and Heidelbaugh N.D. (1981): Ochratoxin A and penicillic acid interaction in mice. *J. Environ. Sci. Health*, **B16**: 557-573.

Stolte, H. and Alt, J., [Editors]. (1980): Research animals and experimental design in nephrology. *Contrib. Nephrol.*, **19**: 1-249.

Stolte, H. and Alt, J. (1982): The choice of animals for nephrotoxic investigations. In: *Nephrotoxicity: Assessment and Pathogenesis.* Bach, P.H., Bonner, F.W., Bridges, J.W. and Lock, E.A. [Editors]. Chichester, Wiley. pp. 102-112.

Svendsen, C. and Skadhauge, E. (1976): Renal Functions in Hens fed Graded Dietary Levels of Ochratoxin A. *Acta Pharmacol. Toxicol.*, **38**: 186-194.

Szczech, G.M., Carlton, W.W. and Tuite, J. (1973a): Ochratoxicosis in beagle dogs. I. Cinical and clinicopathological features. *Vet. Path.*, **10**: 135-154.

Szczech, G.M., Carlton, W.W. and Tuite, J. (1973b): Ochratoxicosis in beagle dogs. II. Pathology. *Vet. Path.*, **10**: 219-231.

Szczech, G.M., Carlton, W.W., Tuite, J. and Caldwell, R.(1973c): Ochratoxin A toxicosis in swine. *Vet. Path.*, **10**: 347-364.

Trump, B.F., Berezesky, I.K., Smith, M.W., Phelps, P.C. and Elliget, K.A. (1989): The relationship between cellular ion deregulation and acute and chronic toxicity. *Toxicol. Appl. Pharmacol.*, **97**: 6-22.

WHO (1979): Environmental Health Criteria 11, Mycotoxins. WHO, Geneva.

WHO (1991): Environmental Health Criteria 119, Principles and Methods for the Assessment of Nephrotoxicity Associated with Exposure to Chemicals. WHO, Geneve.

Summary:

Several mycotoxins are nephrotoxic in animal and likely in human (ochratoxin A, citrinin etc ...). But these are not the only one cause of nephropathy . So specific and selective ways of assessing the lesions caused by mycotoxins are needed which are now laking.

Résumé

Plusieurs mycotoxines sont néphrotoxiques chez l'animal et aussi chez l'Homme vraisemblablement (ochratoxine a , citrinine etc...) . Mais elles ne sont pas les seules causes de néphropathie. Il est donc nécessaire d'avoir des paramètres spécifiques et sélectifs pour évaluer les lésions causées par ces mycotoxines.

Human ochratoxicosis and its pathologies. Eds E.E. Creppy, M. Castegnaro, G. Dirheimer. Colloque INSERM/ John Libbey Eurotext Ltd. © 1993, Vol. 231, pp. 227-234.

Néphropathies interstitielles chroniques. Approches cliniques et étiologiques : ochratoxine A

A. Achour[(1)], M. El-May[(1)], H. Bacha[(2)], M. Hamammi[(3)], K. Maaroufi[(2)] et E.E. Creppy[(4)]

[(1)]*Service de Néphrologie et Hémodialyse CHU Monastir, Tunisie;* [(2)]*Faculté de chirurgie dentaire, Monastir, Tunisie;* [(3)]*Faculté de médecine, Monastir, Tunisie;* [(4)]*Laboratoire de Toxicologie, Université Bordeaux II, France*

I. INTRODUCTION

Les différents types anatomo-cliniques de néphropathies ont commencé à être identifiés à la fin du siècle dernier.

Il existe plusieurs variétés de néphropathies chroniques: les néphropathies glomérulaires chroniques (NGC), les néphropathies interstitielles chroniques (NIC) et les néphropathies vasculaires chroniques. Ces différentes affections sont totalement individualisées sur le plan clinique, histologique, pathogénique et étiologique.

Les NIC ont un tableau clinique bien stéréotypé, un aspect histologique particulier et des étiologies propres. Si elles semblent avoir été bien identifiées, elles n'ont pas livré tous leurs secrets quant à leur étiologie. En effet 50 % des NIC demeurent inexpliquées dont la néphropathie endémique des Balkans (BEN). Une mycotoxine, l'ochratoxine A semble impliquée directement pour une partie des formes inexpliquées (Stephanovic V., Polenakovic M.H. 1991,et Hall P.W. 1992).

Le but de notre étude est d'une part de démontrer s'il existe une corrélation entre cette mycotoxine et l'une des néphropathies en TUNISIE et d'autre part, de voir si cette néphropathie a des ressemblances avec la BEN.

II. ETUDE CLINIQUE :

Rien ne distingue cliniquement de façon indiscutable une NIC d'un autre type histologique de néphropathie chronique. La latence clinique et l'évolution asymptomatique, caractères volontiers attribués aux NIC, sont en réalité le fait de la plupart des néphropathies.

A/ Les manifestations cliniques :

sont dominées par :

- céphalées fréquentes
- asthénie, langue rouge, soif, anémie, polyurie, polydipsie, nyctiurie
- douleur des lombaires
- hypertension artérielle (HTA) absente
- pigmentation pseudoaddisonienne (urochrome)
- absence d'oedème.

B/ Les manifestations biologiques :

- la protéinurie : elle est généralement absente ou minime; elle varie de 0,15g à 0,5g/24h, composée surtout par des protéines de faible poids moléculaire
- la leucocyturie amicrobienne (20-25l/mm3)
- le pouvoir de concentration des urines (abaissée)
- les troubles de l'acidification avec une acidose hyperchlorémique avec $Cl > 110$ mml/l, $HCO_3^- < 22$ mml/l
- le métabolisme du sodium est variable :

. il peut être normal du fait de l'adaptation glomérulo-tubulaire conservée ou bien d'une fuite sodée

. une rétention sodée au terme d'une insuffisance rénale chronique (IRC) avancée.

C/ Signes radiologiques :

Le tableau est dominé par une atrophie des reins se manifestant par une asymétrie anatomique et fonctionnelle. Les caractères morphologiques des reins à l'urographie intra veineuse (UIV) se basent volontiers sur l'inégalité, l'hypotrophie et l'irrégularité des deux reins.

D/ Signes histologiques :

. Au début, avant l'IRC : la biopsie rénale montre une atrophie tubulaire et focale accompagnée d'oedème interstitiel et d'une sclérose sans infiltration cellulaire.

Au stade évolué : l'aspect histologique global est celui d'une atrophie tubulaire globale se noyant dans une large étendue de fibrose.

E/ Evolution :

La NIC évolue toujours vers l'IRC. C'est une évolution progressive très lente (l'IRC est atteinte vers l'âge de 50 à 60 ans). Elle est caractérisée par l'absence de rétention de sel, l'absence d'hypertension artérielle, l'infection urinaire rare et la protéinurie minime.

F/ ETIOLOGIE DES NIC:

L'atteinte anatomique exclusive ou prédominante de l'interstitium rénal définit ces néphropathies. Historiquement, ce terme de NIC fut surtout employé pour la première fois par Spuhler et Zollinger en 1953 chez des horlogers suisses qui prennent quotidiennement des analgésiques.

Elles représentent 30 à 40 % des néphropathies chroniques et relèvent de causes multiples (voir tableau 1).

Tableau 1: Principales étiologies des NIC :

I - Pyélonéphrites chroniques
- 1/ infections urinaires
- 2/ lithiase
- 3/ syndrome obstructif
  - . reflux vésico - urétral
  - . obstruction des voies urinaires

II- Néphropathies des analgésiques :phénacétine(10-20%)

III- Néphropathies métaboliques
- . goutte (hyperuricémie)
- . néphrocalcinose (hypercalcémie)
- . oxalose
- . cystinose
- . intoxication par les métaux lourds (plomb, cadmium.

IV- NIC immunologique : dépôts d'anticorps antimembrane basale tubulaire.

V- NIC héréditaire : néphronophtise

VI- NIC idiopathique.

III- MATERIELS ET METHODES :

Nous avons étudié plusieurs échantillons :
- Un groupe de 20 témoins volontaires sains (G1), un groupe de néphropathes (25 sujets à néphropathies diverses: glomérulaire interstitielle et vasculaire)
- Un groupe de malades en insuffisance rénale traité par l'hémodialyse chronique (83 d'étiologies diverses)
- Un fichier est établi, recensant plusieurs données:

. l'état civil du patient (âge, sexe, origine géographique, profession et ses habitudes)

. les antécédents personnels et familiaux, en insistant surtout sur la prise de médicaments (phénacétine)

. l'état clinique (en rapport avec la néphropathie) en insistant sur les céphalées, lombalgies, hématurie, anémie et prise de la tension artérielle, recherche d'oedème, diurèse

. un examen radiologique (abdomen sans préparation, UIV et/ou échographie rénale) pour apprécier la taille des reins

. un examen histologique (ponction biopsie rénale) pour 2 malades

. une enquête alimentaire précisant la nature de l'alimentation, sa fréquence, les conditions de consommation, la durée et le mode de conservation

. un bilan biologique (GSRh, urée, créatinine, acide urique, NFS, protéine 24h), sédiment urinaire, ionogramme urinaire

. une β2 microglobulie(sang-urine)et un prélèvement destinés à la mise en évidence de l'ochratoxine

. les études statistiques sont basées sur le test T pour série appariée.

IV- RESULTATS ET DISCUSSION

Les étiologies des néphropahies sont multiples et variées, cependant certaines échappent à la classification.

En TUNISIE, les principales causes de l'IRC sont les NGC dans 39,2% des cas, les NIC dans 19,8%, la néphropathie hypertensive dans 14,2%, la néphropathie héréditaire dans 8,2%, la néphropathie diabétique (5,2%) et la néphropathie indéterminée dans 12,5% (tableau 2 (El Matri A. et coll. 1986).

Sur une série personnelle de 1418 malades classés / IRC au CHU de Monastir, la NGC est présente dans 28,3%, la NIC est estimée à 26,1%, les néphropathies indéterminées ont une fréquence de 19,14% (tableau 2).

Tableau 2: Fréquence des néphropathies chroniques en TUNISIE.

	TUNISIE	MONASTIR
NGC	39,2%	28,3%
NIC	19,8%	26,1%
N. vascu.	14,2%	2,46%
Maladie polykystique	8,2%*	9,6%
Diabète	5,2%	9,7%
N. toxique	-	3,3%
N. héréditaire	-	1,4%
Causes indéterminées	12,5%	19,14%
TOTAL	350	1418

* dont néphropathie héréditaire.

Sur les 370 (26,1%) malades classés NIC, la grande partie est d'origine infectieuse et lithiasique (35%), ou bien indéterminée (32%) (voir tableau 3).

Tableau 3: Fréquence des étiologies des NIC.

. Pyélonéphrite chronique	18,7 %
. Lithiase	17,3 %
. Malformation et reflux vésico urétral	8 %
. Goutte	5,3 %
. N. toxique	6,66 %
. indéterminée	32 %

Certaines NIC justifient une enquête étiologique attentive qui demeure parfois négative. Un certain nombre de NIC reste d'étiologies indeterminées, elles

représentent un contingent important de cette catégorie de néphropathie "type BEN" (Hald B. 1991), mais les données chiffrées se font rares.
En fait, cette NIC présente une entité bien précise, stéréotypée sur le plan clinique, biologique, histologique, où aucune étiologie n'est associée (voir tableaux 4 et 5).

Au terme de notre étude préliminaire, 3 questions se posent :
1/ est-il possible que l'ochratoxine A soit à l'origine de ces NIC indéterminées, si on rapproche notre travail de celui réalisé dans les zones de NEB, (BEN), (Petkova-Bocharova et coll. 1988)?.
2/ est-ce que l'OTA est impliquée directement dans la NIC?
3/ les NIC indéterminées existantes dans nos régions sont-elles similaires aux néphropathies dites BEN?

Tableau 4: Signes cliniques selon le type de néphropathie

	GI	GII	GIII
Céphalée	NI	71,42%	42,5%
Anémie	"	57%	40%
Lombalgie	"	57,14%	8%
Hématurie	"	28%	25%
Inf. urine	"	14,28%	0%
Oedème	"	0%	55,5%
HTA	"	28,5%	66,6%
Ages	31,25	46,28	42,47

Tableau 5: Etude biologique selon le type de néphropathies.

	GI	GII	GIII
Urée mmol/l	4,43	13,41	11,55
Créatµmol/l	93,8	178,42	166,5
Urée mmol/l	306	410,7	314,4
Hématocrite%	13,9	9,44	12,55
Na mmol/l	151,79	109,57	66,8**
Leucocyt/ml	<1000	1511	4013 NS
Hématies/ml	<1000	4576,5	4418 NS
Protéinu/24h	0	0,72	4,40

GI: volontaires sains GII:NIC GIII:autres néphropathies

R E F E R E N C E S

1/ STEPHANOVIC V. POLENAKOVIC M.H. 1991
Balkan Nephropathy
An. J. Nephr., 11, 1-11.

2/ HALL (P.W.) 1992
Balkan endemic nephropathy.
More questions than answers
Nephron, 62, 1-5.

3/ EL MATRI A., F. BEN ABDALLAH, CH KECHRID, H. BEN MAIZ, A. KHEDHER, F. BEN MOUSSA, H. BEN AYED. 1986
Néphrologie 7, 109-113.

4/ HALD B. 1991
Ochratoxin A in human blood in european countries. In "Mycotoxins,Endemic Nephropathy and Urinary Tract Tumours". Eds M. Castegnaro, R. Plestina, G. Dirheimer, I.N. Chernozemsky & H. Bartsch. Lyon IARC,Abstract.

5/ PETKOVA-BOCHAROVA T.,CHERNOZEMSKY I.N., CASTEGNARO M. 1988 Ochratoxin A. in human blood in relation to Balkan endemic nephropathy and urinary system tumours in Bulgaria. Food additives and contaminants, Vol 5N°3, pp : 299-301.

Summary :

The chronic nephropathies have several etiologies.

A study of 1418 cases in Tunisia shows 28.3% of Chronic Glomerular Nephropathy, 26.1% of Chronic Interstitial Nephritis (CIN) with 19.1% of unknown cause. In the same time almost 95% of people suffering from nephropathy are OTA positive with clinical and biological signs similar to those observed in BEN.

An attempt is going on to correlate some of these nephropathies with ochratoxin A through the histopathological observation (histology, DNA - adducts etc...) with comparaison to samples from BEN area.

It already appears that a lot of these nephropathies (CIN) could be related to ochratoxicosis.

Résumé

Les étiologies de néphropathies chroniques sont multiples. Il existe plusieurs variétés anatomiques: néphropathies glomérulaires chroniques (NGC), néphropathies interstitielles chroniques (NIC), néphropathies vasculaires chroniques (NVC); cependant, une bonne partie demeure indéterminée.

En TUNISIE les principales étiologies de l'insuffisance rénale chronique (IRC) sont les NGC (39,2%), les NIC (19,8%),les NVC (14,2%) et les N. indéterminées (12,5%).

Dans une série personnelle de 1418 IRC, il y a 28,3% de NGC, 26,1% de NIC et 19,1% d'indéterminées.
Malgré le progrès technique, une grande proportion des NIC demeurent inexpliquées. L'ochratoxine A(OTA) pourrait expliquer une partie des NIC indéterminées.
L'enquête clinique et étiologique des NIC similaires aux BEN dans notre région implique l'OTA dans leurs genèses.

Human ochratoxicosis and its pathologies. Eds E.E. Creppy, M. Castegnaro, G. Dirheimer. Colloque INSERM/ John Libbey Eurotext Ltd. © 1993, Vol. 231, pp. 235-238.

Présence de l'ochratoxine A dans le sang humain et néphropathie en Algérie

A. Khalef[1], M. Benabadji[2], T. Rayan[3] et F. Haddoumi[2]

[1]UGTA, Maison du Peuple, Place du 1er mai, Alger; [2]CHU Thenia, Boumerdes; [3]CHU Parnet, Hussein Dey, Alger, Algérie

Compte-tenu de la fréquence de l'ochratoxicose humaine observée dans la population générale en Algérie (Creppy et coll. 1992, Khalef et coll. 1993) et tout spécialement des taux élevés d'OTA chez presque tous les néphropathes (Creppy et coll. 1992, Khalef et coll. 1993), nous avons émis l'hypothèse de l'existence de néphropathies liées à l'ochratoxicose en Algérie. De tous les néphropathes étudiés à ce jour, la moitié des cas sont d'étiologie inconnue.

Etant donné que tous les néphropathes ont de l'OTA, on peut affirmer sans risque que pour la moitié d'entre-eux dont l'étiologie est connue, il n'y a pas de relation entre l'ochratoxicose et la néphropathie. Ce sont notamment des néphropathies glomérulaires, des uropathies malformatives et des néphropathies d'origine diabétique. La question peut se poser à ce niveau de savoir si tout néphropathe exposé à l'OTA risque d'avoir un taux élevé.

Pour l'autre moitié on constate plusieurs cas de néphropathies familiales non étiquetées, ce qui voudrait probablement dire de même étiologie. Pour ces néphropathes ainsi que pour tous les autres d'étiologie inconnue, il s'agit d'insuffisants rénaux chroniques dialysés (durée moyenne de dialyse 5 ± 1,5 année à raison de 12 heures par semaine). Ces malades qui n'ont pas de syndrome néphrotoxique présentent les signes biologiques suivants:

- protidémie et albuminémie normales
- urémie et créatinémie élevées
- albuminurie et protéinurie élevées.

Ces patients présentent tous de l'OTA dans le sang avant et après hémodialyse ce qui indique entre autres que l'OTA dialyse mal. Chez les personnes présentant un taux supérieur à 8 ng/ml, on constate souvent des complications post-dialyse sous forme de syndrome du canal carpien, polynévrites, arthropathie.

Les signes cliniques sont les suivants : anémie, fatigue, amaigrissement... Ces patients ne présentent pas d'hypertension et ont une diurèse conservée. D'autre part, ils ne présentent pas de phénomènes allergiques particuliers ou d'antécédents familiaux. Ils ne sont pas non plus connus comme abusants d'analgésiques néphrotoxiques.

Des biopsies rénales ont été faites qui vont permettre d'étudier l'histologie du tissu rénal de ces néphropathies afin de les comparer à celles des personnes atteintes de néphropathie endémique des Balkans. De la même façon, il se poursuit actuellement une tentative de détermination du taux des tumeurs rénales parmi les néphropathes ayant de l'OTA dans le sang, dans le but de le comparer à la fréquence de ces tumeurs dans les régions des néphropathies endémiques des Balkans (Petkova-Bocharova et coll. 1988, Delacruz et Bach 1990). En attendant certaines similitudes existent déjà sur les plans biologiques et cliniques entre les néphropathies observées en Algérie et les néphropathies endémiques des Balkans (tableaux 1 et 2).

TABLEAU 1 : SIGNES CLINIQUES, HISTOPATHOLOGIQUES DE LA NEPHROPATHIE ENDEMIQUE DES BALKANS *

<u>Signes cliniques :</u>

Anémie

Asthénie
Céphalée
Polyurie
Anorexie
Lumbago
Insuffisance rénale
Amaigrissement
Sensation de goût amer
Taux élevé d'avortements spontanés

<u>Signes biochimiques:</u>

Urémie
β2 microglobulinémie
Amino-acidurie

<u>Signes hématologiques :</u>

Anémie normochrome

<u>Signes radiologiques :</u>

Contours lisses du rein
Réduction du volume rénal

<u>Signes anatomo-pathologiques :</u>

Fibrose interstitielle
Hyalinisation glomérulaire
Dégénérescence de l'épithélium tubulaire avec desquamation
Perte de la bordure en brosse

<u>Signes histologiques :</u>

Amincissement de la membrane basale

<u>Signes immunologiques :</u>

Augmentation des IgG et IgM (immunoglobulines G et M)
Dépôt d'IgG dans la membrane basale

* Tiré de Austwick, PKC, The practitioner 1981

TABLEAU 2 : SIGNES CLINIQUES ET HISPOPATHOLOGIQUES ABSENTS DANS LA NEPHROPATHIE ENDEMIQUE DES BALKANS *

Hypertension

Modification du fond d'oeil

Modification liquidienne

Athénosclérose

Ostéodystrophie

Cystite

Inflammation glomérulaire

Inflammation tubulaire

Inflammation interstitielle

* Tiré de Austwick, PKC, The practitioner 1981

L'hypothèse de départ selon laquelle il y aurait une relation de cause à effet entre l'ochratoxicose humaine en Algérie et les néphropathies du type NEB, ne peut pas être écartée même s'il manque quelques éléments pour achever de l'étayer.

Il apparait clairement que les signes cliniques non-observés dans la NEB sont aussi absents chez les néphropathes algériens ayant l'OTA dans le sang; De la même façon, la plupart des signes cliniques et biologiques sont retrouvés chez ces néphropathes OTA-positifs dont la maladue est jusqu'à présent d'étiologie inconnue.

BIBLIOGRAPHIE :

Austwick, PKC, 1981, Balkan nephropathy, The practitioner 225, 1031-1038

Creppy E.E., Betbeder A.M., Sanchez D., Gharbi A., Khalef A., Gauret M.F., Counord J., Dirheimer G., Bacha H., Hadidane R., Hammami M., El-May , Achour L., Zidane C. 1992. Human ochratoxicosis in northern Africa, correlation with cases of nephropathy. 14. Mykotoxin-workshop, Justus-Liebig-Universität Giessen, 67-68

Delacruz L. and Bach P.H., 1990, The role of ochratoxin A metabolism and biochemistry in animal and human nephropathy. J. of Biopharm. Sci., 1, 277-304

Khalef A., Zidane C., Charef A., Gharbi A., Tadjerouna M., Betbeder A.M. et Creppy E.E. 1993. Ochratoxicose humaine en Algérie. Publication scientifique INSERM, ed. E.E. Creppy, M. Castegnaro et G. Dirheimer, J. Libbey, Paris

Petkova-Bocharova T., Chernozemsky I.N. and Castegnaro M. 1988. Ochratoxin A in human blood in relation to Balkan endemic nephropathy and urinary system tumours in Bulgaria. Food Addit. Contam., 5, 299-301

SUMMARY

Cases of nephropathy are found in Algeria in people having ochratoxin A in blood in the range of the detected levels in Balkan Endemic Nephropathy (BEN) area. Clinical signs and biological parameters are similar to that observed in BEN patient without nephrotic syndrom and hypertension.

Histopathological studies are now needed to confirm the similarity of this disease with BEN.

Résumé

Des cas de néphropathie ont été détectés en Algérie chez des personnes ayant de l'ochratoxine A dans le sang à des taux comparables à ceux qui sont déterminés dans les zones de Néphropathie Endémique des Balkans (NEB). Les signes cliniques et les paramètres biologiques sont comparables à ceux observés chez les patients souffrant de NEB, en absence de syndrome néphrotique et d'hypertension.

Des études histopathologiques sont maintenant nécessaires pour confirmer la similitude de cette maladie avec la NEB.

Author index

Index des auteurs

Achour A., 111, 227
Alvinerie M., 59
Andrieux M., 147
Ayi-Fanou L., 101

Bach P.H., 33, 43, 217
Bacha H., 111, 227
Bartsch H., 147
Baudrimont I., 177, 189
Benabadji M., 235
Betbeder A.M., 75, 123, 147, 189
Blanc R., 75
Blom M., 67
Boisard F., 75
Bouraïma Y., 101
Brera C., 129

Carstensen B., 209
Castegnero M., 147, 199
Cazenave J.-P., 147
Chambon P., 147
Charef A., 123
Corneli S., 129
Counord J., 147
Creppy E.E., 75, 101, 111, 123, 147, 177, 189, 199, 227

Deberghes P., 75
De Dominicis R., 129
Deffieux G., 75
Delaby J.-F., 75
Dirheimer G., 147, 177, 189, 199
Douet C., 147

Ellouz F., 111
El-May M., 227

Fink-Gremmels J., 67
Fouillet B., 147

Galtier P., 59
Gauret M.-F., 147
Gharbi A., 75, 123, 147, 177, 189
Giacomotto P., 147
Grosse Y., 147, 177, 199
Guiot-Guillain M., 147

Haddoumi F., 235
Hadlock R.M., 141
Hald B., 209
Hamammi M., 111, 227

Kane A., 177, 199
Kawamura O., 159
Khalef A., 123, 235
Kora I., 101
Kuiper-Goodman T., 167

Larrieu G., 59
Laustriat D., 147
Lombaert G.A., 167

Maaroufi K., 111, 127
Maki S., 159
Malcolm S., 167
Manier C., 147
Marquardt R.R., 167
McLean M., 33

McMullen E., 167
Mériaux J., 147
Miraglia M., 129
Moncharmont P., 147
Morton T., 167

Obrecht S., 177, 199
Olsen J.H., 209
Ominski K., 167

Pfohl-Leszkowicz A., 147, 177, 189, 199

Rayan T., 235

Sanni A., 101
Sato S., 159
Setondji J., 101
Speijers G.J.A., 85
Steyn P.S., 3, 51

Tadjerouna M., 123
Thorup I., 209

Ueno Y., 159

Van Egmond H.P., 85
Vezon G., 147

Waller C., 147
Woutersen van Nijnanten F., 67

Zidane C., 123

Colloques INSERM

ISSN 0768-3154

Other *Colloques* published as co-editions by John Libbey Eurotext and INSERM

153 Hormones and Cell Regulation (11th European Symposium). *Hormones et Régulation Cellulaire (11e Symposium Européen).*
Edited by J. Nunez and J.E. Dumont.
ISBN : John Libbey Eurotext 0 86196 104 8
INSERM 2 85598 324 X

158 Biochemistry and Physiopathology of Platelet Membrane. *Biochimie et Physiopathologie de la Membrane Plaquettaire.*
Edited by G. Marguerie and R.F.A. Zwaal.
ISBN : John Libbey Eurotext 0 86196 114 5
INSERM 2 85598 345 2

162 The Inhibitors of Hematopoiesis. *Les Inhibiteurs de l'Hématopoïèse.*
Edited by A. Najman, M. Guignon, N.C. Gorin and J.Y. Mary.
ISBN : John Libbey Eurotext 0 86196 125 0
INSERM 2 85598 340 1

164 Liver Cells and Drugs. *Cellules Hépatiques et Médicaments.*
Edited by A. Guillouzo.
ISBN : John Libbey Eurotext 0 86196 128 5
INSERM 2 85598 341 X

165 Hormones and Cell Regulation (12th European Symposium). *Hormones et Régulation Cellulaire (12e Symposium Européen).*
Edited by J. Nunez, J.E. Dumont and E. Carafoli.
ISBN : John Libbey Eurotext 0 86196 133 1
INSERM 2 85598 347 9

167 Sleep Disorders and Respiration. *Les Evénements Respiratoires du Sommeil.*
Edited by P. Lévi-Valensi and D. Duron.
ISBN : John Libbey Eurotext 0 86196 127 7
INSERM 2 85598 344 4

169 Neo-Adjuvant Chemotherapy. *Chimiothérapie Néo-Adjuvante.*
Edited by C. Jacquillat, M. Weil, D. Khayat.
ISBN : John Libbey Eurotext 0 86196 150 1
INSERM 2 85598 349 5

171 Structure and Functions of the Cytoskeleton. *La Structure et les Fonctions du Cytosquelette.*
Edited by B.A.F. Rousset.
ISBN : John Libbey Eurotext 0 86196 149 8
INSERM 2 85598 351 7

Colloques INSERM
ISSN 0768-3154

172 The Langerhans Cell. *La Cellule de Langerhans.*
Edited by J. Thivolet, D. Schmitt.
ISBN : John Libbey Eurotext 0 86196 181 1
INSERM 2 85598 352 5

173 Cellular and Molecular Aspects of Glucuronidation. *Aspects Cellulaires et Moléculaires de la Glucuronoconjugaison.*
Edited by G. Siest, J. Magdalou, B. Burchell
ISBN : John Libbey Eurotext 0 86196 182 X
INSERM 2 85598 353 3

174 Second Forum on Peptides. *Deuxième Forum Peptides.*
Edited by A. Aubry, M. Marraud, B. Vitoux
ISBN : John Libbey Eurotext 0 86196 151 X
INSERM 2 85598 354 1

176 Hormones and Cell Regulation (13th European Symposium). *Hormones et Régulation Cellulaire (13e Symposium Européen).*
Edited by J. Nunez, J.E. Dumont, R. Denton
ISBN : John Libbey Eurotext 0 86196 183 8
INSERM 2 85598 356 8

179 Lymphokine Receptors Interactions. *Interactions Lymphokines-récepteurs.*
Edited by D. Fradelizi, J. Bertoglio
ISBN : John Libbey Eurotext 0 86196 148 X
INSERM 2 85598 359 2

191 Anticancer Drugs (1st International Interface of Clinical and Laboratory responses to anticancer drugs). *Médicaments anticancéreux (1re Confrontation internationale des réponses cliniques et expérimentales aux médicaments anticancéreux).*
Edited by H. Tapiero, J. Robert, T.J. Lampidis
ISBN : John Libbey Eurotext 0 86196 223 0
INSERM 2 85598 393 2

193 Living in the Cold (2nd International Symposium). *La Vie au Froid (2e Symposium International).*
Edited by A. Malan, B. Canguilhem
ISBN : John Libbey Eurotext 0 86196 234 9
INSERM 2 85598 395 9

Colloques INSERM
ISSN 0768-3154

194 Progress in Hepatitis B Immunization. *La Vaccination contre l'épatite B.*
Edited by P. Coursaget, M.J. Tong
ISBN : John Libbey Eurotext 0 86196 249 4
INSERM 2 85598 396 7

196 Treatment Strategy in Hodgkin's Disease. *Stratégie dans la maladie de Hodgkin.*
Edited by P. Sommers, M. Henry-Amar, J.H. Meezwaldt, P. Carde
ISBN : John Libbey Eurotext 0 86196 226 5
INSERM 2 85598 398 3

198 Hormones and Cell Regulation (14th European Symposium). *Hormones et Régulation Cellulaire (14e Symposium Européen).*
Edited by J. Nunez, J.E. Dumont
ISBN : John Libbey Eurotext 0 86196 229 X
INSERM 2 85598 400 9

199 Placental Communications : Biochemical, Morphological and Cellular Aspects. *Communications placentaires : aspects biochimique, morphologique et cellulaire.*
Edited by L. Cedard, E. Alsat, J.C. Challier, G. Chaouat, A. Malassiné
ISBN : John Libbey Eurotext 0 86196 227 3
INSERM 2 85598 401 7

204 Pharmacologie Clinique : Actualités et Perspectives. (6e Rencontres Nationales de Pharmacologie clinique).
Edited by J.P. Boissel, C. Caulin, M. Teule
ISBN : John Libbey Eurotext 0 86196 225 7
INSERM 2 85598 454 8

205 Recent Trends in Clinical Pharmacology (6th National Meeting of Clinical Pharmacology).
Edited by J.P. Boissel, C. Caulin, M. Teule
ISBN : John Libbey Eurotext 0 86196 256 7
INSERM 2 85598 455 6

206 Platelet Immunology : Fundamental and Clinical Aspects. *Immunologie plaquettaire : aspects fondamentaux et cliniques.*
Edited by C. Kaplan-Gouet, N. Schlegel, Ch. Salmon, J. McGregor
ISBN : John Libbey Eurotext 0 86196 285 0
INSERM 2 85598 439 4

Colloques INSERM

ISSN 0768-3154

207 Thyroperoxidase and Thyroid Autoimmunity. *Thyroperoxydase et auto-immunité thyroïdienne.*
Edited by P. Carayon, T. Ruf
ISBN : John Libbey Eurotext 0 86196 277 X
INSERM 2 85598 440 8

208 Vasopressin. *Vasopressine.*
Edited by S. Jard, R. Jamison
ISBN : John Libbey Eurotext 0 86196 288 5
INSERM 2 85598 441 6

210 Hormones and Cell Regulation (15th European Symposium). *Hormones et Régulation Cellulaire (15e Symposium Européen).*
Edited by J.E. Dumont, J. Nunez, R.J.B. King
ISBN : John Libbey Eurotext 0 86196 279 6
INSERM 2 85598 443 2

211 Medullary Thyroid Carcinoma. *Cancer Médullaire de la Thyroïde.*
Edited by C. Calmettes, J.M. Guliana
ISBN : John Libbey Eurotext 0 86196 287 7
INSERM 2 85598 440 0

212 Cellular and Molecular Biology of the Materno-Fetal Relationship. *Biologie cellulaire et moléculaire de la relation materno-fœtale.*
Edited by G. Chaouat, J. Mowbray
ISBN : John Libbey Eurotext 0 86196 909 1
INSERM 2 85598 445 9

215 Aldosterone. Fundamental Aspects. *Aspects fondamentaux.*
Edited by J.P. Bonvalet, N. Farman, M. Lombes, M.E. Rafestin-Oblin
ISBN : John Libbey Eurotext 0 86196 302 4
INSERM 2 85598 482 3

216 Cellular and Molecular Aspects of Cirrhosis. *Aspects cellulaires et moléculaires de la cirrhose.*
Edited by B. Clément, A. Guillouzo
ISBN : John Libbey Eurotext 0 86196 342 3
INSERM 2 85598 483 1

217 Sleep and Cardiorespiratory Control. *Sommeil et contrôle cardio-respiratoire.*
Edited by C. Gaultier, P. Escourrou, L. Curzi-Dascalora
ISBN : John Libbey Eurotext 0 86196 307 5
INSERM 2 85598 484 X

Colloques INSERM
ISSN 0768-3154

218 Genetic Hypertension. *Hypertension génétique.*
Edited by J. Sassard
ISBN : John Libbey Eurotext 0 86196 313 X
INSERM 2 85598 485 8

219 Human Gene Transfer. *Transfert de gènes chez l'homme.*
Edited by O. Cohen-Haguenauer, M. Boiron
ISBN : John Libbey Eurotext 0 86196 301 6
INSERM 2 85598 497 1

220 Medicine and Change: Historical and Sociological Studies of Medical Innovation. *L'innovation en médecine : études historiques et sociologiques.*
Edited by Ilana Löwy
ISBN : John Libbey Eurotext 2 7420 0010 0
INSERM 5 85598 508 0

221 Structures and Functions of Retinal Proteins. *Structures et fonctions des rétino-protéines.*
Edited by J.L. Rigaud
ISBN : John Libbey Eurotext 0 86196 355 5
INSERM 2 85598 509 9

222 Cellular and Molecular Biology of the Adrenal Cortex. *Biologie cellulaire et moléculaire du cortex surrénal.*
Edited by J.M. Saez, A.C. Brownie, A. Capponi, E.M. Chambaz, F. Mantero
ISBN : John Libbey Eurotext 0 86196 362 8
INSERM 2 85598 510 2

223 Mechanisms and Control of Emesis. *Mécanismes et contrôle du vomissement.*
Edited by A.L. Bianchi, L. Grélot, A.D. Miller, G.L. King
ISBN : John Libbey Eurotext 0 86196 363 6
INSERM 2 85598 511 0

224 High Pressure and Biotechnology. *Haute pression et biotechnologie.*
Edited by C. Balny, R. Hayashi, K. Heremans, P. Masson
ISBN : John Libbey Eurotext 0 86196 363 6
INSERM 2 85598 512 9

Colloques INSERM
ISSN 0768-3154

228 Non-Visual Human-Computer Interactions. *Communication non visuelle homme-ordinateur.*
Edited by D. Burger, J.C. Sperandio
ISBN : John Libbey Eurotext 2 7420 0014 3
INSERM 2 85598 540 4

230 From Research in Oncology to Therapeutic Innovations. *De la recherche oncologique à l'innovation thérapeutique.*
Edited by P. Tambourin, M. Boiron
ISBN : John Libbey Eurotext 2 7420 0016 X
INSERM 2 85598 542 0

L O U I S - J E A N
avenue d'Embrun, 05003 GAP cedex
Tél. : 92.53.17.00
Dépôt légal : 494 — Juin 1993
Imprimé en France